OBERLEY

AGING
Volume 27

FREE RADICALS
IN MOLECULAR BIOLOGY,
AGING, AND DISEASE

Aging Series

Volume 27: Free Radicals in Molecular Biology, Aging, and Disease, *D. Armstrong, R. S. Sohal, R. G. Cutler, and T. F. Slater, editors, 432 pp., 1984*
Volume 26: Nutrition in Gerontology, *J.M. Ordy, D. Harman, and R.B. Alfin-Slater, editors, 352 pp., 1984*
Volume 25: Alcoholism in the Elderly: Social and Biomedical Issues, *J. T. Hartford and T. Samorajski, editors, 304 pp., 1984*
Volume 24: Perspectives on Prevention and Treatment of Cancer in the Elderly, *R. Yancik, P. Carbone, W. B. Patterson, K. Steel, and W. D. Terry, editors, 360 pp., 1983*
Volume 23: Aging Brain and Ergot Alkaloids, *A. Agnoli, G. Crepaldi, P. F. Spano, and M. Trabucchi, editors, 464 pp., 1983*
Volume 22: Aging of the Brain, *S. Algeri, S. Gershon, D. Samuel, and G. Toffano, editors, 400 pp., 1983*
Volume 21: Brain Aging: Neuropathology and Neuropharmacology, *J. Cervós-Navarro and H. I. Sarkander, editors, 1983*
Volume 20: The Aging Brain: Cellular and Molecular Mechanisms of Aging in the Nervous System, *E. Giacobini, G. Filogano, and A. Vernadakis, editors, 284 pp., 1982*
Volume 19: Alzheimer's Disease: A Report of Progress in Research, *S. Corkin, K. L. Davis, J. H. Growdon, E. Usdin, and R. J. Wurtman, editors, 544 pp., 1982*
Volume 18: Neural Aging and Its Implications in Human Neuropathological Pathology, *R. D. Terry, C. L. Bolis, and G. Toffano, editors, 269 pp., 1982*
Volume 17: Brain Neurotransmitters and Receptors in Aging and Age-Related Disorders, *S. J. Enna, T. Samorajski, and B. Beer, editors, 290 pp., 1981*
Volume 16: Clinical Pharmacology and the Aged Patient, *L. F. Jarvik, D. J. Greenblatt, and D. Harman, editors, 256 pp., 1981*
Volume 15: Clinical Aspects of Alzheimer's Disease and Senile Dementia, *N. E. Miller and G. D. Cohen, editors, 350 pp., 1981*
Volume 14: Neuropsychiatric Manifestations of Physical Disease in the Elderly, *A. J. Levenson and R. C. W. Hall, editors, 168 pp., 1981*
Volume 13: Aging of the Brain and Dementia, *L. Amaducci, A. N. Davison, and P. Antuono, editors, 344 pp., 1980*
Volume 12: The Aging Heart: Its Function and Response to Stress, *M. L. Weisfeldt, editor, 335 pp., 1980*
Volume 11: Aging, Immunity, and Arthritic Diseases, *M. M. B. Kay, J. E. Galpin, and T. Makinodan, editors, 275 pp., 1980*
Volume 10: Sensory Systems and Communication in the Elderly, *J. M. Ordy and K. R. Brizzee, editors, 334 pp., 1979*
Volume 9: Neuropsychiatric Side-Effects of Drugs in the Elderly, *A. J. Levenson, editor, 256 pp., 1979*
Volume 8: Physiology and Cell Biology of Aging, *A. Cherkin, C. E. Finch, N. Kharasch, T. Makinodan, F. L. Scott, and B. S. Strehler, editors, 248 pp., 1979*
Volume 7: Alzheimer's Disease: Senile Dementia and Related Disorders, *R. Katzman, R. D. Terry, and K. L. Bick, editors, 452 pp., 1978*
Volume 6: Aging in Muscle, *G. Kaldor and W. J. DiBattista, editors, 240 pp., 1978*
Volume 5: Geriatric Endocrinology, *R. B. Greenblatt, editor, 256 pp., 1978*
Volume 4: The Aging Reproductive System, *E. L. Schneider, editor, 291 pp., 1978*
Volume 3: Neurobiology of Aging, *R. D. Terry and S. Gershon, editors, 452 pp., 1976*
Volume 2: Genesis and Treatment of Psychologic Disorders in the Elderly, *S. Gershon and A. Raskin, editors, 288 pp., 1975*
Volume 1: Clinical, Morphologic, and Neurochemical Aspects in the Aging Central Nervous System, *H. Brody, D. Harman, and J. M. Ordy, editors, 240 pp., 1975*

Aging
Volume 27

Free Radicals in Molecular Biology, Aging, and Disease

Editors

Donald Armstrong, Ed.D., Ph.D.
*Departments of Ophthalmology, Biochemistry, and Molecular Biology
University of Florida College of Medicine
and
Department of Comparative Ophthalmology
University of Florida College of Veterinary Medicine
Gainesville, Florida*

R. S. Sohal, Ph.D.
*Department of Biology
Southern Methodist University
Dallas, Texas*

Richard G. Cutler, Ph.D.
*Gerontology Research Center
National Institutes on Aging
Baltimore City Hospitals
Baltimore, Maryland*

Trevor F. Slater, Ph.D., D.Sc., M.D. (Hon.)
*Department of Biochemistry
Brunel University
Uxbridge, Middlesex
United Kingdom*

Raven Press ■ New York

Raven Press, 1140 Avenue of the Americas, New York, New York 10036

© 1984 by Raven Press Books, Ltd. All rights reserved. This book is protected by copyright. No part of it may be reproduced, stored in a retrieval system, or transmitted, in any form or by any means, electronic, mechanical, photocopying, recording, or otherwise, without the prior written permission of the publisher.

Made in the United States of America

Library of Congress Cataloging in Publication Data
Main entry under title:

Free radicals in molecular biology, aging, and disease.

(Aging; v. 27)
Papers presented at a meeting held in Washington, D.C., on Oct. 6–8, 1983.
Includes bibliographies and index.
1. Aging—Congresses. 2. Free radicals (Chemistry)—Physiological effect—Congresses. 3. Lipofuscins—Congresses. 4. Diseases—Causes and theories of causation—Congresses. 5. Pathology, Molecular—Congresses.
I. Armstrong, Donald, 1931– . II. Series. [DNLM:
1. Aging—congresses. 2. Free Radicals—congresses.
3. Molecular Biology—congresses. W1 AG342E v.27 /
WT 104 F853 1983]
QP86.F727   1984   599'.021   84-16102
ISBN 0-88167-048-0

Papers or parts thereof have been used as camera-ready copy as submitted by the authors whenever possible; when retyped, they have been edited by the editorial staff only to the extent considered necessary for the assistance of an international readership. The views expressed and the general style adopted remain, however, the responsibility of the named authors. Great care has been taken to maintain the accuracy of the information contained in the volume. However, neither Raven Press nor the editors can be held responsible for errors or for any consequences arising from the use of information contained herein.

The use in this book of particular designations of countries or territories does not imply any judgment by the publisher or editors as to the legal status of such countries or territories, of their authorities or institutions or of the delimitation of their boundaries.

Some of the names of products referred to in this book may be registered trademarks or proprietary names, although specific reference to this fact may not be made; however, the use of a name without designation is not to be construed as a representation by the publisher or editors that it is in the public domain. In addition, the mention of specific companies or of their products or proprietary names does not imply any endorsement or recommendation on the part of the publisher or editors.

Authors were themselves responsible for obtaining the necessary permission to reproduce copyright material from other sources. With respect to the publisher's copyright, material appearing in this book prepared by individuals as part of their official duties as government employees is only covered by this copyright to the extent permitted by the appropriate national regulations.

Materials appearing in this book prepared by individuals as part of their official duties as U.S. Government employees are not covered by the above-mentioned copyright.

Second Printing, September 1985

Preface

All organisms go through stages of development, maturity, aging, and death. Historically, scientific observations on the declines associated with aging focused first on such clearly demonstrable end points as increasing mortality with age, expressed in the classic Gompertz Plot, and the progressive decline in vigor, health, sensory, cognitive, and motor functions; these age-related changes are observable at the organismic level. During the last decade, considerable scientific information has accumulated on age-related changes observable at organ, tissue, and cell levels. Recent advances in molecular biology have made it possible to examine whether the overt physical manifestations of aging may be causally related to basic mechanisms of aging at the molecular level. The dramatic increase in knowledge provided by molecular biology may make it possible to focus on basic mechanisms of aging or, at least, begin to study cause and effect sequences in aging from molecular to organismic levels. This volume presents the rapidly emerging knowledge concerning the role of free radicals in molecular biology with particular emphasis on free radical activity in age-pigment accumulation and cell loss during normal aging, and in specific age-related diseases.

Despite significant differences in life spans among species, correlations among life span, body weight, brain weight, and particularly metabolic rate have provided the initial clues that different species may age by similar mechanisms. Correlations among metabolic rate, rate of lipofuscin accumulation, cell loss, and life span have suggested that lipofuscin pigment of postmitotic cells may represent a basic cellular marker for normal aging, as well as for a variety of age-related diseases. Considerable evidence has also emerged recently to suggest a causal or, at least, a close link between age pigments and the free radicals that are generated as a result of univalent reduction of oxygen during aerobic respiration. As yet, evidence for oxygen free-radical damage of cells *in vivo* is considerable but indirect. However, recent clues suggest that abnormal free radical activity with increasing age and in certain diseases may represent a significant cause of cellular breakdown. Age-related changes in molecular oxygen, trace metals, and antioxidant activity have been implicated in the generation of free radicals, lipopigment accumulation, and cell loss. Free radicals are believed to cause peroxidation of polyunsaturated lipids of membranes resulting in lipopigment accumulation, cell loss, and tissue breakdown. Thus, age- and disease-related tissue and cell breakdown may represent a consequence of increasingly abnormal free radical activity. In turn, this cellular breakdown may trigger greater free radical activity and tissue damage.

As there is age-related breakdown of cell structure and function, the topic of free radicals has become increasingly important not only in molecular biology but also in aging, nutrition, pharmacology, pathology, and even in clinical practice. The upsurge of interest in free radical damage of cell structure and function has occurred in researchers of widely different disciplines. As a result, current information on the role of free radicals in aging and disease is widely scattered in the literature.

This volume brings together both the more basic and the most recent information on free radicals in aging and disease. The chapters deal with free radical theories of aging, basic chemistry of free radical reactions, effects on lipopigments, free radicals in age-related diseases, and antioxidant defense mechanisms. The chapters in this volume should make it possible to plan and conduct further research on the causes or mechanisms of aging, ranging from the molecular to the organismic levels. Due to the significant increase in the proportion of the elderly in our population, the search for causes of aging in molecular biology will attract increasing numbers of investigations from many disciplines.

This volume will be of great interest to students, scientists, physicians, nutritionists, pharmacologists, and other individuals who are interested in the rapid progress that will be made in future studies of basic mechanisms of aging and prospects of nutritional and chemical intervention.

J. Mark Ordy
Chairman, AGE Symposia

Acknowledgments

The editors wish to express their appreciation to the following companies for partial support of the symposium at which chapters in this book were presented as papers: Burroughs Wellcome Co.; Hoffmann-LaRoche, Inc.; Johnson and Johnson; Mead Johnson Nutritional Division; Pennwalt Corporation; Promonta Chemische Fabrik, GmbH; Stuart Pharmaceuticals (Div. of ICI Americas); Syntex Corporation; Twin Laboratories, Inc.; Upjohn Company; Warner-Lambert Company; and a departmental grant to the Department of Ophthalmology, University of Florida, from Research to Prevent Blindness, Inc.

Contents

1   Free Radicals and the Origination, Evolution, and Present Status of the Free Radical Theory of Aging
    D. Harman

Free Radicals: Basic Reactions

13 Free Radicals in Autoxidation and in Aging
    W. A. Pryor

43 Biological Sites and Mechanisms of Free Radical Production
    B. A. Freeman

53 Free Radical Mechanisms in Lipid Peroxidation and Prostaglandins
    J. F. Mead

67 Free Radicals in Melanin Formation, Structure, and Reactions
    R. C. Sealy

77 Superoxide Dismutase: An Antioxidant Defense Enzyme
    H. M. Hassan

87 Free Radical Injury During Inflammation
    R. F. Del Maestro

103 Free Radical-Induced Changes in the Surface Morphology of Isolated Hepatocytes
    M. T. Smith, H. Thor, S. A. Jewell, G. Bellomo, M. S. Sandy, and S. Orrenius

Free Radicals: In Aging

119 Metabolic Rate, Free Radicals, and Aging
    R. S. Sohal

129 Free Radical Involvement in the Formation of Lipopigments
    D. Armstrong

143 Possible Involvement of Iron and Oxygen Free Radicals in Aspects of Aging in Brain
    R. A. Floyd, M. M. Zaleska, and H. J. Harmon

163 Potential Role of Autoxidation in Age Changes of the Retina and Retinal Pigment Epithelium of the Eye
M.L. Katz, W.G. Robison, Jr., and E.A. Dratz

181 Dietary Restriction and the Aging Process
R. Weindruch

203 Adrenocortical Cultures as Model Systems for Investigating Cellular Aging
P.J. Hornsby, K.A. Aldern, and S.E. Harris

Free Radicals: Effects of Antioxidants

223 Effects of Antioxidants on Neuronal Lipofuscin Pigment
K. Nandy

235 Antioxidants and Longevity
R.G. Cutler

Free Radicals: In Disease

267 Clinical Laboratory Tests as Indicators of Free-Radical Reactions
T.L. Dormandy and D.G. Wickens

275 Electron Spin Resonance Studies of Cancer: Experimental Results and Conceptual Implications
H.M. Swartz

293 Free Radicals, Lipid Peroxidation, and Cancer
T.F. Slater, K.H. Cheeseman, and K. Proudfoot

307 Oxy-Radical Production in Alloxan-Induced Diabetes: An Example of an *In Vivo* Metal-Catalyzed Haber-Weiss Reaction
G. Cohen

317 Free Radicals and Damage to Ocular Tissues
R.D. Wiegand, J.G. Jose, L.M. Rapp, and R.E. Anderson

355 Free Radicals in Inflammatory Disease
M.V. Torrielli and M.U. Dianzani

381 Free Radicals and Lung Injury
C.T. Bishop, B.A. Freeman, and J.D. Crapo

391 Cytotoxicity and Somatic Mutation Induced in Mammalian Cells in Culture by Hyperoxia and Ionizing Radiation: Effect of Superoxide Dismutase and Catalase Inhibitors
S. Lesko, L. Trpis, and S. Yang

397 Closing Remarks
T. F. Slater

399 *Subject Index*

Contributors

Kathy A. Aldern, B.S.
Division of Endocrinology and Metabolism
Department of Medicine
University of California at San Diego
La Jolla, California 92093

Robert E. Anderson, M.D., Ph.D.
Department of Ophthalmology
Cullen Eye Institute
Baylor College of Medicine
Houston, Texas 77030

Donald Armstrong, Ed.D., Ph.D.
Department of Ophthalmology
Box J-284
University of Florida College of Medicine
J. Hillis Miller Health Center
Gainesville, Florida 32610

Giorgio Bellomo, M.D.
Department of Forensic Medicine
Karolinska Institute
Box 60400
Stockholm 104 01, Sweden

Clark T. Bishop, M.D.
Division of Allergy, Critical Care and
 Respiratory Medicine
Department of Medicine
Box 3177
Duke University Medical Center
Durham, North Carolina 27710

Kevin H. Cheeseman, Ph.D.
Department of Biochemistry
School of Biological Sciences
Brunel University
Uxbridge, Middlesex UB8 3PH,
 United Kingdom

Gerald Cohen, M.D., M.I. Biol., Ph.D.
Department of Neurology
Mount Sinai School of Medicine of the
 City University of New York
One Gustave Levy Place
New York, New York 10029

James D. Crapo, M.D.
Division of Allergy, Critical Care and
 Respiratory Medicine
Department of Medicine
Box 3177
Duke University Medical Center
Durham, North Carolina 27710

Richard G. Cutler, Ph.D.
Gerontology Research Center
National Institute on Aging
Baltimore City Hospital
Baltimore, Maryland 21224

Rolando F. Del Maestro, M.D., Ph.D.
Brain Research Laboratory
Department of Clinical Neurological
 Sciences
Victoria Hospital
University of Western Ontario
London, Ontario, Canada NGA 4G5

M. U. Dianzani, M.D.
Institute of General Pathology
University of Turin
C.so Raffaello, 30
10125 Torino, Italy

Thomas L. Dormandy, M.D., Ph.D.
Department of Chemical Pathology
Whittington Hospital
London N19 5NF, United Kingdom

CONTRIBUTORS

Edward A. Dratz, Ph.D.
Division of Natural Sciences
University of California
Santa Cruz, California 95064

Robert A. Floyd, Ph.D.
Biomembrane Research Laboratory
Oklahoma Medical Research Foundation
825 North East 13th Street
Oklahoma City, Oklahoma 73104

Bruce A. Freeman, Ph.D.
Division of Allergy, Critical Care and
 Respiratory Medicine
Department of Medicine
Box 3177
Duke University Medical Center
Durham, North Carolina 27710

Denham Harman, M.D., Ph.D.
Departments of Medicine and
 Biochemistry
University of Nebraska Medical Center
Omaha, Nebraska 68105

H. James Harmon, Ph.D.
Department of Zoology
Oklahoma State University
Stillwater, Oklahoma 74078

Sandra E. Harris, M.S.
Department of Medicine
University of California at San Diego
La Jolla, California 92093

Hosni M. Hassan, Ph.D.
Departments of Food Science and
 Microbiology
North Carolina State University
Raleigh, North Carolina 27650

Peter J. Hornsby, Ph.D.
Division of Endocrinology and Metaboli.
Department of Medicine
University of California at San Diego
La Jolla, California 92093

Sarah A. Jewell, B.S.
Department of Forensic Medicine
Karolinska Institute
Box 60400
Stockholm 104 01, Sweden

Jule G. Jose, O.D., Ph.D.
College of Optometry
University of Houston
Houston, Texas 77002

Martin L. Katz, Ph.D.
National Eye Institute
Laboratory of Vision Research
National Institutes of Health
Bethesda, Maryland 20205

Stephen A. Lesko, Ph.D.
Division of Biophysics
School of Hygiene and Public Health
The Johns Hopkins University
615 North Wolfe Street
Baltimore, Maryland 21205

James F. Mead, Ph.D.
Laboratory of Biomedical and
 Environmental Sciences
Department of Biological Chemistry
School of Medicine and School of Public
 Health—Nutrition
University of California at Los Angeles
900 Veteran Avenue
Los Angeles, California 90024

Kalidas Nandy, M.D., Ph.D.
Geriatric Research, Education and
 Clinical Center
Edith Nourse Rogers Memorial Veterans
 Administration Hospital
Bedford, Massachusetts 01730

Sten Orrenius, M.D.
Department of Forensic Medicine
Karolinska Institute
Box 60400
Stockholm 104 01, Sweden

Karen Proudfoot, B.Sc.
Department of Biochemistry
School of Biological Sciences
Brunel University
Uxbridge, Middlesex UB8 3PH,
 United Kingdom

William A. Pryor, Ph.D.
Departments of Chemistry and
  Biochemistry, and Division of
  Toxicology
Louisiana State University
Baton Rouge, Louisiana 70803-1804

Laurence M. Rapp, Ph.D.
Department of Biochemistry
Cullen Eye Institute
Baylor College of Medicine
Houston, Texas 77030

W. Gerald Robison, Jr., Ph.D.
National Eye Institute
Laboratory of Vision Research
National Institutes of Health
Bethesda, Maryland 20205

Martha S. Sandy, B.S.
Department of Biomedical and
  Environmental Health Sciences
School of Public Health
University of California
Berkeley, California 94720

Roger C. Sealy, Ph.D.
Radiation Biophysics Section and National
  Biomedical ESR Center
Medical College of Wisconsin
8701 Watertown Plank Road
Milwaukee, Wisconsin 53226

Trevor F. Slater, Ph.D., D.Sc., M.D. (Hon.)
Department of Biochemistry
School of Biological Sciences
Brunel University
Uxbridge, Middlesex UB8 3PH,
  United Kingdom

Martyn T. Smith, Ph.D.
Department of Biomedical and
  Environmental Health Sciences
School of Public Health
University of California
Berkeley, California 94720

Rajinder S. Sohal, Ph.D.
Department of Biology
Southern Methodist University
Dallas, Texas 75275

Harold M. Swartz, M.D., Ph.D.
University of Illinois College of Medicine
  at Urbana-Champaign
Urbana, Illinois 61801

Hjördis Thor
Department of Forensic Medicine
Karolinska Institute
Box 60400
Stockholm 104 01, Sweden

Mario V. Torrielli, M.D.
Institute of General Pathology
University of Turin
C.so Raffaello, 30
10125 Torino, Italy

Ludmilla Trpis, B.S.
Division of Biophysics
School of Hygiene and Public Health
The Johns Hopkins University
615 North Wolfe Street
Baltimore, Maryland 21205

Richard Weindruch, Ph.D.
Department of Pathology
University of California at Los Angeles
  School of Medicine
Los Angeles, California 90024

David G. Wickens, B.Sc.
Department of Chemical Pathology
Whittington Hospital
London N19 5NF, United Kingdom

Rex D. Wiegand, Ph.D.
Department of Ophthalmology
Cullen Eye Institute
Baylor College of Medicine
Houston, Texas 77030

Shu-Uin Yang, M.S.
Division of Biophysics
School of Hygiene and Public Health
The Johns Hopkins University
615 North Wolfe Street
Baltimore, Maryland 21205

Malgorzata M. Zaleska, Ph.D.
Oklahoma Medical Research Foundation
825 North East 13th Street
Oklahoma City, Oklahoma 73104

Free Radicals and the Origination, Evolution, and Present Status of the Free Radical Theory of Aging

Denham Harman

University of Nebraska College of Medicine, Omaha, Nebraska 68105

The term "radical" comes from the early days of chemistry. Lavoisier (33) in his theory of acids designated the element or group of elements which combined with oxygen in the acid a "radical". This theory was soon discarded but the word "radical" continued to be used to signify a group of elements which retained their identity through a series of reactions, e.g., a methyl radical (CH_3). Today a "radical" is defined as an atom or group of atoms with an unpaired electron, e.g., $Cl\cdot$, $HC_3\cdot$, or $HO\cdot$.

Numerous "radicals" were discovered in the early 1800's including the cacodyl radical, C_2H_6As. In 1849 it was found that heating zinc with ethyl iodide in a sealed tube produced a gas which was thought to be "free ethyl". When measurement of molecular weight by the method of vapour densities was established, it was soon realized that groups such as methyl or ethyl did not persist in the free state but combined to form dimers.

The first stable free radical found, triphenylmethyl, was prepared in benzene in 1900. In 1911 tetraphenylhydrazine was shown to disociate into two identical stable free radicals.

By the 1920's free radicals were being proposed as intermediates in gas-phase reactions. In the 1930's the body of knowledge of free radical reactions began to increase rapidly as a result of studies of halogenation, oxidation, formation of polymers from vinyl chloride, addition of hydrogen bromide and of bisulfite to olefins, etc. My introduction to free radical chemistry was a course in photochemistry in 1939. Subsequently I spent most of the period 1943-1949 studying free radical reactions involving compounds of sulfur and phosphorus, as well as the reaction of O_2 with organic compounds. By the 1950's free radical chemistry was a well-established field of chemistry (36, 43).

Prodded in part by interest in the biological properties of some of the compounds I had synthesized, I entered medical school in the fall of 1949. Five years later, in July 1954, I was fortunate to have time to address the question of why all living things age and die. I had first become interested in this problem in December of 1945 after reading the article, "Tomorrow You Will be Younger", by William L. Laurence, science editor of the New York Times (34). This article was concerned with the work of Dr. Alexander A. Bogomolets of Kiev. An English translation of Dr. Bogomolet's book, The Prolongation of Life, was published in 1947 (2). The free radical theory of aging was formulated (9,10) in the first part of November, 1954, after four frustrating months in the library. This theory assumes that there is a single basic cause of aging, modified by genetic and environmental factors, and postulates that free radical reactions are involved in aging and disease. By chance this theory was put forth at the time when the rate of increase in average life expectancy at birth in the United states began to decrease; it is now 73.7 years and progressing slowly toward a plateau value of 74-76 years.

EARLY STUDIES

Direct validation of the free radical theory of aging did not seem feasible. Hence attempts were started to obtain support for the theory indirectly by determining if dietary changes designed to lower adverse endogenous free radical reactions would increase the life span.

The first study, published in 1957 (12) was encouraging. Addition of one of several free radical reaction inhibitors to the diet at a level of 0.5 - 1.0% by weight throughout life increased the average life span of C3H female mice and of AKR male mice. This study was fortunate as the effect would probably not have been observed at lower or higher dietary concentrations.

Studies prompted by the free radical hypothesis were summarized twice in the 1960's, in 1962 (14) and 1969 (18). Knowledge of free radical reactions in biological systems was still so meager in 1962 that information such as the following was included to support the probable presence of $HO\cdot$ and $HO_2\cdot$: "Investigation of the action of xanthine oxidase, with either sulfite oxidation or luminal used to detect free radicals, indicate that $HO\cdot$ and/or $HO_2\cdot$ are formed. The organic free radicals, detected by ESR arising in several dehydrogenase systems, in the action of peroxidase and H_2O_2 on a number of substrates, and in illuminated choroplast preparations, would be expected to react, to a greater or lesser degree, depending on the availability of oxygen and the resonance stability of the free radical, with oxygen with the formation of radicals such as $RO_2\cdot$ and $HO_2\cdot$." All of the available information was summarized with the statement, "The foregoing, taken as a whole, strongly indicates that chemically active free radicals of the nature of $HO\cdot$ and $HO_2\cdot$ are produced in living things in the course of normal metabolism".

By the early 1970's it had become evident that dietary measures designed to decrease endogenous free radical reaction levels in mice tended to increase the average life expectancy by as much as 20-30 percent, but had little, if any, effect on maximum life span. Consideration of such data resulted in the suggestion (22) in 1972, expanded in 1983 (27), that the mitochondria might serve as the "biologic clock"; slowing mitochondrial degradation may increase the maximum life span.

PRESENT SUPPORT

Support for the free radical theory of aging, summarized in part again in 1981 (26), now includes (27,28): 1) studies on the origin of life and evolution, 2) life span experiments in which adverse free radical reactions were expected to be lowered by dietary manipulations, 3) the plausible explanations it provides for aging phenomena, and 4) the growing number of studies which implicate free radical reactions in the pathogenesis of specific diseases.

ORIGIN OF LIFE AND EVOLUTION

Life apparently arose spontaneously (26) about 3.5 billion years ago from amino acids, nucleotides and other basic chemicals of living things produced from the simple, reduced, components of the primitive atmosphere by free radical reactions, which were initiated mainly by ionizing radiation from the sun. It seems likely that evolution was made possible initially by the constant presence of ionizing radiation, which served on the one hand to provide compounds in the environment necessary for the survival and growth of the first protocells and on the other to produce more-or-less random changes throughout the cells.

From the beginning, the evolution of more complex cells apparently occurred through the gradual selection and development of: a) defenses against deleterious chemical reactions (e.g., free radical reactions), and b) means to repair or replace cellular components (e.g., DNA, RNA, proteins) that were rendered defective by such adverse reactions.

Defenses that now help limit free radical damage include antioxidants, such as tocopherols and carotens, heme-containing peroxidases, the selenium-containing glutathione peroxidase, superoxide dismutases, and elevated serum uric acid levels.

The first process evolved to help restore altered DNA to its original form was probably excision repair, followed later by recombinational repair (28). The evolution of germ cells (cells capable of encoding for themselves as well as somatic cells, cells with similar basic functions on which are superimposed differences allowing for the growth and development of multicellular organisms suitable for the continuation and evolution of the germ cells) has been accompanied by the development of more complex systems, e.g., meiosis, for restoring the "purity" of the DNA in the zygote to the degree necessary for the formation of a "normal" member of the species while still allowing changes in DNA that might lead to further evolution of the germ cells and of the somatic cells encoded in them. The foregoing implies that evolution, driven almost entirely by energy from the sun, is the selection and improvement of measures to increase the probability of survival of germ cell DNA.

The original basic pattern of evolution does not appear to have changed except that the sun-initiated free radical reactions, which were essential for the origination and early evolution of the protocells, have largely been replaced by those derived from enzymatic and non-enzymatic reactions. Enzymatic reactions serving as sources of free radicals include those involved in the respiratory chain, in phagocytoses, in prostaglandin synthesis, and in the cytochrome P-450 system. Free radicals also arise in the non-enzymatic reactions of oxygen with organic compounds as well as those initiated by ionizing radiation. Changes attributed to free radical reactions include: 1) accumulative oxidative alterations in the long-lived molecules collagen, elastin, and chromosomal material; 2) breakdown of mucopolysaccarides through oxidative degradation; 3) accumulation of metabolically inert material such as ceroid and age pigment through oxidative polymerization reactions involving lipids, particularly polyunsaturated lipids, and proteins; 4) changes in membrane characteristics of such elements as mitochondria and lysosomes because of lipid peroxidation; and 5) arteriolocapillary fibrosis secondary to vessel injury by products resulting from peroxidation of serum and vessel-wall components.

The somatic and germ cells resulting from the "readout" of the zygote DNA all age and die except for those germ cells which are rejuvenated by entering into zygote formation. Presumably the cells age and die in much the same manner as the early protocells except that now the adverse free radical reactions, such as those listed above, arise largely from within the cells rather than as a result of ionizing radiation from the sun. Collectively the defenses that have evolved against changes that could have deleterious effects on germ cells, and hence on evolution, have permitted the life spans of multicellular organisms to increase by enabling somatic cells, particularly those critical to the existence of the organism as a whole (e.g., cells of the respiratory center or of the myocardium) to function longer.

DIETARY MANIPULATIONS

Dietary (23) manipulations expected to lower the rate of production of free radical reaction damage are also in accord with the possibility that more-or-less random damage produced by free radical reactions constitute the basic aging process; these include: 1) minimizing dietary components - such as copper and polyunsaturates, which tend to increase free radical reaction levels, 2) adding to the diet compounds able to inhibit free radical reaction-induced damage, e.g., 2-mercaptoethylamine (2-MEA), α-tocopherol, butylated hydroxytoluene (BHT), and 1,2-dihydro-6-ethoxy-2,2,4-trimethlquinoline (ethoxyquin).

For example, dietary antioxidants increase the life span (26) of mice, rats, fruit flies, nematodes, and rotifers, as well as the "life span" of neurospora. In the case of mice, addition of 1.0% (wt/wt) 2-mercaptoethylamine to the diet of male LAF_1 mice (17), starting shortly after weaning, increased the average life span of 30%; this increase is equivalent to raising the human life span from 73 to 95 years. Corresponding increases produced by 0.5% ethoxyquin in the diet of male and female C3H mice (4), were 18.1% and 20.0%, respectively. Although it has been relatively easy to increase the average life span of mice, the increases were not accompanied by any certain extension of maximum life spans.

AGING PHENOMENA

The free radical theory of aging is also supported by the plausible explanations it provides for aging phenomena, including the:

(a) Inverse relationship between the average life spans of mammalian species and their basal metabolic rates (27).
(b) Observation that antioxidants which increase the average life span of mice, depress body weight and fail to increase maximum life span (27).
(c) Clustering of degenerative diseases in the terminal part of the life span (27).
(d) Exponential nature of the mortality curve (27).
(e) Beneficial effect of caloric restriction on life span and degenerative diseases (27).
(f) Increase in autoimmune manifestations with age (25).
(g) Greater longevity of females (24).

Thus, for example, the greater longevity of females may be due, at least in part, to the greater protection of female embryos from free radical damage during a period (about 48 hours in the mouse) of both high mitotic and metabolic activity just prior to the random inactivation of one of the two functioning female X chromosomes in the late blastocyst state of development. The X chromosome codes for glucose-6-phosphate dehydrogenase, a key enzyme in the production of NADPH. NADPH acts to maintain glutathione in the reduced state. Glutathione, the major cellular sulphydryl compound, serves to minimize free radical damage to the organism by acting as a free radical reaction inhibitor and as a hydrogen donor for glutathione peroxidase in the reduction of hydrogen peroxide and hydroperoxides.

The increase in autoimmune manifestation with age is probably largely due to a disproportionate decrease in the radiosensitive T-suppressor cell function owing to the increasing levels of free radical reactions associated with advancing age.

Compounds that increase average life span, such as 2-MEA and ethoxyquin, tend to depress body weight and fail to increase maximum life. These effects are most likely mainly caused by decreased ATP production resulting from the interaction of the antioxidants with free radicals in the respiratory chain.

PATHOGENESIS OF THE "FREE RADICAL" DISEASES

The free radical theory of aging postulates that free radical reactions, arising largely in the course of normal metabolism, are responsible for the progressive accumulation of the changes with time associated with or responsible for the ever-increasing likelihood of disease and death which accompanies advancing age. The ubiquitous free radical reactions would be expected to produce the progressive changes throughout the body in a more-or-less common pattern, i.e., the normal aging pattern, on which are superimposed patterns that differ from individual to individual owing to genetic differences - it is reasonable to expect that genes responsible for free radical reactions and those associated with defenses against free radicals have evolved unevenly - as well as to the differences in the environment that modulate free radical reaction damage. The superimposed patterns of change may become progressively more discernable with time and in some individuals eventually be recognized as diseases. In accord with the foregoing there are a growing number of diseases in which free radical reactions are now believed to be implicated. These "free radical" diseases may be classified into three broad groups in which a given disorder is mainly due to: 1) genetics, 2) a combination of genetic and environmental factors, and 3) largely to environmental influences. The autosomal recessive disorders, Fanconi's anemia (30) and the Bloom syndrome (6) are examples of the first group. These diseases are apparently caused by genetically determined inadequate protection from oxygen radicals. The second group is represented by systemic lupus erythematosis (SLE). The basic defect in SLE (25) seems to be a genetically determined increased sensitivity of the nuclear DNA of one or more cell types to free radical damage. Whether or not this defect is expressed, and to what degree, would seen to be determined by both genetic and environmental factors, e.g., the amount of polyunsaturated fats in the diet, that affect the level of endogenous free radical reactions. Cancer and cardiovascular diseases are examples of group 3. Unlike SLE, where a genetically determined defect must exist in an individual if the disorder is to be made manifest, in the majority of cases of cardiovascular disease and in many cancer cases there does not seem to be a significant degree of genetic susceptibility. Disease development depends in large part on environmental influences, e.g., the presence of carcinogens in the case of cancer and of increased intake of dietary lipids in atherosclerosis.

In addition, more subtle free radical induced change may cause death. Thus, excessive cellular accumulation of lipofuscin or melanin may kill and/or interfere with the function of cells critical to the organism as a whole, such as some groups in the CNS. The foregoing may account, at least in part, for the observation that autopsies on individuals 85 years and older fail to find a cause of death in about one-third of the cases (32).

The above discussion indicates that the relationship between aging and diseases in which free radical reactions are involved is a direct one. Modulation of the normal distribution of deleterious free radical reaction-induced changes throughout the body by genetic and environmental differences between individuals, results in patterns of changes in some sufficiently different from the normal aging pattern to be recognized as disease. The growing number of such "free radical diseases" includes (28) the two major causes of death, cancer and atherosclerosis as well as the other common degenerative diseases, essential hypertension, senile dementia of the Alzheimer type, amyloidosis, immune deficiency of age, osteoarthritis, senile macular degeneration and Parkinson's disease. Additional disorders in which free radical reactions are implicated (in addition to Fanconi's anemia, the Bloom syndrome and SLE mentioned above) include Batten's disease, insulin dependent diabetes mellitus, preeclamsia, Huntington's chorea, xeroderma pigmentosum, and ataxia telangiectasia. The latter two diseases, both autosomal recessive disorders, are due to defective repair of DNA, presumably damaged, at least in part, by endogenous free radical reactions.

SPECIFIC "FREE RADICAL" DISEASES

Some of the data implicating free radical reactions in cancer and atherosclerosis, the two leading causes of death are presented briefly below.

Cancer

Cancer initiation and promotion (40) is associated with chromosomal defects (47) and oncogene activation (8). Ionizing radiation is a "complete carcinogen", being both an initiator and promoter (40). It is a reasonable possibility that some endogenous free radical reactions, like those initiated by ionizing radiation, will result in tumor formation (10,44) by serving as a continuous source of tumor initiators and promoters. Support for this possibility includes studies (28) of the effect on cancer induction of dietary fat (5,20), antioxidants (13,21,46) and increasing age (45). The parallelism between cancer incidence and age is probably due, at least in part, to the increasing level of endogenous free radical reactions with age (27) coupled with the apparently progressively diminishing capacity of the immune system to eliminate the altered cells.

Such studies indicate that the incidence of human cancer may be lowered by antioxidant supplemented diets whose components have been selected to minimize adverse free radical reactions in the cells and tissues.

Atherosclerosis

Atherosclerosis is the major cause of death in developed countries. Atherosclerosis research has been concerned largely with lipids and their metabolism (38) and with thrombosis (1), and to a lesser extent with the role of the arterial wall

(1,11,15,16,41). Intervention studies have not been promising (37).

Atherosclerotic lesions tend to form in areas of the vascular tree subject to injury (15,16) - usually mechanical injury. The lesions can be initiated and enchanced by substances capable of irritating the arterial wall (11,15,16). A possible constant source of irritating compounds is the reaction of molecular oxygen with the polyunsaturated compounds present in serum and arterial-wall lipids (11). The readily oxidized polyunsaturated substances comprise about 30 percent of the total fatty acids present, mainly as esters, in the lipids of both serum (39) and atherosclerotic plaques (3). Hence the oxidation products, including peroxides and compounds of higher molecular weight formed through oxidative polymerization, as well as substances arising from the reaction of intermediate lipid-free radicals with proteins and other substances, may be produced in amounts large enough to significantly contribute, directly or indirectly, to atherogenesis. In addition, oxy-radicals arising from leucocytes (35) may also serve as an important source of vessel wall irritants.

Many observations (23,28) now support the possibility that lipid peroxidation is involved in atherogenesis including the recent observation (29) that oxidized serum low density lipoproteins, the lipoprotein fraction positively associated with atherosclerosis is significantly more toxic to cells in culture than the serum high density lipoprotein, the fractions that seemingly serves to "protect" from atherosclerosis.

Taken as a whole, studies on the pathogenesis of atherosclerosis are compatible with the possibility that the disease is basically due to free radical reactions largely involving dietary-derived lipids in the arterial wall and serum to yield peroxides and other substances. The peroxides should increase platelet aggregation and, along with the other peroxidation products, serve to initiate and sustain an inflammatory reaction in the vessel wall that in turn interacts with serum derived lipids. Increases in serum lipid levels due to diet or other causes, e.g., LDL receptor deficiency (7), should enhance the inflammatory response, as should advancing age because of the accompanying higher levels of endogenous free radical reactions. Assuming that the foregoing suggested pathogenesis of atherosclerosis is correct, attempts to modify the degree of atherosclerosis by diet during the relatively short time periods of experimental studies might be expected to be difficult. The rapid increase in hydroperoxides in atherosclerotic plaques after death (31) attests to the ready peroxidizability of plaque components while the fact that the hydroperoxide content is essentially zero during life (31) suggests that glutathione peroxidase along with glutathione and other reducing agents serve to keep free radical reaction chain lengths short. Hence, the effects of diets formulated to increase the rate of peroxidation whould be masked by the normal high rate of endogenous peroxidation while attempts to decrease free radical chain lengths should have little effect

since they are already short. In agreement with the foregoing
are the inconclusive results obtained in a 20 month study (19)
designed to modulate the rate of atherogenesis in minipigs by
diets formulated to change the level of free radical reactions
in the serum (the diets contained varying amounts of polyunsaturated fats and of copper sulfate). Likewise in agreement
is the association between ischemic heart disease and low serum
selenium concentrations (42).

The above discussion suggests that the incidence of clinical
cardiovascular disease may be decreased by a low fat diet supplemented with vitamin E and possibly other antioxidants.

COMMENT

In the 29 years since the free radical theory was first proposed, in November 1954, considerable data have been accumulated
which demonstrate that free radical reactions do contribute to
the degradation of biological systems, and a beginning has been
made with the aid of the rapidly increasing general knowledge
of free radical reactions in biology to determine the mechanisms
by which the changes are produced.

Today it seems very likely that the assumption that there
is a basic cause of aging is correct and that the sum of the
deleterious free radical reactions going on continuously throughout the cells and tissues is the aging process or a major contributor to it.

Although all will not agree with this assessment, it is apparent that the free radical theory of aging can serve as a useful
guide for efforts to increase the healthy life span. This is
not to say that other approaches to increasing the functional
life span may not be found in the future, but only that it now
seems feasible to augment it to some degree by interfering with
what appears to be the natural process which leads to death.
Future studies directed to this end should include, in particular,
a search for free radical reaction inhibitors that have only
a slight depressing effect on ATP production at levels which
significantly inhibit adverse free radical reactions elsewhere
in the body.

SUMMARY

The term "radical" comes from the early days of chemistry.
In the 1920's "radicals" were first proposed as intermediates,
i.e., free radicals, in some chemical reaction. By the 1950's
free radical chemistry had become a well established field of
chemistry. In 1954 this class of chemical reaciton was implicated in aging and disease. The first direct experimental support
of the free radical theory of aging was published in 1957. In
1972 the hypothesis was extended by the suggestion that the mitochondria might serve as the "biologic clock". A summary paper
was published in 1981. By this time it had become clear, beyond
reasonable doubt, that the sum of the deleterious free radical

reactions going on continuously throughout the cells and tissues, was the aging process or a major contributor to it.

The relationship between aging and diseases involving free radical reactions seem to be a direct one. Such diseases can be attributed to modulation of the normal pattern of distribution of deleterious changes, produced by free radical reactions throughout the body, by genetic and environmental differences. The number of diseases in which free radical reactions are recognized to be involved is increasing and includes the two major causes of death, cancer, and atherosclerosis.

It is reasonable to expect on the basis of present data alone that the healthy life span can be increased by 5-10 or more years through a judicious selection of diets and antioxidant supplements.

REFERENCES

1. Benditt, E. P., and Gown, A. M. (1980): Atheroma: the artery wall and the environment. In: International Review of Experimental Pathology, Vol. 21, edited by Richter, G. W., and Eptstein, M. A., pp. 55-118. New York, Academic Press.
2. Bogomolets, A. A. (1947): The Prolongation of Life. Duell, Sloan, and Pearce, New York.
3. Bottcher, C. J. F., Boelsum-Van Houte, E., Ter Haar Romeny-Wachter, C. C., Woodford, F. P., and Van Gent, C. M. (1960): Lipid and fatty acid composition of coronary and cerebral arteries at different stages of atherosclerosis. Lancet, 2:1162-1166.
4. Comfort, A. (1971): Effect of ethoxyquin on the Longevity of C3H mice. Nature, 229:254-255.
5. Editorial (1982): Obesity: The cancer connection. Lancet, 1:1223-1224.
6. Emerit, I., and Cerutti P. (1981): Clastogenic activity from Bloom syndrome fibroblast cultures. Proc. Natl. Acad. Sci., U.S.A., 78:1868-1872.
7. Goldstein, J. L., Kita, T., and Brown, M. S. (1983): Defective lipoprotein receptors and atherosclerosis. New Engl. J. Med., 309:288-296.
8. Hamlyn, P., and Sikora, K. (1983): Oncogenes. Lancet, 1:326-330.
9. Harman, D. (1955): Aging: A theory based on free radical and radiation chemistry. Univ. Cal. Rad. Lab. Report No. 3078, July 14.
10. Harman, D. (1956): Aging: A theory based on free radical and radiation chemistry. J. Gerontol., 11:298-300.
11. Harman, D. (1957): Atherosclerosis: hypothesis concerning the initiating steps in pathogenesis. J. Gerontol., 12:199-202.
12. Harman, D. (1957): Prolongation of the normal life span by radiation protection chemicals. J. Gerontol., 12:257-263.

13. Harman, D. (1961): Prolongation of the normal lifespan and inhibition of spontaneous cancer by antioxidants. J. Gerontol., 16:247-254.
14. Harman, D. (1962): Role of free radical in mutation, cancer, aging, and the maintenance of life. Rad. Res., 16: 753-763.
15. Harman, D. (1962): Atherosclerosis: Inhibiting effect of an antihistaminic drug, chlorpheniramine. Circulation Res., 11:277-282.
16. Harman, D. (1962): Atherosclerosis: effect of the rate of growth. Circulation Res., 10:851-852.
17. Harman, D. (1968): Free radical theory of aging: Effect of free radical reaction inhibitors on the mortality rate of male LAF_1 mice. J. Gerontol., 23:476-482.
18. Harman, D. (1969): Prolongation of life: Role of free radical reactions in aging. J. Amer. Geriatrics Soc., 17: 721-735.
19. Harman, D. (1970): Atherogenesis in minipigs: Effect of dietary fat unsaturation and of copper. In: Atherosclerosis: Proceedings of the Second International Symposium, edited by R. J. Jones, pp. 472-475. Springer-Verlag, New York.
20. Harman, D. (1971): Free radical theory of aging: Effect of the amount and degree of unsaturation of dietary fat on mortality rate. J. Gerontol., 26:451-457.
21. Harman, D. (1972): Free radical theory of aging: Effect of vitamin E on tumor incidence. Gerontologist, 12(3): Part 2, 33 (abstr.)
22. Harman, D. (1972): The biologic clock: The mitochondria? J. Amer. Geriatrics Soc., 20:145-147.
23. Harman, D. (1978): Free radical theory of aging: Nutritional implications. Age, 1:145-152.
24. Harman, D. (1979): Free radical theory of aging: Beneficial effect of adding antioxidants to the maternal mouse diet on life span of offspring; possible explanation of the sex differences in longevity. Age, 2:109-122.
25. Harman, D. (1980): Free radical theory of aging: Beneficial effect of antioxidants on the life span of male NAB mice; role of free radical reactions in the deterioration of the immune system with age and in the pathogenesis of systemic lupus erythematosus. Age, 3:64-73.
26. Harman, D. (1981): The aging process. Proc. Natl. Acad. Sci., U.S.A., 78:7124-7128.
27. Harman, D. (1983): Free radical theory of aging: Consequences of mitochondrial aging. Age, 6:86-94.
28. Harman, D. (1984): Role of free radicals in aging and disease. In: Relations Between Normal Aging and Disease, edited by H. A. Johnson. Raven Press, New York (In press).
29. Hessler, J. R., Morel, D. W., Lewis, L. J., and Chisolm, G. M. (1983): Lipoprotein oxidation and lipoprotein-induced cytotoxicity. Atherosclerosis, 3:215-222.

30. Joenje, H., Arwert, F., Eriksson, A. W., deKoning, H., and Oostra, A. B. (1981): Oxygen-dependence of chromosomal aberrations in Fanconi's anemia. Nature, 290:142-143.
31. Johnson, R. J. (1966): Atherosclerosis: The possible role of peroxide. Ph.D. Thesis, University of Nebraska College of Medicine, Omaha, Nebraska.
32. Kohn, R. R. (1982): Cause of death in very old people. JAMA, 247:2793-2797.
33. Lavoisier, A. L. (1965): Traite' Elementaire de Chimie, Vol. 1, Paris, Cuchet, 1789, p. 293. Translation by R. Kerr, p. 66. Reprinted, Dover Press, New York.
34. Laurence, W. L. (1946): Tomorrow you may be younger. Ladies Home J., 62:22.
35. Ludwig, P. W., Hunninghake, D. B., and Hoidal, J. R. (1982): Increased leucocyte oxidative metabolism in hyperlipoproteinanemia. Lancet, 2:348-350.
36. Lundberg, W. O. editor (1961): Autoxidation and Antioxidants Vols. 1 and 2. Interscience Publ., New York.
37. Mann, G. V. (1977): Diet-heart: End of an era. New Engl. J. Med., 297:644-650.
38. Miller, N. E., and Lewis, B., editors (1981): Lipoproteins, Atherosclerosis, and Coronary Heart Disease, Elsevier/North Holland, New York.
39. Patil, V. S., and Magar, N. G. (1960): Effect of dietary fat intake and age on polyunsaturated fatty acids in human blood serum. Biochem. J., 76:417-420.
40. Pitot, H. C. (1982): The natural history of neoplastic development: The relation of experimental models to human cancer. Cancer, 49:1206-1211.
41. Ross, R. (1981): Atheorsclerosis: A problem of the biology of arterial wall cells and their interactions with blood components. Atherosclerosis, 1:293-311.
42. Salonen, J. T., Alfthan, G., Huttunen, J. K., Pikkarainen, J., and Puska, P. (1982): Association between cardiovascular death and myocardial infarction and serum selenium in a matched-pair longitudinal study. Lancet, 1:175-179.
43. Steacie, E. W. R. (1946): Atomic and Free Radical Reactions: The Kinetics of Gas-Phase Reactions Involving Atoms and Organic Radicals. Reinhold Publ. Corp., New York.
44. Totter, J. R. (1980): Spontaneous cancer and its possible relationship to oxygen metabolism. Proc. Natl. Acad. Sci. U.S.A., 77:1763-1767.
45. Upton, A. C. (1977): Pathobiology. In: The Biology of Aging, edited by Finch, C. E. and Hayflick, L., pp. 513-535. Van Nostrand Reinhold, New York.
46. Wattenberg, L. L. (1980): Inhibition of chemical carcinogenesis by antioxidants. In: Carcinogenesis, Vol. 5, Modifers of Chemical Carcinogenesis, edited by Slaga, T. J., Raven Press, New York.
47. Yunis, J. J. (1983): The chromosomal basis of human neoplasia. Science, 221:227-236.

Free Radicals in Autoxidation and in Aging

William A. Pryor

Departments of Chemistry and Biochemistry, Louisiana State University, Baton Rouge, Louisiana 70803

PART I. Kinetics of the autoxidation of linoleic acid in SDS micelles: Calculations of radical concentrations, kinetic chain lengths, and the effects of vitamin E.

PART II. The role of radicals in chronic human diseases and in aging.

My task in this presentation will be two-fold. Firstly, I have been asked to tell you something about the nature of free radicals and the reactions that they undergo. My aim here will be to set the stage: I will describe the nature of the actors that the following speakers will enrole in life-and-death dramas of biological significance. Secondly, I will present a hypothesis for the life-limiting role that radicals can exert in mammals, and particularly in humans.

PART I

ELECTRON-PAIR CHEMICAL BONDS, THE DEFINITION OF FREE RADICALS, ELECTRON SPIN RESONANCE, AND OTHER "ODD" TOPICS

Consider the molecule R-H (where R is a part of an organic molecule) and H is one hydrogen atom that is shown explicitly. The line between R and H represents the chemical bond, and (almost) all chemical bonds consist of two electrons. "Why two?" The reason for this is that electrons have "spin".

Electrons prefer to occur in *pairs* such that one electron has its spin pointing "up" and one "down" ("antiparallel"), so a pair of electrons has no *net* spin. This is energetically favorable, and therefore electrons *pair up* in bonds and there are two electrons in all chemical bonds.* Thus, a free radical with its unpaired electron is odd, in many senses of the word.

---
*This statement, like *all* statements that contain the word "all", does have exceptions.

Two-electron bonds can break in two ways: homolytically (that is symmetrically) with one electron going with each partner, as shown in eq 1.

$$R-H \longrightarrow R\cdot + H\cdot \qquad (1)$$

(The "dots" represent electrons from the bond that broke.) Bonds also can break heterolytically, with both electrons in the bond going with one partner:

$$R-H \longrightarrow R:^- + H^+ \qquad (2)$$

Species formed by homolytic processes (e.g., eq 1 above) are called "free radicals" and have an unpaired electron; because of this odd electron, free radicals have a *net* spin and spin is a vector with an associated direction. When a free radical is placed in a magnetic field, the spin vector for the odd electron and its associated spin magnetic moment can be aligned either with or against the field, and these two alignments differ slightly in energy. In a field of about 3600 gauss, this transition occurs at a wavelength of 3 cm, or about 1 cal for most organic radicals. When this transition occurs, the energy change due to the electron "flipping" its spin direction in the magnetic field can be measured by an electron spin resonance (ESR) spectrometer. Since only free radicals have odd electrons, ESR detects free radicals specifically and with great sensitivity.

Most small free radicals are quite reactive, and with this background we can understand why. Electrons want to pair, so a free radical, with its odd number of electrons, tends to collide with other molecules and to try to pair up its electron. As long as an odd number of electrons goes into a reaction, an odd number of electrons, and therefore another radical, must come out of the reaction. So whenever a free radical reacts with a non-radical, other free radicals are produced. These reactions are called propagation reactions, and radicals keep reacting until two free radicals find one another and collide--that's called a termination--and a 2-electron bond is remade.

THE REACTIVITY OF FREE RADICALS

I have said that most radicals are quite reactive; just how reactive are they? Table I compares the halflife of the hydroxyl radical with that of the hydroxyl ion, and the contrast is quite remarkable. The hydroxyl radical abstracts hydrogen atoms from a moderately good donor like linoleate with a rate constant of 10^9 liter mole^{-1} sec^{-1} (M^{-1} sec^{-1}).

(Hydroxyl radicals are so reactive that they react even with some poorer donors with about this same rate constant.) I have assumed a concentration of substrate of 1 M in Table I, so the apparent, pseudo-unimolecular rate constant for the HO·/linoleate reaction is then 10^9 sec^{-1}, giving a half life of 7×10^{-10} sec. (All the calculations in Table I are for reaction at about 37°C.) In contrast, the hydroxyl ion reacts with typical substrates (such as isopropyl bromide) with rate constants of about 10^{-5} M^{-1} sec^{-1}, giving a pseudo-unimolecular rate constant (again for 1 M substrate) of 10^5 sec^{-1} and a half life of about 7×10^4 sec. Thus, the hydroxyl radical is about 10^{14} times more reactive than is the hydroxyl ion, a truly mind-boggling difference!

However, not all free radicals are as reactive as is the hydroxyl radical; in fact some have reasonably long lifetimes. For example, Table I shows that the alkoxyl radical would react with 1 M linoleate with a halflife of about 7×10^{-7} sec, still quite short lived but 10^3 less so than is the hydroxyl radical. The peroxyl radical under the same conditions would have a lifetime of 7×10^{-3} sec, or 10^7 times longer than the hydroxyl radical.

Thus, while the hydroxyl radical is so reactive that it diffuses only one or two molecular diameters before it reacts, the peroxyl radical can be expected to diffuse freely in a biological mileau. Obviously, it is this type of radical that might cause extensive biological damage even some distance from the site at which it is generated.

Some radicals are even longer lived. For example, the radical in cigarette tar is a quinone-hydroquinone charge transfer complex and has a lifetime of days (72). Despite this long lifetime, we have shown that tar binds to DNA <u>in vitro</u> and might cause biological damage (82).*

So the moral is: Free radicals, and especially small free radicals, are generally reactive. But their reactivity varies widely, over many powers of ten. Some have only a fleeting existence and react very close to where they are produced; some can diffuse long distances or even live for days and still react with appropriate substrates and cause biological damage. Thus, radicals are reactive, but they are not monotonously so and they differ remarkably in their lifetimes and reactivities.

---

*We do not yet know if the paramagnetic groups are responsible for the binding or other as yet unidentified functional groups.

Table I. Lifetimes of the hydroxyl radical, the hydroxyl ion, the alkoxyl radical, and the peroxyl radical.

Species	Halflife at 37°C for reaction with a typical substrate (1M).	Ref
HO·	7×10^{-10} sec	a
HO⁻	7×10^{4} sec	b
RO·	7×10^{-7} sec	c
ROO·	7×10^{-3} sec	d
Cigarette tar radical	2 days	e

Bimolecular rate constants (in M^{-1} sec^{-1}) and the model substrate used in the halflife calculation are: (a) 10^9; linoleate, (b) 10^{-5}; isopropyl bromide, (c) 10^6; cyclohexene, (d) 10^2; linoleate. (e) See ref. 72. The rate constants for RO· and ROO· are from ref 43.

AUTOXIDATION AS AN EXAMPLE OF A FREE RADICAL CHAIN REACTION. USE OF LINOLEIC ACID MICELLES AS A TEST SYSTEM FOR ANTIOXIDANTS.

Let's consider autoxidation as an example of a system in which radical reactions occur. To start the reaction, an *initiation* process must form free radicals from even-electron species. (We will consider how radicals might form in vivo below.) For example, consider the decomposition of an azo compound into two alkyl radicals as shown in eq 3.

$$R-N=N-R \longrightarrow 2R· + N_2 \quad (3)$$

These R· radicals rapidly react with oxygen in an aerated system, and a peroxyl radical is formed, eq 4. This peroxyl

$$R· + O_2 \longrightarrow ROO· \quad (4)$$

radical can then abstract reactive hydrogen atoms from the substance undergoing autoxidation.

In order to make my discussion specific, and so I can give actual rate constants and concentrations of free radicals, I will illustrate the situation that occurs when linoleic acid undergoes autoxidation in SDS micelles. This is a system my group is currently using to assess relative antioxidant potentials of vitamin E, BHT, and new synthetic antioxidants and anti-inflammatory agents (70). To perform these calculations, I have leaned on texts (64,98) and excellent reviews of autoxidation by Mill and Hendry (57) and Howard (43), publications by Ingold and coworkers (2,15) and our own data (70). When linoleic acid undergoes autoxidation, the

ROO· radicals formed in eq 4 attack linoleic acid as shown in eq 5*. The linoleyl radical, L·, is of course, the

$$ROO· + LH \xrightarrow{} ROOH + L· \quad (5)$$

conjugated dienyl radical. Equations 3-5 constitute the *initiation sequence*.

The carbon-centered radical, L·, reacts with oxygen to become peroxidized, eq 6.

$$L· + O_2 \xrightarrow{k_o} LOO· \quad (6)$$

These peroxyl radicals then attack another molecule of linoleic acid to abstract an allylic hydrogen and produce the conjugated diene hydroperoxide, LOOH, eq 7.

$$LOO· + LH \xrightarrow{k_p} LOOH + L· \quad (7)$$

Equations 6 and 7 are the *propagation sequence*: note that they constitute a chain. Each primordial radical R· initiates a chain of reactions 6-7, and the ratio of the number of product molecules produced (LOOH) to primordial radicals that initiate the chain is called the kinetic chain length, ν.

Equations like 6 and 7 are called *propagation reactions* since they propagate the chain; as long as equations like these occur, the number of radicals is conserved and the reaction will keep going until the substrate is used up. However, radical chains are stopped by reactions called terminations. In the absence of an antioxidant, termination occurs by collision of any two of the radicals involved.

$$2LOO· \xrightarrow{k_t} \text{non-radical products (NRP)} \quad (8)$$
$$LOO· + L· \xrightarrow{} NRP \quad (9)$$
$$2L· \xrightarrow{} NRP \quad (10)$$

For autoxidations conducted under 1 atmosphere of air, the ratio of peroxyl to alkyl radicals is very high, so eq 8 is the only important termination (p. 291 in ref 64).

THE RATE CONSTANTS AND CONCENTRATIONS OF FREE RADICALS IN THE AUTOXIDATION OF LINOLEIC ACID IN MICELLES

A steady-state analysis can now be performed for the autoxidation of linoleic acid. The rate of initiation is given by eqs 3-5 and can be simply written as R_i. At the steady-state,

---

*We will abbreviate linoleic acid as LH, where the H is one of the reactive, doubly allylic hydrogens.

the rate of initiation must equal the rate at which termination occurs; otherwise the process either would stop or would continue to increase in rate and eventually explode! Thus, we can write eq 11. (The factor of two occurs since each termination destroys

$$R_i = 2k_t [LOO\cdot]^2 \qquad (11)$$

two free radicals, and rate constants are written on a per-radical basis by convention.) Using eq 11 and the values of the rate constants given in Table II, we can solve for the concentration* of the main chain-carrying species, the peroxyl radical. In our studies, $R_i = 3 \times 10^{-7}$ M s^{-1}, a typical value for an in vitro autoxidation. Therefore, we obtain 2×10^{-7} M as the steady-state concentration of the peroxyl radical, eq 12.

$$[LOO\cdot] = (R_i/2k_t)^{0.5} = 2 \times 10^{-7} \text{ M} \qquad (12)$$

This value for the concentration of LOO· is just at the borderline of detectability of most ESR spectrometers; thus, only in special cases can a standing concentration of peroxyl radicals be observed in autoxidations (75). (Oxygen, if present, also broadens the signal and makes it more difficult to observe peroxyl radicals.)

We also can now calculate the concentration of L·, the carbon-centered radicals. Another steady-state relationship is that the two chain reactions must occur at the same rate; that is, the chain consists of eq 6 and eq 7 occurring alternatively, so each time one occurs the other then follows. That means that the material passed through these steps must be equal and their rates must be equal.

---

*Our system uses 14% linoleic acid in an SDS micelle. This percentage of linoleic acid roughly equals that found in a biological membrane. However, micelles are only about 30Å in diameter whereas cells are many orders of magnitude larger. The advantage of micelles is that diffusion into them is so rapid that initiator and inhibitors can be injected into the bulk aqueous phase and they equilibrate into the linoleic micelle layer too fast for any delay to be observed in our kinetic traces. (See Figure 1 below and ref. 97a.)

In this article I have calculated concentrations of linoleic acid, vitamin E, the initiator, and O_2 *in the micelle*, since this is where the autoxidation occurs. Thus, the concentrations quoted in the text are 100 times larger than the values that would be obtained if the solution were assumed to be a single homogeneous phase and average concentrations were used. Similarly, therefore, the value of R_i I have used is 100 times larger than the value calculated from the rate of disappearance of the initiator averaged over the entire solution.

Table II. Rate constants adopted for kinetic calculations on the autoxidation of linoleic acid at 37°C in SDS micelles.

eq	Symbol	Value	Notes
6	k_o	2×10^9 M^{-1} sec^{-1}	Taken from ref 49a. I have neglected the reverse of reaction 6.
7	k_p	62 $M^{-1} sec^{-1}$	These are the values for homogeneous solution at 30°C (p. 92 of ref 43). The values at 37°C are expected to be similar within the accuracy of these illustrative calculations. However, k_t in a micelle may be smaller than the homogeneous value since diffusion from the micelle may be rate limiting (42a).
8	$2k_t$	9×10^6 $M^{-1} sec^{-1}$	
16	$\dfrac{k_p}{(2k_t)^{0.5}}$	2×10^{-2} $M^{-0.5}$ $sec^{-0.5}$	
3-5	R_i	3×10^{-7} $M\ sec^{-1}$	The initiation rate in the micelle, for the system described here (and in 70).
7	[LH]	0.62 M	Linoleic acid molarity in the micelle, assuming the reagent associates entirely with the lipid phase (70).
17	[InH]	5×10^{-4} M	Ditto for α-tocopherol (70).
17	k_{inh}	2×10^5 M^{-1} sec^{-1}	For α-tocopherol as the inhibitor; data obtained in ref 70.

$$k_p[LOO\cdot][LH] = k_o[L\cdot][O_2] \tag{13}$$

The value of k_o is 2×10^9 (49a), so the $L\cdot$ concentration is given by eq 14, where 0.63 M is the concentration of linoleic acid in our micelles and 1×10^{-3} M is taken as the initial oxygen concentration in the oil phase of the micelle.

$$[L\cdot] = \frac{(62)(2 \times 10^{-7})(0.63)}{(2 \times 10^9)(10^{-3})} = 4 \times 10^{-12}\ M \tag{14}$$

We see that the $LOO\cdot$ is much greater than the $L\cdot$ concentration.

We also can calculate the kinetic chain length; this is equal to the rate of the autoxidation process (and thus to the rate of either propagation step) divided by the rate of primary radical production, eq 15.

$$v = \frac{k_p[LOO\cdot][LH]}{R_i} = \frac{(62)(2 \times 10^{-7})(0.63)}{3 \times 10^{-7}} = 26 \qquad (15)$$

Thus, 26 molecules of linoleic acid undergo autoxidation when a single free radical is introduced into this model membrane system.* That much damage might well be enough to destroy the membrane and produce cell lysis and death; however, we must remember that in the real system, the polyunsaturated fatty acids (PUFA) would be protected by antioxidants such as vitamin E.

THE OXIDIZABILITY OF OLEFINS

It is useful to have an expression for the oxidizability of a given olefin. The rate of oxidation is equal to the rate of use of oxygen. If the chain length is long, oxygen is mainly used in eq 6, and since eqs 6 and 7 occur at equal rates (eq 13), we can write:

$$\frac{-d[O_2]}{dt} \equiv R_{oxi} \cong k_p[LH][LOO\cdot]$$

If we substitute the expression for [LOO·] taken from eq 12, we get eq 16.

$$R_{oxi} \cong \frac{k_p}{(2k_t)^{0.5}}[LH](R_i)^{1/2} \qquad (16)$$

The term $k_p/(2k_t)^{0.5}$ in eq 16 expresses the rate of product formation (at a given rate of initiation and substrate concentration) and is therefore called the "oxidizability" of the olefin. Table III gives some values of this parameter for simple olefins, dienes and polyenes. Two things should be pointed out: The first is the square root dependence in eq 16 on the rate of initiation and on the value of k_t; this square root dependence is typical for free radical processes and results from the fact that radicals are produced in a first order reaction (eq 3 in our scheme) and destroyed in a reaction that is second order (eq 8). Secondly, note that it is very easy to obtain values of the oxidizability; it only requires measuring the rate of oxygen usage or product production at a known rate of radical formation (initiation). However, obtaining the individual values of k_p and k_t takes

---

*Remember these calculations are only approximate and meant to be merely illustrative.

considerably more effort. I will not discuss the ways in which this can be done, since it has been reviewed (4,43,57,64,98). Table III also shows the values of k_p and k_t; note that the oxidizability often does not parallel values of k_p, since k_t may vary.

TABLE III. Values of $k_p/(2k_t)^{0.5}$ as well as k_p and $2k_t$ for several olefins, dienes, and polyenes at 30°C (43). (Rate constants in M^{-1} sec^{-1}.)

Substrate	$k_p/(2k_t)^{0.5}$ x 10^3	k_p	$2k_t$ x 10^{-6}
Styrene	6	–	–
1,4-Dihydronaphthalene	35	–	–
1-Octene	0.062	1.	260.
1,4-Pentadiene	0.42	14.	1080.
3-Heptene	0.54	1.4	6.4
Methyl oleate	0.89	0.92	1.1
Methyl linoleate	21.	62.	8.8
Methyl linolenate	39.	234.	36.

INHIBITED AUTOXIDATION

Inhibitors such as α-tocopherol are effective antioxidants because they rapidly trap peroxyl radicals to give a stabilized radical that does not continue the chain, eq 17. In fact, in the case of vitamin E,

$$LOO \cdot + InH \xrightarrow{k_{inh}} LOOH + In \cdot \qquad (17)$$

the inhibitor radical that is produced (In·) reacts with a second peroxyl radical to form non-radical products (NRP).

$$LOO \cdot + In \cdot \longrightarrow NRP \qquad (18)$$

(Inhibitors like this are said to have a stoichiometric factor of 2; that is 2 radicals are stopped per molecule of inhibitor.) We can calculate the fraction of the peroxyl radicals that undergo reaction 7 and continue the chain versus those that react with tocopherol, eq 17, ultimately to terminate the autoxidation. In this calculation (eq 19), the concentration of peroxyl radicals cancels out.* We will use the

---
*The factor of 2 occurs in eq 19 since each inhibitor stops 2 kinetic chains.

concentration of tocopherol that we use in our micelle studies
and the value of k_{inh} that we measure for tocopherol (70).

$$\frac{2R_{17}}{R_7} = \frac{2k_{inh}}{k_p} \frac{[LOO\cdot][InH]}{[LOO\cdot][LH]} = \frac{(2)(2 \times 10^5)(5 \times 10^{-4})}{(62)(0.63)} = 5 \quad (19)$$

Thus, even when there is much more linoleic acid than
tocopherol, 5/6 = 83% of the peroxyl radicals are stopped by
the inhibitor and the chain reaction is greatly slowed, as
shown in Figure 1.

We can now calculate the concentration of the peroxyl
radicals in this inhibited autoxidation. The principal
termination reaction is now reaction with vitamin E, rather
than reaction 8 as had been previously true. Therefore, we can
write eqs 20 and 21, giving the new concentration of peroxyl
radicals as 2×10^{-9}. Thus, the inhibitor acts to keep

$$R_i = R_{inh} = nk_{inh}[LOO\cdot][Inh] \quad (20)$$

$$[LOO\cdot] = \frac{3 \times 10^{-7}}{(2)(2 \times 10^5)(5 \times 10^{-4})} = 2 \times 10^{-9} \text{ M} \quad (21)$$

the peroxyl radical concentration about 100-fold lower than it
was in the uninhibited autoxidations. (Compare eqs 12 and
21.) We also can now calculate the new kinetic chain length.
The formula is given in eq 15, and we only need to supply the
new concentration of the peroxyl radical. This is done in eq 22.

$$\nu = \frac{(62)(2 \times 10^{-9})(0.63)}{(3 \times 10^{-7})} = 0.3 \quad (22)$$

Thus, when the inhibitor is present the chain length is very
small.

THE KINETICS OF THE INHIBITED AUTOXIDATION

I have used a value of k_{inh}, the rate constant for
reaction 17, of 2×10^5 M^{-1} sec^{-1} in the
calculations above. How was this value determined? Figure 1
shows a plot of oxygen concentration (determined using an
oxygen electrode) versus time when linoleic acid undergoes
autoxidation in SDS micelles at 37°C and with α-tocopherol
as the inhibitor. First let me describe the advantages of this
system for studying autoxidation and then let me describe the
data that are acquired in order to calculate k_{inh}.

Ingold and his coworkers have described an autoxidation system in which styrene is oxidized to styrene polyperoxide in chlorobenzene as a solvent (15). This obviously is a far cry from a biological lipid bilayer system, but Ingold has argued convincingly of the merits of this system. (It gives a single product, has a high value of k_p and the reversal of reaction 17 can be neglected, and the rate constants are well characterized.) If a system is studied that is a close model for in vivo autoxidation (such as red blood cells, a classical model system), initiators and inhibitors cannot be injected into the bulk aqueous phase and produce an instantaneous response, since diffusion into the bilayer from the aqueous phase is too slow. [Even an egg lecithin bilayer vesicle system gives this problem (2).] Our system, on the other hand, is an extremely useful halfway house.

In our system, the rate of autoxidation of linoleic acid is essentially zero in the absence of the initiator. (Notice the flatness of the oxygen trace at the far left in Figure 1 before the initiator is added.) Our system produces classical inhibition kinetics. Initiator can be injected into the bulk aqueous phase and the autoxidation starts instantly, avoiding the problems that plagued Barclay and Ingold (2). When the vitamin E is injected, it also produces an instantaneous effect. The rate of autoxidation before E is added, R_{oxi}, is also observed after all the E has been used. (See Figure 1.) The two quantities that we need to measure to obtain a value of k_{inh} are shown on this plot; they are τ, the length of the inhibition period, and R_{inh}, the rate of autoxidation in the presence of the inhibitor.*

We can derive the necessary equations as follows. We can combine eq 20 and the definition of the rate during the inhibition period, given in eq 23, to give eq 24. Notice that

$$R_{inh} = k_p [LH][LOO\cdot] \qquad (23)$$

$$R_{inh} = k_p [LH] \left[\frac{R_i}{n\, k_{inh} [InH]} \right] \qquad (24)$$

the rate of autoxidation during the inhibition period, R_{inh}, shows a first power dependence on R_i, and not the square root

---

*Because the chain lengths are not large in inhibited runs and one O_2 is used in eq 4 in the initiation sequence, the long-chain approximation used in eq 16 is no longer correct, and R_{inh} is defined as $(R_{oxi} - R_i)$, where $R_{oxi} = -dO_2/dt$.

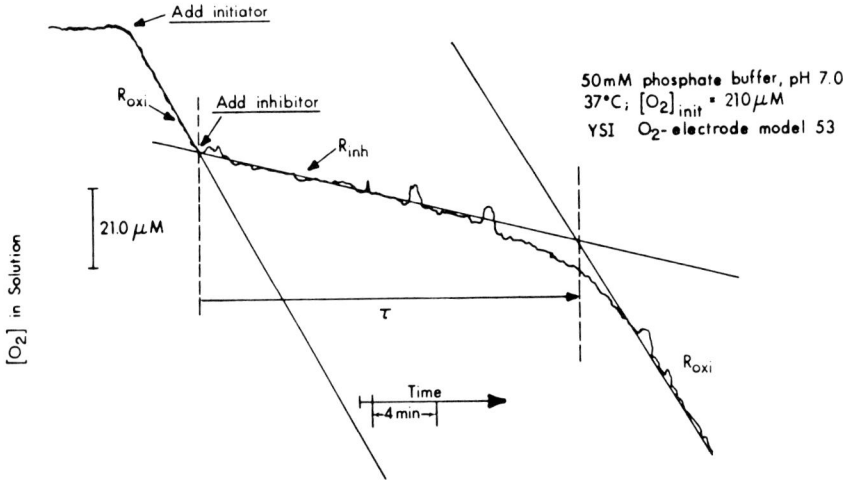

FIGURE 1. The autoxidation of linoleic acid in SDS micelles. The initiator is first injected into the bulk buffer phase and then the antioxidant is injected. The oxygen-electrode trace shown is for alpha-tocopherol as the antioxidant (70).

dependence that was observed in eq 16; this is because the LOO· radicals are scavenged by the inhibitor, InH, rather than undergoing a bimolecular termination reaction.

We can simplify eq 24 as follows. The lag time, τ, is defined as shown in eq 25.

$$\tau \equiv \frac{n[\text{InH}]}{R_i} \quad \text{(in sec)} \tag{25}$$

That is, τ is the ratio of the total number of radicals that are scavenged (n [InH]) divided by the rate at which radicals are being produced, R_i. (This definition perhaps is not obvious at first glance, but notice that it does have the correct units, seconds, since the numerator is in mole/liter and the denominator is a rate in moles/liter-sec.) If eq 25 is substituted into eq 24, we obtain the final equation, eq 26.

$$R_{inh} = \frac{k_p[\text{LH}]}{k_{inh} \tau} \tag{26}$$

Since both the rate during the inhibition period, R_{inh}, and τ can be directly measured from traces like Figure 1, and since the linoleic acid concentration in the micelle can be calculated from the dimensions of the micelle and the total amount of the linoleic acid that is added, a knowledge of k_p allows the calculation of k_{inh}. Alternatively, the ratio of k_{inh}/k_p can be obtained. (This inhibitor parameter is equivalent to the chain transfer constant C that is obtained in the polymer field when the rate of reaction of the polymeric radical with the transfer agent is obtained relative to k_p.)

Our group is measuring values of k_{inh} for a series of natural and synthetic antioxidants and for non-steroidal anti-inflammatory drugs using this new test system, and these data will be published at a later time (70). In this chapter, I have merely indicated the value we obtain for α-tocopherol, $k_{inh} = 2 \times 10^5$ M^{-1} sec^{-1}, as indicative of the data that can be obtained using this method.

Our value for k_{inh} is smaller than that of Niki et al. (60) who obtain a value of 51×10^4 M^{-1} sec^{-1} in homogeneous solution in tert-butanol using linoleate as the substrate and Burton and Ingold (15) who report 235×10^4 M^{-1} sec^{-1} using styrene as the substrate in homogeneous solution with chlorobenzene as the solvent. These differences should not be overinterpreted until we have determined the values of k_p and k_t in our system (by the rotating sector method) and can compare k_{inh}/k_p in our system with the comparable values in the homogeneous solutions.

WAYS IN WHICH FREE RADICALS CAN BE PRODUCED IN VIVO

Theory tells us that free radicals can be produced in a limited number of ways. From the biological perspective, the most interesting possibilities are those listed in Table IV.

Table IV. Ways of producing free radicals in vivo.

    Irradiation
       Ionizing radiation
       Light

    Bond homolysis
       Uncatalyzed
       Catalyzed
          Non-enzymatic catalysis
          Enzyme-catalyzed reactions

Radicals can be produced by irradiation, as shown in Table IV. Ionizing radiation is well known to produce free radicals in living systems, but I will not review that subject here. To that subject belongs the massive literature of radiation chemistry and radiation biology. The subject of light-initiated free radical reactions also is quite complex, since light can either excite molecules that transfer their energy to oxygen or the light can directly excite oxygen to produce singlet oxygen. Again this subject has its own literature, and I will not review it here in detail. I do, however, wish to make the point that in mammals, one tissue is particularly at risk for light-induced free radical reactions: that tissue is, of course, the eye. A number of disease processes of the eye appear to involve free radicals and to be initiated by light. For example, the action spectrum for the autoxidation of rod outer segments corresponds to the absorption spectrum of rhodopsin (101). In eye tissue, GSH is oxidized (102) and thiobarbituric acid reactive materials (TBARM) are produced (8). Cataract lenses give more TBARM than do normal lenses (8); interestingly the superoxide dismutase (SOD) levels in the lens decrease with age in rat (27). It has recently been suggested that the principal human antiprotease, α-1-protease inhibitor (αlPI), is oxidized in cataract lens (13), suggesting the involvement of oxidizing species and perhaps oxy-radicals, in cataract development. Retrolential fibroplasia, a disease that blinds some premature human infants, also appears to involve free radical reactions (47).

Uncatalyzed Homolysis

I have argued for some years (65,66,68) that the most important routes to produce free radicals *in vivo* are enzyme-catalyzed reactions and the action of xenobiotics, topics that I will review below. I believe that spontaneous unimolecular homolysis of lipid hydroperoxides usually is not fast enough to be important. The bond dissociation energy of RO-OH is 42 kcal/mole, and is independent of the nature of the R group (7). This is so strong a bond that uncatalyzed homolysis would not reach appreciable rates until temperatures near 110°C (6). Nevertheless, there is an old report in the literature that the decomposition of ethyl linoleate hydroperoxide has an activation energy of less than 20 kcal/mole and is fast at 25°C (32). This fast rate must result either from an appreciable chain-induced decomposition, eq 27 (where M· is any free radical), or from some very special

$$M\cdot + ROOH \longrightarrow MH + ROO\cdot \tag{27}$$

interaction of the double bonds in the molecule with the hydroperoxide function, a possibility that I have already discussed (pp 9-10 in ref 65). I believe the former is the more reasonable possibility, but the true unimolecular decomposition rate for any lipid or PUFA hydroperoxide has yet to be measured.

Enzymatically Catalyzed Radical Production

It is beginning to be clear that an important source of free radicals in vivo involves the enzymatic production of superoxide and the reaction of superoxide, by mechanisms that are not yet entirely known, to produce the hydroxyl radical. An extensive literature on this subject exists, predominantly arguing that a metal-catalyzed cycle exists by which superoxide can react to produce the hydroxyl radical (38,54), as exemplified in outline form in eqs 28-29.

$$O_2^{-}\cdot + Fe^{3+} \longrightarrow O_2 + Fe^{2+} \tag{28}$$

$$Fe^{2+} + HOOH \longrightarrow Fe^{3+} + HO\cdot + HO^{-} \tag{29}$$

We have suggested that lipid hydroperoxides can react with superoxide to produce alkoxyl radicals (96), but this mechanism also appears to be complex (9).

$$O_2^{-}\cdot + LOOH \longrightarrow LO\cdot + \text{(other unknown products)} \tag{30}$$

Aust and his group believe that a species, which they call "perferryl", is produced from the reaction of superoxide with metal-ions, and that this species can abstract hydrogen atoms and initiate autoxidation (1). While it is likely that such metal-dioxygen complexes exist, their detailed chemistry remains to be elucidated.

Initiation of Free Radical Production by Xenobiotics

I believe radical production by xenobiotics is one of the most important mechanisms in vivo for the production of free radicals that ultimately cause pathological damage (16,51,52,65,66,97). Xenobiotics can produce free radicals by three distinct mechanisms: (i) Some are themselves free radicals, and consequently react directly to form biopolymer-radicals; (ii) some, while not themselves free radicals, are so reactive that they cause free radical formation in the cell; (iii) and some require enzymatic activation to produce free radicals.

The nitrogen oxides (NO and NO_2) are examples of important toxins that are themselves free radicals. Nitrogen dioxide in smoggy air is known to react with PUFA both in vitro and in vivo to cause free radical production (68,74).

Cigarette smoke is another interesting system. Cigarette smoke consists of both a gas phase and a tar phase (77,81). We have shown that *gas-phase* cigarette smoke contains high concentrations of NO_x as well as reactive organic oxy-radicals, and that these can cause biological damage

(18,26,69,71,79,80). We have implicated gas-phase cigarette smoke in the types of damage that lead to lipid peroxidation (18,26,69,77) and to the oxidation of αlPI, a reaction involved in the etiology of emphysema (26,71). We have shown that the *tar-phase* radical, while less reactive, binds to DNA in vitro, and we have suggested ways in which the paramagnetic species in tar could be implicated in tumor development (74,82).

Ozone is an example of a toxin that, while not a free radical itself, is so reactive that it reacts with biological materials to produce free radicals (76,78). We have shown that ozone reacts with methyl linoleate to form free radicals that can be detected by the ESR spin-trap method (76). We have elucidated the mechanism of this process using both small olefins (78) and cumene as a model of an olefin with an allylic hydrogen (75).

Some xenobiotics require enzymatic activation to produce free radicals. Typical of this group is carbon tetrachloride, a toxin that is metabolized at a cytochrome P-450 site and that leads to the trichloromethyl free radical (85,87,90). Some of the quinone drugs that are used in cancer therapy also are reduced enzymatically to produce free radicals (52). For quinoid species like paraquat and adriamycin, a redox cycle exists in which superoxide is produced. (Q is used as a symbol for a generalized quinone.)

$$Q + e \longrightarrow Q^{-\cdot} \tag{31}$$

$$Q^{-\cdot} + O_2 \longrightarrow Q + O_2^{-\cdot} \tag{32}$$

If oxygen is not present, a more complex cycle occurs where the semiquinone radical-ion itself either reacts with H_2O_2 or an iron-catalyzed cycle occurs (12,103). That is, either eq 33 occurs or eqs 34-35.

$$Q^{-\cdot} + H_2O_2 \longrightarrow Q + HO^- + HO\cdot \tag{33}$$

$$Q^{-\cdot} + Fe^{3+} \longrightarrow Q + Fe^{2+} \tag{34}$$

$$Fe^{2+} + H_2O_2 \longrightarrow Fe^{3+} + HO^- + HO\cdot \tag{35}$$

PART II

AGING

"All the rivers run into the sea;
  yet the sea is not full.."
Ecclesiastes

"Every day you get older,
  it's a law."
*Butch Cassidy in "Butch
  Cassidy & the Sun
  Dance Kid"*

Since this is a conference on radical involvement in aging, I cannot let this opportunity pass without some remarks on this subject! My hypothesis to explain how radicals affect aging can be stated simply: Radicals do not affect the genetically-determined maximum lifespan for each species, but radical-mediated reactions contribute importantly to those diseases that are most important in determining the extent to which a population lives to its maximum lifespan. I believe that aging is best defined as the "increased susceptibility to life-limiting degradative processes", and thus free radical reactions, which can contribute to a generalized "wear-and-tear", are an important factor in this time-dependent degradation (66,68,92).

Figure 2 shows types of survivorship curves for populations of mammals. Curve A shows a first-order decay; this is the type of aging that occurs if deaths are caused by random events (as for songbirds in the wild or radioactive atoms). Curve B1 shows a population in which a mean lifespan is beginning to become evident. As health care improves, the mean lifespan of the population increases (curves B2 and B3) and more of the population lives closer to the fixed maximum lifespan for that species (68). Finally, the ultimate care produces curve C, in which most of the population lives to the maximum lifespan and all of the individuals die in a narrow band of ages close to that maximum. Curve D only can be achieved if a substantial increase in the maximum lifespan can be realized (20,68,92,93).

If we examine the survivorship curves for human populations either estimated for stone age man to the present (20) or taken from the accurate population records from 1930 up to today (93), we see that the progress that has been made is exclusively of the type from curve B1 to curve B3; there is no evidence for any qualitative jump to a new curve of the type of C. (See Figure 3.) Thus, better control of the degradative processes is the only type of control of aging that we have been able to achieve, and it has made great improvements in the lifespan and health of our human population.

Fries and Crapo (34) have recently made a critical survey of the chronic diseases that most limit human lifespan. Their list is given in Table V, and it is interesting that most of these processes appear to involve a free radical component. Furthermore, the evidence for radical involvement in these diseases becomes greater each year as our tools become more sophisticated. For example, whereas 15 or 20 years ago radical involvement in cancer might be speculated upon, now there are firm data. I will briefly review the evidence for radical involvement in some of these processes, outlining the possibilities and citing references that give further details.

Table V. Chronic diseases of humans (p. 83 of ref 34)

Disease	Are Radicals Involved?
Emphysema	Oxidation of an anti-protease, partly by radicals or peroxides, appears dominant in smoker's emphysema
Atherosclerosis	Good evidence for some involvement
Cancer	Good evidence for substantial involvement
Osteoarthritis Cirrhosis Diabetes	Evidence beginning to emerge for some involvement

Emphysema

An overwhelming percentage of the persons who suffer from emphysema are smokers, and there now is very strong evidence that emphysema is caused by the inactivation of alPI in lung lavage fluid by oxidants in smoke (44). Smoke causes pulmonary alveolar macrophages (PAM) to be activated, so smokers lungs contain higher concentrations of superoxide and hydrogen peroxide than do those of non-smokers. These oxidants are known to inactivate alPI (44). In addition, free radicals in smoke such as nitrogen dioxide react with these oxidants to form even more strongly oxidizing materials that inactivate alPI (18,26,69,71). Furthermore, we have some evidence that other species in gas-phase smoke, perhaps the organic free radicals themselves (71), inactivate alPI. Thus, emphysema represents a major chronic disease in which radical involvement is important. The hope remains that this insight ultimately will lead to strategies for protection of the lung against this type of oxidative damage.

FREE RADICALS IN AUTOXIDATION AND AGING

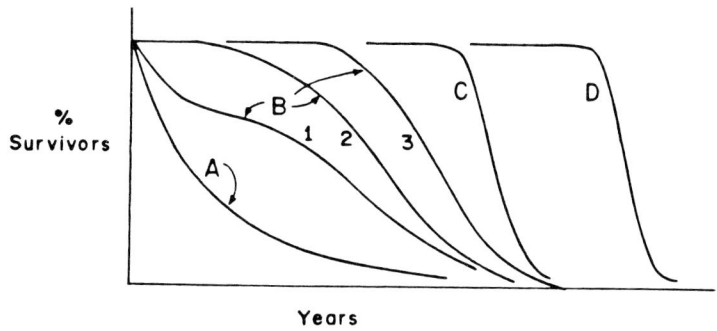

FIGURE 2. Survivorship curves. From W.A. Pryor "Vitamin E" (B. Lubin and L.J. Machlin, editors) New York Academy of Sciences, Volume 393 pp 1-23, 1982. Redrawn from a figure in A. Comfort, "The Biology of Senescence." Holt, Rinehart, and Winston, Inc., New York, 1956.

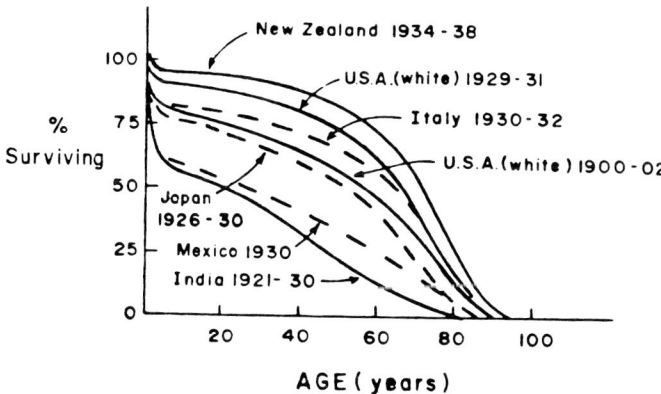

FIGURE 3. Human survivorship curves. From W.A. Pryor "Vitamin E" (B. Lubin and L.J. Machlin, editors) New York Academy of Sciences, Volume 393 pp 1-23, 1982. Redrawn from a figure in A. Comfort, "The Biology of Senescence." Holt, Rinehart, and Winston, Inc., New York, 1956.

Atherosclerosis

The arachidonic acid cascade produces hydroperoxide-containing products such as 15-HPETE. These hydroperoxides are reduced to alcohols by a peroxidase that is associated with prostaglandin cyclooxygenase activity (35,51). In this process, an oxidant is formed that causes inactivation of the enzyme systems; prostacyclin (PGI) synthetase but not thromboxane (TXA) synthetase is inhibited (39,58). Thus, high hydroperoxide levels may lead to a high TXA/PGI ratio, leading to diseases associated with hypertension and atherosclerosis. This peroxide-mediated mechanism for vessel-wall injury also can be initiated by endotoxin (104). Interestingly, cigarette smoke alters the metabolism of arachidonic acid in rat aortas, platelets and lungs in a manner that results in increased TXA and decreased PGI, also creating a condition that favors the development of cardiovascular disorders (49). Cigarette smoke contains a variety of oxidizing species that could be involved in these effects (18,69,79).

There is increasing evidence for the involvement of superoxide radicals and other oxy-radicals in ischemic injury of tissue (14,53,61) including myocardial injury (36). It appears that antioxidants and radical scavengers may play a role in reducing tissue damage during the reprofusion phase of surgical procedures or in shock (62).

Cancer

The evidence for radical involvement in some types of chemically-induced carcinogenesis is now quite strong (16,21,29,30,45,51,52,97). Many chemicals are metabolized to carcinogens by processes that do not involve free radicals; a well-understood example is the oxidation of benzo(a)pyrene to the diol epoxide. However, the oxidation of B(a)P and other polynuclear aromatic hydrocarbons (PAH) to carcinogenic compounds can involve radicals, and this statement, while once controversial, now has ample evidence (16,30,97). While the involvement of these radical-mediated PAH products in cancer may be less clear than the fact that they do occur, I believe that such evidence does now exist (16,63). In addition, the phenomenon of promotion of tumor growth appears to involve radicals and radical precursors, although the mechanisms are not at all clear as yet (28,89).

Originally it was hoped that ESR could be used as a technique for detecting early stages of tumor development. However, this hope, to date at least, has not been realized (94). There is some very recent evidence, however, that some tumors do possess a highly anomalous ESR signal (5).

Another line of evidence that free radicals may be involved in chemical carcinogenesis is that anti-oxidants often act as anti-cancer compounds. Many types of chemical carcinogens, in a variety of organs and in many animal species are protected against by an extensive series of anti-oxidants (88,97,99). While it is not yet clear that all of these compounds act as anti-cancer compounds *because* of their anti-oxidant properties, it seems reasonable that some of them do. In the mouse-skin test, where peroxides act as promoters, anti-oxidants again show protective effects (42).

Systems that produce superoxide, such as human neutrophils, give a positive Ames test (100), again indicating a connection between radical-forming activity and mutagenesis and/or carcinogenesis.

Osteoarthritis, Cirrhosis and Diabetes

Although evidence for free radical involvement in these diseases is less strong, there are some indications that radicals might be implicated (10,19,37).

Diseases of the Eye

In the section above entitled "Ways in Which Free Radicals are Produced In Vivo" I reviewed evidence for singlet oxygen and/or free radical involvement in some of the pathologies observed in the eye, including cataract and retrolental fibroplasia. There now is considerable evidence that vitamin E protects against the lipid peroxidation that appears to accompany retrolental fibroplasia in premature infants. In this context, it is interesting that vitamin E deficiency in two species of monkeys produces a massive degeneration of the rod outer segments (ROS), as revealed by light microscopy, that has been attributed to lipid peroxidation (41).

The Wear and Tear Theory

The "wear and tear" theory of aging postulates that random processes produce deleterious changes in biomolecules with time, and that these changes contribute to cellular aging. Most forms of this theory propose that changes gradually occur in enzymes, that these "bad" enzymes then lead to "bad" DNA, which leads to bad RNA, and finally inaccurate enzymes. This spiral eventually produces an error catastrophe and death (48,93). Since plastics, paper, and rubber gradually decay in air due to autoxidation, it seems reasonable to postulate that tissue would as well. Of course a living system, unlike inert materials, possesses impressive repair mechanisms to overcome this damage. Nevertheless, it is reasonable to presume that random oxidative damage would accumulate over the life of an organism, and in some cases, overwhelm the repair mechanisms. The evidence for the accumulation of inaccurate enzymes with time is still rather controversial (48). However, there does appear to be no doubt that random errors do accumulate in biomolecules with time (50), and it seems reasonable that free radical reactions would contribute to this. For example, I have already reviewed the evidence for the oxidation of the methionine residues at the active site of αlPI by radicals. This oxidation also has been postulated to occur in any protein or enzyme containing methionine residues (13), again suggesting a connection between free radicals and the random processes associated with aging.

The finding, originally discovered by Harman (40), that antioxidants extend the mean life span of small mammals is impressive evidence for radical involvement in some of the processes that contribute to aging. The correlations of enzyme levels (such as superoxide dismutase) that protect against radical damage and the lifespan of species also is evidence for the same processes (22).

Acknowledgement. Research in the areas discussed here is supported by the National Institute of Health, the National Foundation for Cancer Research, and the National Science Foundation. I also wish to acknowledge many helpful suggestions from Drs. Laurence Castle and Daniel F. Church.

REFERENCES

1. Aust, S.D., and Svingen, B.A. (1982): In: Free Radicals in Biology Volume V, edited by W.A. Pryor, pp 1-25, Academic Press, New York.

2. Barclay, L.R.C., and Ingold, K.U. (1981): J. Amer. Chem. Soc., 103:6478-6485.

3. Bateman, L., and Gee, G. (1951): Trans. Faraday Soc., 47:155-164.

4. Bateman, L. (1954): Quart. Rev., 8:147-172.

5. Benedetto, C. (1982): In: Free Radicals, Lipid Peroxidation, and Cancer, edited by D.C.H. McBrien and T.F. Slater, pp. 27-54, Academic Press, New York.

6. Benson, S.W. (1964): J. Chem. Phys., 40:1007-1013.

7. Benson, S.W. (1976): In: Thermochemical Kinetics. John Wiley and Sons, New York.

8. Bhuyan, K.C., and Bhuyan, D.K. (1983): (in press).

9. Bielski, B., and Thomas, M.J. (1984): (in press).

10. Blake, D.R., Hall, N.D., Bacon, P.A., Dieppe, P.A., Halliwell, B., and Gutteridge, J.M.C. (1981): Lancet, 1142-1144.

11. Bodaness, R.S., Zigler, J.S. (1983): Biochem. Biophys. Res. Com., 113: 592-597.

12. Borg, D.G., Schaich, K.M., Elmore, J.J., and Bell, J.J. (1978): Photochem. Photobiol., 28: 887-908.

13. Brod, N., and Weissbach, H. (1983): Arch. Biochem. Biophys., 223:271-281.

14. Bulkley, G.B. (1983): Surgery 94:407-414.

15. Burton, G.W., and Ingold, K.U. (1981): J. Amer. Chem. Soc., 103:6472-6477.

16. Cavalieri, E.L., and Rogan, E.G. (1984): In: Free Radicals in Biology Volume VI, edited by W.A. Pryor, Chapter 10, Academic Press, New York.

18. Church, D.F., Crank, G., Chopard, C., Govindan, C.K., and Pryor, W.A. (1982): Fed. Proc., 41:2346.

19. Cohen, G., and Greenwald, R.A. (1983): Oxy-Radicals in The Scavenger Systems, Molecular Aspects, Volume I, Elsevier, New York.

20. Comfort, A. (1979): In: The Biology of Senescence 3rd Edition, p 6, Elsevier, New York.

21. Cornwell, D.G., and Marisaki, N. (1984): In: Free Radicals in Biology Volume VI, edited by W.A. Pryor, Chapter 4, Academic Press, New York.

22. Cutler, R.G. (1984): In: Free Radicals in Biology Volume VI, edited by W.A. Pryor, Chapter 11, Academic Press, New York.

23. Delamere, N.A., Paterson, C.A., and Cotton, T.R. (1983): Exp. Eye Res., 37:45-53.

24. Dix, T.A., and Marnett, L.J. (1983): Science, 221:77-79.

25. Docampo, R., and Moreno, S.N.J. (1984): In Free Radicals in Biology Volume VI, edited by W.A. Pryor, Chapter 8, Academic Press, New York.

26. Dooley, M.M., and Pryor, W.A. (1982): Biochem. Biophys. Research Comm., 106:981-987.

27. Dovrat, A., and Gershon, D. (1981): Exper. Eye Res., 33:651-661.

28. Emerit, I., and Cerutti, P.A. (1982): Proc. Nat. Acad. Sci., 79:7509-7513.

29. Floyd, R.A. (1980): In: Free Radicals in Biology Volume V, edited by W.A. Pryor, Academic Press, New York, pp 187-206.

30. Floyd, R.A., editor (1982): In: Free Radicals in Cancer, Marcel Dekker, New York.

31. Foote, C.S. (1976): In: Free Radicals in Biology Volume II, edited by W.A. Pryor, pp 135-156, Academic Press, New York.

32. Franks, F., Gent, N., Roberts, B. (1965): J. Applied Chem., 15:243-249.

33. Fridovich, I. (1976): In: Free Radicals in Biology Volume I, edited by W.A. Pryor, pp 239-271, Academic Press, New York.

34. Fries, J.F. and Crapo, L.M. (1981): Vitality and Aging. W.H. Freeman and Company, San Francisco.

35. Gale, P.H., and Egan, R.W. (1984): In: Free Radicals in Biology Volume VI, edited by W.A. Pryor, Chapter, 1, Academic Press, New York.

36. Gardner, T.J., Stewart, J.R., Casale, A.S., Downey, J.M., and Chambers, D.E. (1983): Surgery 94:423-432.

37. Greenwald, R.A. and Cohen, G. (1983): Oxy-Radicals in Their Scavengers Systems, Volume II, Cellular and Molecular Aspects, Elsevier, New York.

38. Gutteridge, J.M.C., Rowley, D.A., and Halliwell, B. (1981): Biochem. J., 199:263-265.

39. Ham, E.A., Egan, R.W., Soderman, D.D., Gale, P.H., and Kuehl, F.A. (1979): J. Biol. Chem., 254:2191-2194.

40. Harman, D. (1982): In: Free Radicals in Biology Volume V, edited by W.A. Pryor, pp 255-271, Academic Press, New York.

41. Hayes, K.C. (1974): Invest. Ophthal., 13:499-510.

42. Heckler, E., Fusenig, N.E., Kunz, W., Marks, F., and Thielmann, H.W., editors (1982): Cocarcinogenesis and Biological Effects of Tumor Promoters, Raven Press, New York.

42a. Henglein, A., and Proske, T. (1978): J. Am. Chem. Soc., 100:3706-3709.

43. Howard, J.A. (1972): In: Advances in Free-Radical Chemistry, Volume IV, edited by G.H. Williams, pp. 49-174. Academic Press, New York.

44. Janoff, A., Carp, H., Laurent, P., and Raju, L. (1983): Am. Rev. Respir. Dis., 127:S31-S38.

45. Kalyanaraman, B., and Sivarajah, K. (1984): In: Free Radicals in Biology Volume VI, edited by W.A. Pryor, Chapter 5, Academic Press, New York.

46. Kohn, R.R. (1971): In: Principles of Mammalian Aging, pp 107-108, Prentice Hall Inc., Englewood Cliffs, New Jersey.

47. Kretzer, F.L., Hittner, H.M., Johnson, A.T., Mehta, R.S., and Godio, L.B. (1982): In: Vitamin E, edited by B. Lubin, and L.J. Machlin, New York Academy of Sciences, Volume 393, pp 145-166. New York.

48. Laughrea, M. (1982): Exp. Gerontol., 17:305-317.

49. Lubawy, W.C., Valentovic, M.A., Atkinson, J.E., and Gairola, G.C. (1983): Life Sci., 33:577-584.

49a. Maillard, B., Ingold, K.U., and Scaiano, J.C. (1983): J. Am. Chem. Soc., 105:5095-5099.

50. Man, E.H., Sandhouse, M.E., Burg, J., and Fisher, G.H. (1983): Science, 220:1407-1408.

51. Marnett, L. (1984): In: Free Radicals in Biology Volume VI, edited by W.A. Pryor, Chapter 3, Academic Press, New York.

52. Mason, R.P. (1982): In: Free Radicals In Biology Volume V, edited by W.A. Pryor, pp 161-196, Academic Press, New York.

53. McCord, J.M. (1983): Surgery, 94:412-414.

54. McCord, J.M., and Day, E.D. (1978): FEBS Lett., 86:139-142.

55. Mead, J.F. (1976): In: Free Radicals in Biology Volume I, edited by W.A. Pryor, pp 51-67, Academic Press, New York.

56. Menzel, D.B. (1976): In: Free Radicals in Biology Volume II, edited by W.A. Pryor, pp 181-200, Academic Press, New York.

57. Mill, T., and Hendry, D.G. (1980): In: Chemical Kinetics, edited by C.H. Bamford, and C.F.H. Tipper, pp. 1-83. Elsevier Scientific Publishing Co., New York.

58. Moncada, S., and Vane, J.R. (1978): Brit. Med. Bull., 34: 129-134.

59. Mudd, J.B. (1976): In: Free Radicals in Biology Volume II, edited by W.A. Pryor, pp 203-210, Academic Press, New York.

60. Niki, E., Yamamoto, Y., and Kamiya, Y. (1984): In: Oxygen Radicals in Chemistry and Biology, edited by W. Bors et al., Walter de Gruyter and Co., Berlin, Germany.

61. Parks, D.A., Bulkley, G.B., and Granger, D.N. (1983): Surgery 94:415-422.

62. Parks, D.A., Bulkley, G.B., and Granger, D.N. (1983): Surgery 94:428-432.

63. Preston, B.D., Miller, J.A., and Miller, E.C. (1983): J. Biol. Chem. 258:8304-8311.

64. Pryor, W.A., (1966): Free Radicals, McGraw Hill Book Co., New York.

65. Pryor, W.A. (1976): In: Free Radicals in Biology Volume I, edited by W.A. Pryor, pp 1-43. Academic Press, New York.

66. Pryor, W.A. (1977): In: Medicinal Chemistry Volume V, edited by J. Mathieu, pp 331-359, Elsevier, Amsterdam.

67. Pryor, W.A. (1978): Photochem. Photobiol., 28:787-801.

68. Pryor, W.A. (1982): New York Acad. Sci., 393:1-30.

69. Pryor, W.A., Chopard, C., Tamura, M., and Church, D.F. (1982): Fed. Proc., 41:2346.

70. Pryor, W.A., Church, D.F., and Castle, L., (1984): (to be submitted).

71. Pryor, W.A., Dooley, M.M., and Church, D.F. (1984): (in press).

72. Pryor, W.A., Hales, B.J., Premovic, P.I., and Church D.F. (1983): Science, 220:425-427.

73. Pryor, W.A., and Lightsey, J.W. (1981): Science, 214:435-437.

74. Pryor, W.A., Lightsey, J.V., and Church, D.F. (1982): J. Amer. Chem. Soc., 104:6685-6692.

75. Pryor, W.A., Ohto, N., and Church, D.F. (1983): J. Amer. Chem. Soc., 105:3614-3622.

76. Pryor, W.A., Prier, D.G., and Church, D.F. (1981): Environ. Res., 24:42-52.

77. Pryor, W.A., Prier, D.G., and Church, D.F. (1983): Environmental Health Perspect., 47:345-355.

78. Pryor, W.A., Prier, D.G., and Church, D.F. (1983): J. Amer. Chem. Soc., 105:2883-2888.

79. Pryor, W.A., Tamura, M., and Church, D.F. (1984): J. Amer. Chem. Soc., (in press).

80. Pryor, W.A., Tamura, M., Dooley, M.M., Premovic, P., Hales, B.J., and Church, D.F. (1983): In: Oxy Radicals and Their Scavenger Systems: Cellular and Medical Aspects, edited by R. Greenwald, and G. Cohen, pp 185-192, Elsevier, New York,

81. Pryor, W.A., Terauchi, K., and Davis, W.H. (1983): Environ. Health Perspec., 16:161-175.

82. Pryor, W.A., Uehara, K., and Church, D.F. (1984): In: Oxygen Radicals in Chemistry and Biology, edited by W. Bors et al. Walter de Gruyter & Co., Berlin, Germany.

83. Pryor, W.A., Uehara, K., and Church, D.F. (1984): To be submitted.

84. Recknagel, R.O. (1983): Life Sciences, 33: 401-408.

85. Recknagel, R.O., Glende, E.A., and Hruszkewycz, A.M. (1977): In: Free Radicals in Biology Volume III, edited by W.A. Pryor, pp 97-130, Academic Press, New York.

86. Renneberg, R., Capdevila, J., Chacos, N., Estabrook, R.W. and Prough, R.A. (1981): Biochem. Pharm., 30:843-848.

87. Reynolds, E.S., and Moslen, M.T. (1980): In: Free Radicals in Biology Volume V, edited by W.A. Pryor, pp 49-90, Academic Press, New York.

88. Shamberger, R.J., Baughman, F.F., Kalchert, S.L., Willis, C.E., and Hoffman, G.C. (1973): Proc. Natl. Acad. Sci. 70:1461-1463.

89. Slaga, T.J., Klein-Szanto, A.J.P., Triplett, L.L., Yotti, L.P., and Trosko, J.E. (1981): Science, 213:1023-1025.

90. Slater, T.F. (1972): Free Radical Mechanisms in Tissue Injury, Pion Ltd., London.

91. Spector, A., and Garner, W.H. (1981): Exp. Eye. Res., 33:673-681.

92. Strehler, B.L. (1962): Time Cells and Aging 3rd Edition. Academic Press, New York.

93. Strehler, B.L. (1977): In: Time Cells and Aging 2nd Edition, p 25, Academic Press, New York.

94. Swartz, H.M. (1979): In: CIBA Symposium Number 67 (New Series) "Submolecular Biology and Cancer", Elsevier Publisher, New York.

95. Tappel, A.L. (1980): In: Free Radicals in Biology Volume V, edited by W.A. Pryor, pp 2-44, Academic Press, New York.

96. Thomas, M.J., Mehl, C.S., and Pryor, W.A. (1982): J. Biol. Chem., 257:8343-8348.

97. Ts'o, P.O.P., Caspary, W.J., and Lorentzen, R.J. (1977): In: Free Radicals in Biology Volume III, edited by W.A. Pryor, pp 251-300, Academic Press, New York.

97a. Turro, N.J., Zimmt, M.B., and Gould, I.R. (1983): J. Am. Chem. Soc. 105:6347-6349.

98. Walling, C., (1957): Free Radicals in Solution, John Wiley and Sons, New York.

99. Wattenberg, L.W. (1972): J. Natl. Cancer Insti. 48:1425-1430.

100. Weitzman, S.A., and Stossel, T.P. (1981): Science, 212:546-547.

101. Williams, T.P., and Howell, W.L. (1983): Invest. Opth. Vision Sci., 24:285-287.

102. Winkler, B.S., and Giblin, F.J. (1983): Exper. Eye Res., 36:287-297

103. Winterbourne, C., and Sutton, H. (1983): In: Oxy Radicals and Their Scavenger Systems, edited by R.A. Greenwald and G. Cohen, pp 105-112, Vol. II, Elsevier, New York.

104. Yoshikawa, T., Murakami, M., Furukawa, Y., Kato, H., Takemura, S., and Kondo, M. (1983): Thromb Haemostas, 49: 214-216.

Free Radicals in Molecular Biology, Aging, and Disease, edited by D. Armstrong et al. Raven Press, New York © 1984.

Biological Sites and Mechanisms of Free Radical Production

Bruce A. Freeman

Department of Medicine, Duke University Medical Center, Durham, North Carolina 27710

Molecules having an odd number of electrons can be classified as free radicals. Both organic (ie., quinones) and inorganic molecules (ie., oxygen) can exist as free radical species having varying degrees of reactivity and toxicity. In cells, free radicals can be derived during normal metabolic processes, following exposure to ionizing radiation and after exposure to drugs or xenobiotics which can be metabolized to radicals in situ. Because of the ubiquity of molecular oxygen and its ability to readily accept electrons, oxygen-centered free radicals are often primary or secondary mediators of cellular free radical reactions.

The catalytic activity of many cellular enzymes, electron transport processes and autoxidation of cell components yields free radical intermediates. Free radicals are also produced by activated phagocytes for microbicidal purposes. Many anthracyclic antineoplastic agents such as adriamycin, daunorubicin, doxorubicin, and other antibiotics that depend on quinoid groups or bound metals for activity are able to generate oxygen radicals. Many of the chemotherapeutic effects and cytotoxic side effects of these drugs have been ascribed to their ability to reduce oxygen to O_2^-, H_2O_2 and hydroxyl radical (OH·). Irradiation of organisms with electromagnetic radiation (x-rays and γ-rays) and particulate radication (electrons, protons, neutrons, deuterons, and α and β particles) generate primary radicals by transferring their energy to cellular components such as water. These primary radicals, including e^-_{aq}, OH· and H·, can then undergo secondary reactions with dissolved O_2 or with cellular solutes. In addition, a wide variety of environmental agents including dietary substances, photochemical air pollutants, hyperoxia, pesticides, tobacco smoke, solvents, anesthetics, and the general class of aromatic hydrocarbons also generate free radicals. These xenobiotics either already exist as free radicals or are converted to radical species by intracellular metabolic and detoxification processes. This chapter will discuss cellular sites of free radical generation and the processes by which radical species arise. Free radical generation by xenobiotics has recently been reviewed extensively (29-32) and will not be discussed herein.

Plasma Membrane

Phagocytic cells such as polymorphonuclear leukocytes and monocytes generate partially reduced oxygen species as a defense against invading microorganisms. A membrane-bound NADPH oxidase produces O_2^- and H_2O_2 directly and OH· by secondary reactions during the metabolic burst experienced during phagocytosis or exposure to chemotactic stimuli (5). Purification efforts have shown cell fractions enriched with this enzyme to contain granule and plasma membrane marker enzymes and to have a molecular weight of 300,000 or less. Another phagocytic cell and monocyte enzyme, myeloperoxidase, generates oxidized halides using H_2O_2 and halides, including I^-, Br^- and Cl^-, as substrates. This enzyme has a molecular weight of 150,000 and is also important in bactericidal killing processes. Myeloperoxidase-induced killing occurs via halogenation of target cells, or as a result of formation of chloramines, labile halide derivatives, hypochlorous acid or singlet oxygen.

Free radical production by microsomal and plasma membrane-associated enzymes such as lipoxygenase and cyclooxygenase is of current interest because of many recent discoveries regarding arachidonic acid metabolism. Arachidonic acid is the predominant substrate for these enzymes which is converted into biologically potent products. These products include prostaglandins, thromboxanes, leukotrienes, and slow-reacting substances of anaphylaxis. The enzymatic oxidation of arachidonic acid by membrane-bound cyclooxygenase involves free radical intermediates, one of which has been shown by electron spin resonance to be a carbon-centered free radical. This radical was produced by cyclooxygenase-mediated abstraction of one of the methylene hydrogens of arachidonic acid, producing a fatty acid free radical species (32). Also, during cyclooxygenase-catalyzed arachidonic acid metabolism, an oxygen-centered radical which can be scavenged by methional or phenol is produced during the breakdown of the hydroperoxide on PPG_2 (16). This radical has been proposed to be OH· (16), possibly hemoprotein derived (38). The nature of this free radical is controversial and has recently been proposed to be a cyclooxygenase hemoprotein radical, distinct from oxygen-centered free radicals such as OH· (28). In any event, the formation of a free radical intermediate could account for the irreversible oxidative self deactivation of cyclooxygenase during catalysis which can be prevented by free radical scavengers (38). The production of hydroxyl radical or another radical species during prostaglandin synthesis could lead to feedback regulation of cyclooxygenase, modulate both the rate and extent of prostaglandin biosynthesis, and participate in secondary messenger and cytotoxic effects after prostaglandin synthesis. Thus, it appears that the biosynthesis of prostaglandins and thromboxanes results in hemoprotein-, oxygen- and carbon-centered free radicals capable of reacting with the biosynthetic enzymes themselves and other cell components. All of these effects are in addition to the demonstrated chemotactic, vasoactive, and platelet-aggregating effects of prostaglandins.

It is important to note that, because of the proximity of arachidonic acid-metabolizing enzymes and other plasma membrane free radical sources to cell and organelle surfaces, metabolic byproducts can affect cytotoxic membrane and extracellular components, depending on product solubility and byproduct diffusion distances. The reactivity of by-products has a major influence on diffusion distances. For example, OH· has a high and indiscriminate reactivity, so this free radical is not likely to diffuse away from cellular sites of production -- it will react nearby. Less reactive free radicals may be capable of reacting distally from sites of generation. Superoxide, much less reactive with cell components than OH·, could potentially diffuse further way from sites of generation, were it not for the high concentration of superoxide dismutase in cells (estimated to be 3×10^{-5} M in liver cells (22), which maintains O_2^- at 10^{-11} to 10^{-12} M (49). Unless O_2^- becomes protonated, this charged molecule will not diffuse through nonpolar microenvironments of a cell. Hydrogen peroxide, must less reactive than OH· and O_2^-, is maintained under normal conditions at concentrations of 10^{-7} to 10^{-9} M by intracellular catalase and peroxidases (41). Hydrogen peroxide has been shown to diffuse across mitochondrial membranes (47), peroxisomal membranes (12), and across the plasma membrane (45), thus potentially exerting toxic effects at a distance from its site of generation.

Soluble Components of the Cytosol

Many cell components, capable of undergoing oxidation-reduction reactions, are quantitatively important contributors to intracellular free radical production. These include thiols (4), hydroquinones (36), catecholamines (37), flavins (6), and tetrahydropterins (18). In all cases, O_2^- is the primary radical formed by the reduction of dioxygen by these molecules. Also, chelated Fe(III) can be reduced to Fe(II) by thiols, ascorbate, and a host of other reductants. Fe(II) can then autoxidize, producing O_2^- (35). Hydrogen peroxide is a secondary product of one-electron autoxidations, via spontaneous or enzymatically catalyzed dismutation of O_2^-:

$$O_2^- + O_2^- + 2H^+ \longrightarrow H_2O_2 + O_2$$

The spontaneous dismutation of O_2^- has a rate constant at pH 7.4 of approximately 2×10^5 M^{-1} sec^{-1}, whereas the reaction catalyzed by superoxide dismutase is about 10^4 times faster, having a rate constant of 2×10^9 M^{-1} sec^{-1} (22). Thus, cellular processes that yield O_2^- will also produce H_2O_2 as an O_2^- dismutation byproduct.

Hydroxyl radical production by cells has also been reported (40). Formation of OH· and H_2O_2 accounts for many of the effects of O_2^- generating systems, since hydroxyl radical scavengers, in addition to superoxide dismutase and catalase, can protect free radical targets in many in vitro and in vivo test systems. The generation of the potent oxidant OH· seems to require not only O_2^- or H_2O_2, but a transition metal such as iron (35,102). These substances can react by what has been termed an iron-catalyzed Haber-Weiss reaction:

$$Fe(III) + O_2^- \longrightarrow Fe + O_2 \text{ II}$$

$$Fe(II) + H_2O_2 \longrightarrow Fe(III) + OH^- + OH\cdot$$

Thus, O_2^- reduces Fe(III), which in turn reduces H_2O_2 to form OH·. Reductants such as ascorbate can also reduce Fe(III), implying that a source of peroxides in the presence of transition metals can generate OH· in the absence of O_2^- (50).

Numerous enzymes generate free radicals during their catalytic cycling. Xanthine oxidase, probably the most studied free radical-producing enzyme, generates O_2^- during the reduction of oxygen to H_2O_2 (21). Interestingly, human xanthine oxidase serves in vivo as an NAD^+-dependent dehydrogenase and produces no free radical intermediates. Proteolytic modification of xanthine oxidase during purification or during in vivo ischemia converts the enzyme from the dehydrogenase form to the O_2^- producing oxidase form (43,44). Aldehyde oxidase, which is unique but structurally similar to xanthine oxidase, shares many of the same substrates and also generates O_2^- (43). Dihydroorotate dehydrogenase (1), flavoprotein dehydrogenase (33), and tryptophan dioxygenase (25) also utilize O_2^- during their catalytic cycle, deduced either from the observation that superoxide dismutase will inhibit enzyme activity or from electron spin resonance measurement of free radical intermediates during enzyme catalysis.

Studies of the enzyme sources of cellular free radical production have shown that modulation of enzyme activities, cofactor availability, substrate concentration, and oxygen tension can combine to affect rates of intracellular radical production. Thus, certain cellular metabolic states such as hyperoxia, ischemia, or antibiotic therapy can favor free radical production in excess of basal rates.

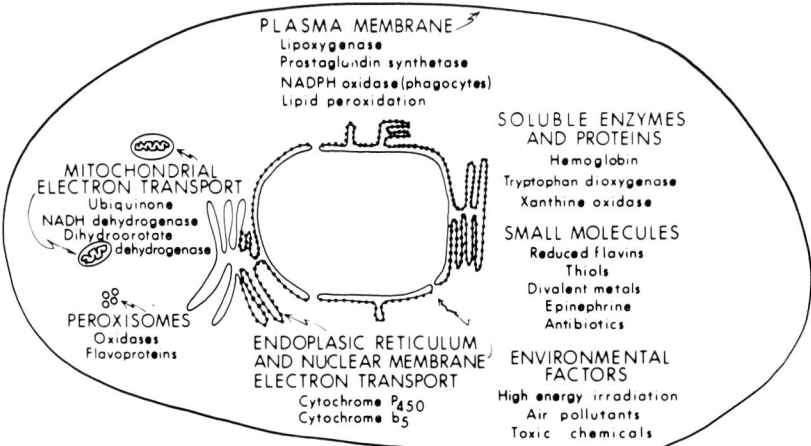

Figure 1. Cellular sources of free radicals. Structural components, subcellular organelles, cytoplasmic contents and xenobiotics all contribute to biological production of free radical species.

Mitochondria

Reduction of O_2 to H_2O by mitochondrial cytochrome c oxidase involves a four-electron transfer with no free radical intermediates, thus this enzyme is not a source of mitochondrial O_2^- production (12). Superoxide radical generation by mitochondria is greatest when respiratory chain carriers located on the inner mitochondrial membrane are highly reduced (48). Hence, endogenous factors that influence mitochondrial radical production are those that regulate respiration. This includes the availability of NAD-linked substrates, succinate, ADP (which serves as a phosphate acceptor and relieves an ion gradient-established increase in respiratory chain reduction) and oxygen. If oxygen is present in concentrations that limit its reduction to H_2O by cytochrome c oxidase (1 to 3 mm Hg, (26)), increased respiratory chain reduction and an accumulation of reduced cofactors in cells may enhance O_2^- production by electron transport components of ischemic cells.

Isolated mitochondria and submitochondrial particles have been used to measure sites and rates of respiratory chain O_2^- and H_2O_2 production. Intact mitochondria can be used to predict the rate of extramitochondrial H_2O_2 release into the cytosol. Rates of intramitochondrial O_2^- and H_2O_2 generation can only be estimated, using inner mitochondrial membrane fragments (submitochondrial particles) washed free of enzymes such as superoxide dismutase, catalase, and glutathione peroxidase. These enzymes, which metabolize O_2^- and H_2O_2, have rate constants orders of magnitude greater than the probes used to quantify O_2^- and H_2O_2. This necessitates removal or inhibition of these enzymes in biological preparations in which O_2^- and H_2O_2 are to be measured. Site-specific respiratory chain inhibitors, such as antimycin A, cyanide, and rotenone, which were used in elucidating the steps of mitochondrial electron transport (11), also assisted in identification of O_2^- producing respiratory chain components.

Studies of mitochondria isolated from bovine and rat heart (15,46), isolated tumor cells (15), porcine lung and rat lung (46,47) showed that the ubiquinone-cytochrome b region is a major site of O_2^- production. The data suggested that O_2^- generation in this region is due to autoxidation of ubisemiquinone. NADH dehydrogenase and dihydroorotate dehydrogenase are also autoxidizable electron carriers responsible for a portion of mitochondrial O_2^- production (20,46). Most, if not all, mitochondrial H_2O_2 is derived from O_2^- via dismutation (43). There are species-to-species and organ-to-organ differences in mitochondrial respiratory chain component concentrations, which can contribute to differences in major sites and specific activities of O_2^- and H_2O_2 production by mitochondria isoalted from different sources (47). Recent reviews have extensively discussed mitochondrial O_2^- and H_2O_2 generation (7,12,19).

Intact mitochondria can release H_2O_2 into the cytoplasm (8,47). It is controversial whether O_2^- can escape intramitochondrial superoxide dismutase and wreak havoc in the cytosol (39,47). Superoxide and H_2O_2 production normally accounts for 1 to 2 percent of mitochondrial oxygen consumption under reduced conditions. The intramitochondrial concentration of O_2^- has been estimated to be 8×10^{-12} M (49), and even though this calculation was based on a series of assumptions, it shows that mitochondrial superoxide dismutase maintained intramitochondrial O_2^- at very low steady state concentrations. Thus, very little if any O_2^- will enter the cytoplasm from mitochondria.

Endoplasmic Reticulum and Nuclear Membrane

Intramitochondrially derived free radicals must escape organelle antioxidant defenses to initiate cytosol damage. Free radicals produced by the endoplasmic reticulum and nuclear membrane can undergo both intraorganelle and cytosolic reactions, unless scavenged by antioxidant defenses. In the case of nuclear membrane-derived radicals, DNA would be particularly susceptible to free radical damage.

According to the concept of membrane spatial continuity, the endoplasmic reticulum and nuclear membrane share many of the same elements. Both of these intracellular membranes contain the cytochromes P_{450} and b_5 that can oxidize unsaturated fatty acids (10), xenobiotics (13) and reduce dioxygen (3), among other substrates. NADH or NADPH are required cofactors for these reactions. Flavoprotein-containing cytochrome reductases which provide the electrons for the cytochrome P_{450}- and b_5-mediated reactions are also capable of autoxidizing to produce O_2^- and H_2O_2 (2). Microsomal and nuclear membrane cytochromes can directly form O_2^- by one-electron transfer or will form H_2O_2 by dissociation of peroxy-cytochrome complexes. Substrates for microsomal cytochrome P_{450} oxidative reactions have been shown to either stimulate or inhibit H_2O_2 formation. This will be in part due to endogenous ratios of both substrate and cofactor (24).

Other sources of microsomal oxygen radical production include flavin-containing oxidases. Hydroxyl radicals are generated by rat liver microsomes, both in the absence (14) and presence (23) of cytochrome P_{450} substrates, and require the presence of NAD(P)H. Addition of azide increased OH· production by these systems. Azide inhibits catalase, thereby preventing H_2O_2 metabolism. This suggested that the microsomal OH· was derived from an iron-catalyzed Haber-Weiss reaction where H_2O_2 could have served as OH· precursor.

Peroxisomes

Peroxisomes are potent sources of cellular H_2O_2 because of high intraorganelle concentrations of oxidases, none of which have been shown to generate O_2^- as an immediate precursor to H_2O_2 (34). Some peroxisomal H_2O_2-generating enzymes include D-amino acid oxidase, urate oxidase, L-α-hydroxy acid oxidase, and fatty acyl-CoA oxidase. Peroxisomal catalase is the enzyme that normally metabolizes most of the H_2O_2 generated by peroxisomal oxidases. The proportion of peroxisomal H_2O_2 that can diffuse out of peroxisomes into the cytoplasm ranges from a calculated 2 percent (42) to a measured 11 to 42 percent, depending on the substrates used for determinations (9). Chance, Sies and Boveris (12) proposed that the discrepancy between these calculated and observed rates of extraperoxisomal H_2O_2 diffusion may be due to increased peroxisomal membrane permeability after isolation for in vitro measurements. Peroxisomal metabolism of H_2O_2 infused into intact hepatocyte suspensions readily occurs (27), showing the ability of H_2O_2 to diffuse across at least two barrier membranes and through the cytoplasm, to reach peroxisomes.

Summary

Precise definition of cellular sources and mechanisms of free radical generation is difficult, with most observations gleaned from in vitro studies of disrupted cell preparations. This is due to both technological limitations of biochemical assays for free radical species and the tremendous scavenging abilities of endogenous cellular free radical defenses. Care must be taken when extending observations derived from in vitro models to intact cell or in vivo systems, because cell disruption may cause rearrangement of autoxidizable components. Also, most in vitro studies of free radical generation do not employ cofactor concentrations approximating those found in vivo. This may lead to inaccurate estimation of in vivo rates of free radical generation. Finally, in an organism, a wide variability in cellular content of free radical-producing components exists, both from organ to organ and within cell types comprising a specific organ. Different environmental factors, respiratory chain compositions or isoenzymes in the cells of an organism also leads to varying specific activities of free radical generation. Thus, each cell type or organ must be separately considered when studying or predicting free radical reactions.

References

1. Aleman, V., and Handler, P. (1967): J. Biol. Chem., 242:4087-4092.
2. Archakov, A.I., Bachmanova, G.I., Isotov, M.V., and Kuznetsova, G.P. (1980): In: Microsomes, Drug Oxidations and Chemical Carcinogenesis, edited by R.W. Estabrook, pp. 289-302, Academic Press, New York.
3. Aust, S.D., Roerig, D.L., and Pederson, T.C. (1972): Biochem. Biophys. Res. Commun., 47:1133-1137.
4. Baccanari, D.P. (1978): Arch. Biochem. Biophys. Res. Commun., 47:1133-1137.
5. Baehner, R.L., Boxer, L.A. and Ingraham, L.M. (1982): In: Free Radicals in Biology, Vol. 5, edited by W.A. Pryor, pp. 91-127, Academic Press, New York.
6. Ballou, D., Palmer, G. and Massey, V. (1969): Biochem. Biophys. Res. Commun., 36:898-904.
7. Boveris, A. (1977): Adv. Exp. Biol. Med., 78:67-98.
8. Boveris, A., and Chance, B. (1973): Biochem. J., 134:707-719.
9. Boveris, A., Oshino, N. and Chance, B. (1972): Biochem. J., 128:619-627.
10. Capdevila, J., Parkhill, L., Chacos, N., Okita, R., and Masters, B.S.S. (1981): Biochem. Biophys. Res. Commun., 101:1357-1361.
11. Chance, B., Boveris, A., and Oshino, N. (1977): In: Alcohol and Aldehyde Metabolizing Systems, edited by R.G. Thurman, J.R. Williamson, H.R. Drott, and Chance, B., pp 261-276, Academic Press, New York.
12. Chance, B., Sies, H., and Boveris, A. (1979): Physiol. Rev., 59:527-672.

13. Chignell, C.G. (1979): In: Spin Labeling, Vol. 2, edited by L.J. Berliner, pp. 223-242, Academic Press, New York.
14. Cohen, G. and Cederbaum, A. (1980): In: Microsomes, Drug Oxidations and Chemical Carcinogenesis, edited by R.W. Estabrook, pp. 307-326, Academic Press, New York.
15. Dionisi, O., Galeotti, T., Terranova, T., and Azzi, A. (1975): Biochem. Biophys. Acta, 403:292-300.
16. Egan, R.W., Paxton, J., and Kuehl, F.A. (1976): J. Biol. Chem., 251:7329-7337.
17. Estabrook, R.W., and Werringloer, J. (1976): In: Drug Metabolism Concepts, edited by M.J. Donald and R.G. Robert, pp. 1-36, American Chemical Society, Washington, D.C.
18. Fisher, D.B. and Kaufman, S. (1973)L J. Biol. Chem., 248:4300-4308.
19. Forman, H.J. and Boveris, A. (1982): In: Free Radicals in Biology, edited by W. Pryor, pp. 65-92, Academic Press, New York.
20. Forman, H.J. and Kennedy, J. (1976): Arch. Biochem. Biophys., 173:219-226.
21. Fridovich, I. (1970): J. Biol. Chem., 245:4035-4042.
22. Fridovich, I. (1975): Annu. Rev. Biochem., 44:147-162.
23. Hawco, F., Hulett, L., and O'Brien, P.J. (1980): In: Microsomes, Drug Oxidations and Chemical Carcinogenesis, edited by R.W. Estabrook, pp. 419-437, Academic Press, New York.
24. Hildebrandt, A.G., Heinemeyer, G. and Roots, I. (1982): Arch. Biochem. Biophys., 216:455-465.
25. Hirata, F. and Hayaishi, O. (1971): J. Biol. Chem. 246:7825-7832.
26. Jobsis, F.F. and LaManna, J.D. (1978): In: Extrapulmonary Manifestations of Respiratory Disease, edited by E.G. Robin, pp. 63-87, Marcel Dekker, New York.
27. Jones, D.P., Eklow, L., Thor, H. and Orrenius, S. (1981): Arch. Biochem. Biophys., 210:505-515.
28. Kalyanaraman, B., Mason, R.P., Tainer, B. and Eling, T.E. (1982): J. Biol. Chem., 257:4764-4770.
29. Mason, R.P. (1979): In: Reviews in Biochemical Toxicology, edited by Hodgson, E., Bend, J.R. and Philpot, R.M., pp. 151-182, Elsevier, North Holland.
30. Mason, R.P. (1982): In: Free Radicals in Biology, Vol. 5, edited by W.A. Pryor, pp. 161-180, Academic Press, New York.
31. Mason, R.P. and Chignell, C.F. (1982) Pharmacol. Rev., 33:189-204.
32. Mason, R.P., Kalyanaraman, B., Tainer, B.E. and Eling, T.E. (1980) J. Biol. Chem., 256:5019-5026.
33. Massey, V., Strickland, S., Mayhew, S.G., Howell, L.G., Engel, P.C., Matthews, R.G.,

Schuman, M. and Sullivan, P.A. (1969) Biochem. Biophys. Res. Commun., 36:898-904.
34. Masters, C., and Holmes, R. (1977): Physiol. Rev., 57:816-857.
35. McCord, J.M. and Day, E.D. (1978) FEBS Lett., 86:139-144.
36. McCord, J.M. and Fridovich, I. (1970): J. Biol. Chem., 245:1374-1381.
37. Misra, H.P. and Fridovich, I. (1972): J. Biol. Chem., 247:2170-2176.
38. Nobuchika, O., Ohki, S., Yamamato, S. and Hayaishi, O. (1978): J. Biol. Chem., 253:5061-5070.
39. Nohl, H. and Hegner, D. (1978): Eur. J. Biochem., 82:563-567.
40. Nohl, H., Jordan, W. and Hegner, D. (1981): FEBS Lett. 123:241-246.
41. Oshino, N. Chance, B., Sies, H. and Bucher, T. (1973): Arch. Biochem. Biophys., 154:117-124.
42. Poole, B. (1975): J. Theor. Biol., 51:149-162.
43. Rajagopalan, K.V. (1980): in: Enzymatic Basis of Detoxication, Vol. 1, edited by Jakoby, W., pp. 295-318, Academic Press, New York.
44. Roy, R.S. and McCord, J.M. (1982): Fed. Proc., 41:767 (abstract).
45. Schroy, C.B. and Baiglow, J.E. (1981): Biochem. Pharmacol., 30:3207-3214.
46. Turrens, J.F. and Boveris, A. (1980): Biochem. J., 191:421-427.
47. Turrens, J.F., Freeman, B.A. and Crapo, J.D. (1982): Arch. Biochem. Biophys., 217:411-419.
48. Turrens, J.F., Freeman, B.A., Levitt, J.G. and Crapo, J.D. (1982): Arch. Biochem. Biophys. 217:401-410.
49. Tyler, D.D. (1975): Biochem. J., 147:793-800.
50. Winterbourn, C.C. (1979): Biochem. J., 182:625-629.

Free Radical Mechanisms in Lipid Peroxidation and Prostaglandins

James F. Mead

Department of Biological Chemistry, School of Medicine; School of Public Health—Nutrition; and Laboratory of Biomedical and Environmental Sciences, University of California at Los Angeles, Los Angeles, California 90024

The mechanisms involved in the chemistry of lipid peroxidation follow directly from those discussed from a theoretical viewpoint in the chapter by Pryor. However, the details are subject to modification by the nature of the lipids as hydrophobic and amphipathic substances usually insoluble in water and, in this medium, tending to form micelles or other aggregates minimizing exposure of the hydrocarbon chains to the aqueous environment. These tendencies are shared by and reflected in the lipid bilayers of most cellular membranes with additional complications stemming from the many lipid and non-lipid components of the membranes obviously capable of taking part in or modifying the reactions of the lipids.

Nevertheless, a great deal of research has been carried out on the mechanisms of lipid peroxidation in the neat state or in moderate to high concentration in aprotic solvents and the information gained from these studies has served as a starting point for the more complex membrane systems. The original ideas of the scientists of the British Rubber Producers Research Society are still largely valid for the initial reactions (e.g. 5,21). These ideas are shown, slightly modernized, in Fig. 1 for linoleic acid. As indicated, a radical initiator, which can be one of a wide variety of substances including some of the products of the reaction, abstracts the hydrogen atom from the diallylic carbon at position 11 and the resulting radical center, delocalized between carbons 9 and 13, readily reacts with molecular oxygen to yield the 9 or 13 peroxy radicals, ROO·, which can in turn act as initiators with the starting linoleic acid or some other hydrogen donor. The peroxy radicals are in this way converted to the 9 and 13 hydroperoxides. Oleic acid gives, as products of peroxidation, the 8, 9, 10 and 11 hydroperoxides while trienoic, tetraenoic and high polyunsaturated fatty acids yield the positional isomers that would be predicted from their double bond positions.

As shown in Fig. 1, termination reactions involve radical addition, usually resulting in the production of dimers. The first reaction is unlikely since the alkyl radical reacts with oxygen more rapidly than it adds to a second radical, particularly in the presence of partial pressures of oxygen necessary for the initiation and propagation reactions. The third reaction is unlikely at the low oxygen pressures usually present in living tissues.

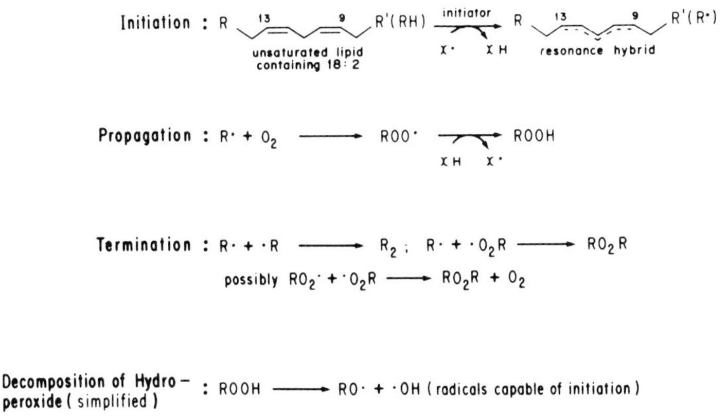

Initiation : $R \underset{\text{unsaturated lipid containing 18:2}}{\overset{13\qquad 9}{\diagup\!=\!\diagdown\!=\!\diagup}} R'(RH) \xrightarrow[X\cdot\quad XH]{\text{initiator}} R \underset{\text{resonance hybrid}}{\overset{13\qquad 9}{\diagup\!\cdots\!\diagdown\!\cdots\!\diagup}} R'(R\cdot)$

Propagation : $R\cdot + O_2 \longrightarrow ROO\cdot \underset{XH\quad X\cdot}{\rightleftharpoons} ROOH$

Termination : $R\cdot + \cdot R \longrightarrow R_2$; $R\cdot + \cdot O_2R \longrightarrow RO_2R$
possibly $RO_2\cdot + \cdot O_2R \longrightarrow RO_2R + O_2$

Decomposition of Hydroperoxide (simplified) : $ROOH \longrightarrow RO\cdot + \cdot OH$ (radicals capable of initiation)

FIG. 1. INITIAL REACTIONS IN THE AUTOXIDATION OF A DIENE (E.G. LINOLEIC ACID). R· = ALKYL RADICAL: RO· = ALKOXYL RADICAL; XH IS ANY HYDROGEN DONOR, POSSIBLY THE STARTING SUBSTANCE, RH.

Although in the initial stages of the reaction the principal products are, as shown, the hydroperoxides, secondary reactions become increasingly important as the reaction progresses.

First, the cis,trans isomeric hydroperoxides shown as products in Fig. 1 can isomerize following reversible loss and addition of oxygen from the peroxy radicals as shown in Fig. 2 for the peroxy-9-octadecadienoates (11,13).

Second, the peroxy radicals and hydroperoxides undergo a number of important reactions leading to a wide variety of secondary products.

They can undergo homolysis, usually assisted by a transition metal ion (Eq. 1):

$$ROOH + M^{n+} \to RO\cdot + OH^- + M^{(n+1)+}$$

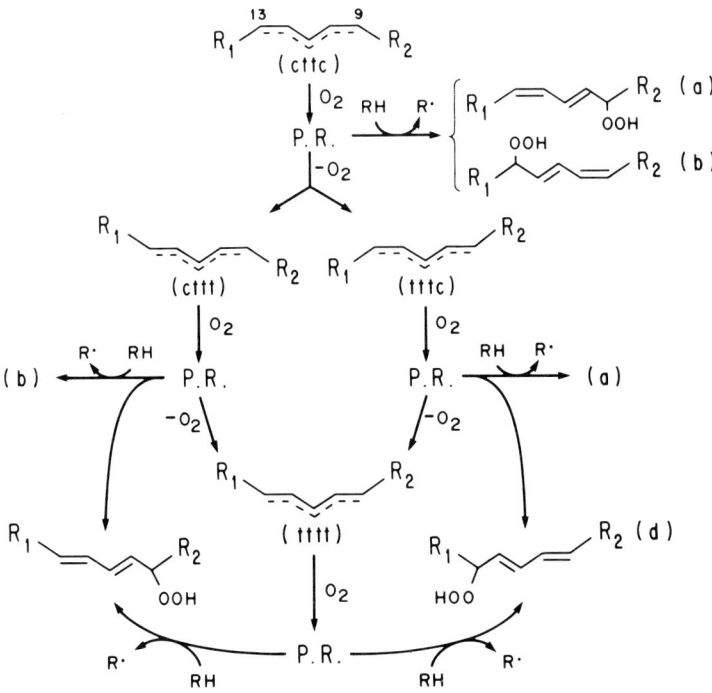

FIG. 2. FORMATION OF GEOMETRIC ISOMERS OF LINOLEATE HYDROPEROXIDES (FROM PORTER ET AL. (1980). c=cis; t = trans; thus cttc: cis,trans,trans,cis. P.R. = peroxyl radical. RH is a hydrogen-containing substance, possibly the starting fatty acid.

This reaction results in the formation of alkoxy radicals which react with any organic substance in their vicinity. The alkoxy radicals can also undergo several well-recognized reactions leading to products, some of which may be used as measures of lipid peroxidation. One of these is the so-called β-scission, in which the c-c bonds on either side of the oxygen-bearing carbon are split (Eq. 2):

$$R_1-\underset{\underset{O\cdot}{|}}{C}H-R_2 \begin{array}{c} \nearrow R_1\cdot + R_2CHO \\ \searrow R_1CHO + R_2\cdot \end{array}$$

This reaction can result, in the case of the ω6 and ω3 fatty acids, in the production of pentane and ethane, respectively, which can serve as measures of the extent of the peroxidation reaction (4,15,20) (Figs. 3 and 4). As can be seen from the equations in Figs. 3 and 4, the other product is usually an unsaturated aldehyde which, when produced as a result of rancidity in unsaturated oils, contributes to their typically unpleasant odors.

It is obvious that scission does not necessarily take place at the positions shown in Figs. 3 and 4 but, particularly in the case of the more highly unsaturated acids, may occur at several sites and result in a wide variety of carbonyl and hydrocarbon products. Moreover, the fate of the radical moiety is more likely to be reaction with oxygen rather than with a hydrogen donor, thus leading to formation of peroxy radicals which initiate new chains and are converted to hydroperoxides. These, in turn, can ultimately undergo β-scission giving rise to an overwhelming array of products.

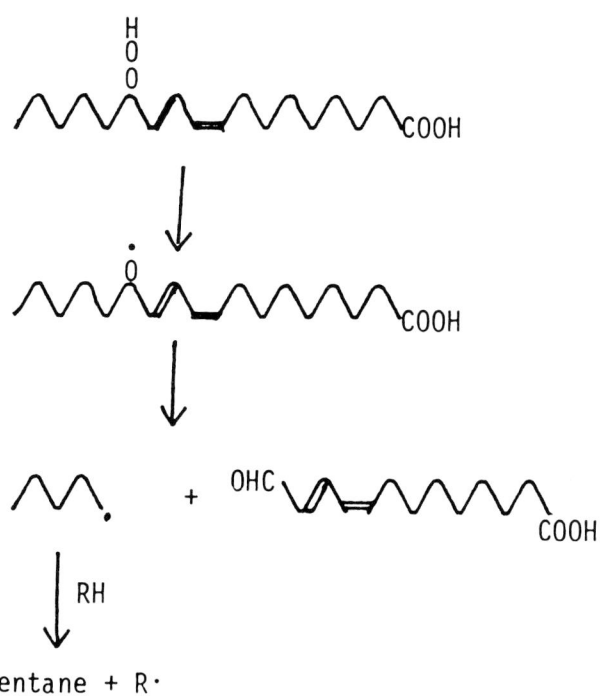

FIG. 3. MECHANISM OF PRODUCTION OF PENTANE FROM 13-HYDROPEROXYLINOLEIC ACID.

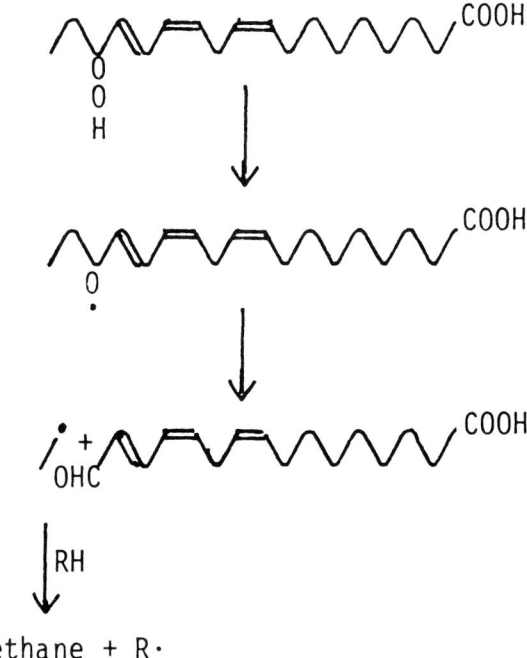

FIG. 4. MECHANISM OF PRODUCTION OF ETHANE FROM 16-HYDROPEROXYLINOLENIC ACID.

If these were not sufficient, all the other possible reactions discussed by Pryor also contribute to the products isolated from peroxidized lipids (6). These reactions include formation of epoxides and other cyclic products by intramolecular addition of peroxy radicals to neighboring double bonds, further reaction of the secondarily formed radicals with additional molecules of the initiating species yielding di- and trihydroxy compounds and formation of ketones and hydroxyketones by decomposition of hydroperoxides. Polymers are also formed by radical addition as, for example, the termination reactions shown in Fig. 1. Altogether, the mixture resulting eventually from long-term peroxidation of bulk lipids is one of great complexity which has not yet been completely resolved in any case (6,7).

It has been a matter of faith that the lipid bilayer present in most biomembranes should be a fertile field for a radical-initiated and -propagated chain reaction. The ordered arrangement of polyunsaturated fatty acids in the bilayer with chains more or less parallel and unsaturated centers more or less in a plane would seem to be the ideal site for such a reaction (Fig. 5). However, a little reflection on the nature of membranes

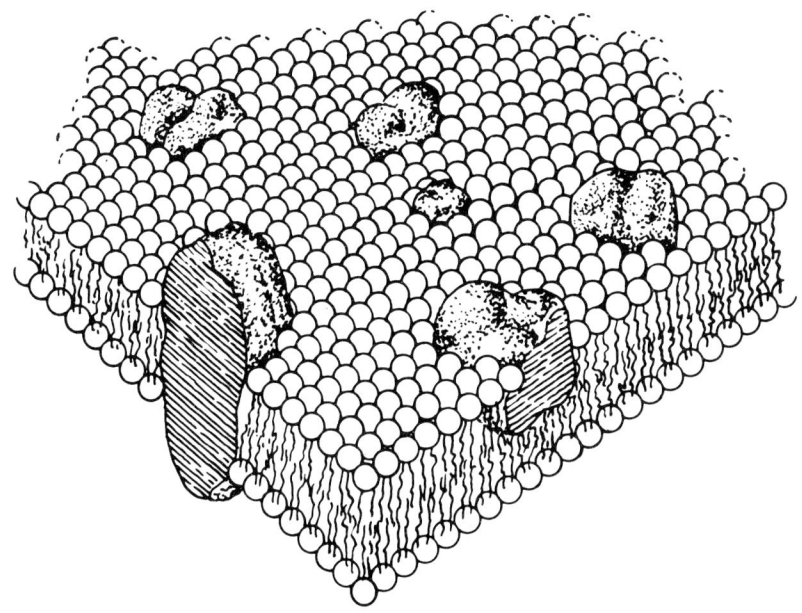

FIG. 5. SCHEMATIC VIEW OF THE LIPID BILAYER MEMBRANE FROM S.J. SINGER AND G.L. NICHOLSON, SCIENCE 175:720 [1972], COPYRIGHT [c] 1972 BY THE AMERICAN ASSOCIATION FOR THE ADVANCEMENT OF SCIENCE.)

reveals that this cannot be true. In addition to the antioxidants, such as the tocopherols, normally present in the membrane, many other substances, including membrane proteins with various functions, would react readily with the oxy radicals limiting their range. Furthermore, very few membrane lipids contain two unsaturated fatty acids and the saturated acids would serve to damp the radical propagation. It is not surprising, therefore, to find that the peroxidation of lipids in both membrane models and biomembranes is at least quantitatively different from that in bulk phase lipid. Thus, it has been reported that in a rudimentary model membrane, fatty acids adsorbed on silica (Fig. 6), the autoxidation reaction shows different kinetics and products (23). Kinetically the reaction appears to be first order (unlike bulk-phase autoxidation) and the major products are the isomeric epoxy-octadecadienoic acids rather than the expected hydroperoxides. Moreover, when intermolecular transfers are inhibited by inclusion of saturated fatty acids in the monolayer, intramolecular reactions give rise to hydroxyepoxides and other similar products. Possible mechanisms for these reactions are

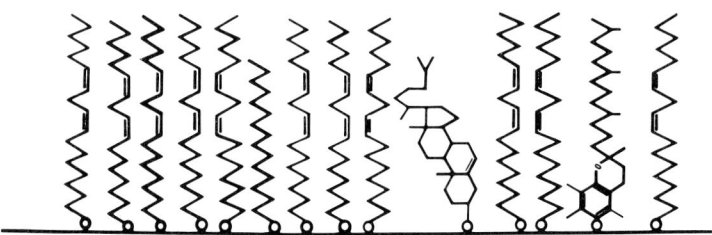

FIG. 6. SCHEMATIC REPRESENTATION OF FATTY ACIDS AND OTHER MEMBRANE COMPONENTS (CHOLESTEROL, α-TOCOPHEROL) ADSORBED ON SILICA.

depicted in Figs. 7, 8 and 8a.

A more complex model membrane system, the mixed saturated-soybean phosphatidyl choline liposome, on autoxidation also gave rise to major products that are not found in large amounts in bulk-phase autoxidation and are somewhat similar to those from the monolayer system. These included epoxides, hydroxyepoxides and di- and trihydroxy fatty acid-containing lipids in addition to the usual hydroperoxides (25). The same products were formed during autoxidation of a suspension of membranes from Acholeplasma laidlawii, indicating that, at least for this relatively simple membrane system, the autoxidation reaction follows closely the course shown by the model systems (24).

These studies and those of several other groups (12) have shown that the autoxidation reaction in lipid membranes does not follow closely the mechanism predicted from the chain reaction as it occurs in bulk phase lipids. Moreover, it is of interest that some of the products identified, such as the fatty acid epoxides and those of cholesterol, if it is present, are potential or known mutagens and carcinogens. Some of the in vivo effects of membrane peroxidation might be explained on this basis.

That some of these products are indeed formed in vivo has been demonstrated by Sevanian et al. (16) who isolated fatty acid and cholesterol epoxides in the rat lung tissue and washings, particularly after a 24-hour exposure to 6 PPM NO_2. Although these substances are potentially harmful to the organism with long-term exposure, they do not necessarily explain the extensive cell damage brought about by short-term exposure to radical initiation such as ozone and nitrogen oxides. Two types of mechanism can be postulated to account for such damage. A direct attack of

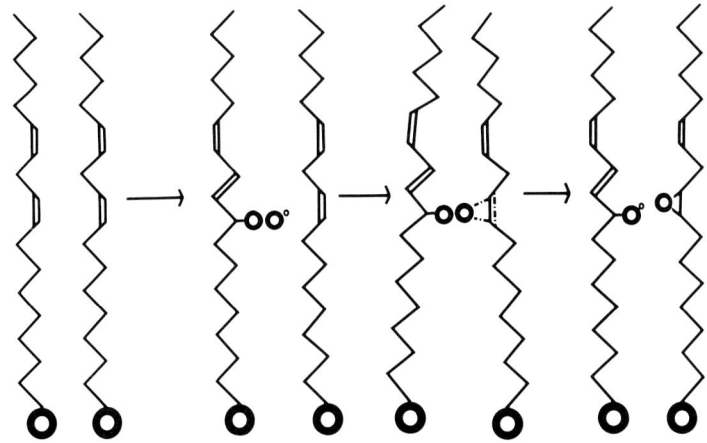

FIG. 7. PROPOSED MECHANISM FOR FORMATION OF EPOXIDES IN A MONOLAYER OF LINOLEIC ACID ADSORBED ON SILICA.

reactive radicals on the membrane constituents would result from the formation of the radicals in the immediate vicinity of the constituents or, possibly, by the formation of more stable radicals, such as $O_2^-\cdot$, that might travel greater distances before being converted to more active species (as $HO_2\cdot$) at susceptible sites (Fig. 5) (1). That direct reaction of this type occurs is evident from the formation of several tissue pigments such as lipofuscin (aging pigment) or ceroid in susceptible tissues. Lipofuscin, which increases in tissues such as heart, liver and brain as a function of time and metabolic rate is found to consist of peroxidized lipid and denatured membrane protein, the combination of which forms an intractable mass ultimately sequestered in the lysosomes (19).

In addition to the direct reaction, an indirect type of cell damage can also occur as the result of the formation of oxidized fatty acids in the phospholipids of the lipid bilayer. Although these substances are not particularly damaging by themselves, their presence in the bilayer lipids may trigger an abortive protective mechanism. It has been shown that inclusion of an oxidized fatty acid (such as 9,10-epoxy-octadecenoic acid) in the 2-position of phosphatidyl choline in place of its precursor oleic acid results in a two-fold activation of phospholipase A_2 (17) and a many-fold activation of phospholipase C (9). These reactions are illustrated in Fig. 9. That the subsequent increased rate of hydrolysis of membrane phospholipids can result in membrane destruction and cell death has been shown by Shier and Du Bourdieu (18), who reported that CA^{2+} + ionaphore A23187 have a similar effect stemming from an increased rate of hydrolysis

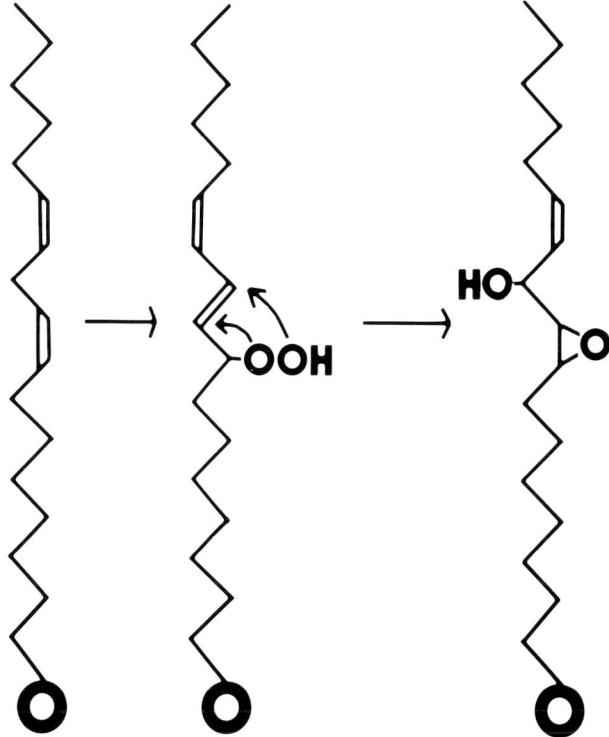

FIG. 8. PROPOSED MECHANISM FOR FORMATION OF HYDROXYEPOXIDES IN A MONOLAYER OF LINOLEIC ACID PLUS SATURATED FATTY ACID.

brought about by this means of phopholipase activation.

Thus, whether or not the very damaging products of lipid peroxidation seen in bulk-phase autoxidation are formed in appreciable amounts in vivo, nevertheless, magnification of the effect of the much less damaging oxidized fatty acids through phospholipase activation could produce the cell damage seen as a result of a peroxidation reaction.

The picture painted so far is one of attack on the cellular defense mechanisms by externally initiated peroxidation. Cells are also under attack by internally-generated oxy-radicals produced in the very reactions used by the cell to generate energy. Despite the many defense systems such as sequestration of susceptible structures from damaging substances, presence of antioxidants and radical traps and enzymes specifically designed to remove radicals and oxidizing agents, some of the oxy-radicals

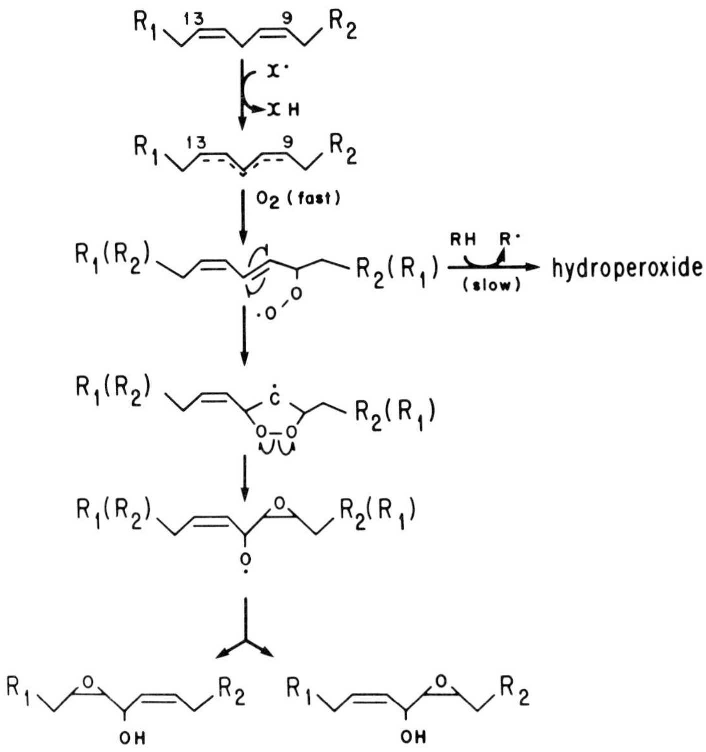

FIG. 8a. PROPOSED MECHANISM FOR THE FORMATION OF HYDROXYEPOXIDES IN BULK PHASE FATTY ACIDS.

produced during the reduction of oxygen in metabolic oxidations undoubtedly escape and create some damage. In particular, the many flavin-mediated oxidations result in production of peroxides. Moreover, a major family of messengers produced in all cells in response to a great variety of stimuli is formed in reactions almost precisely like those discussed above. These, of course, are the prostanoids and leukotrienes and related substances. The major difference between the uncontrolled mechanisms of autoxidation and those leading to eicosenoid formation is that the latter are enzymatically controlled and directed to the formation of specific products. Indeed, some of these products are formed during non-enzymatic peroxidation of the fatty acid precursors of the eicosenoids such as arachidonic acid (14). In Fig. 10 is presented a simplified chart of the major pathways of arachidonic acid oxidation via the cyclooxygenase and lipoxygenase pathways. Hemler and Lands (8) have proposed a mechanism for the formation

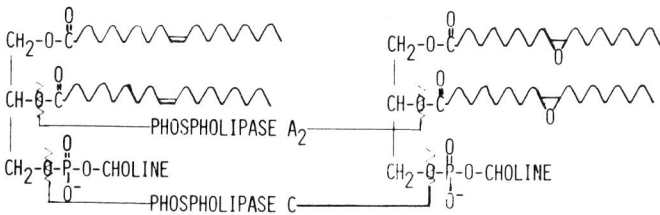

PC (N MOL/MG LIPASE/MIN)   PC EPOXIDE (N MOL/MG LIPASE/MIN)
51.0 ± 6.9   95.0 ± 10.4

FIG. 9. PHOSPHOLIPASES A_2 (C. ADAMANTHEUS) AND C (C. WELCHII) ACTIVITIES AGAINST 1-PALMITOYL, 2-9(10)EPOXYSTEAROYLGLYCERYLPHOSPHORYLCHOLINE AND 1-PALMITOYL, 2-OLEOYLGLYCERYLPHOSPHORYLCHOLINE.

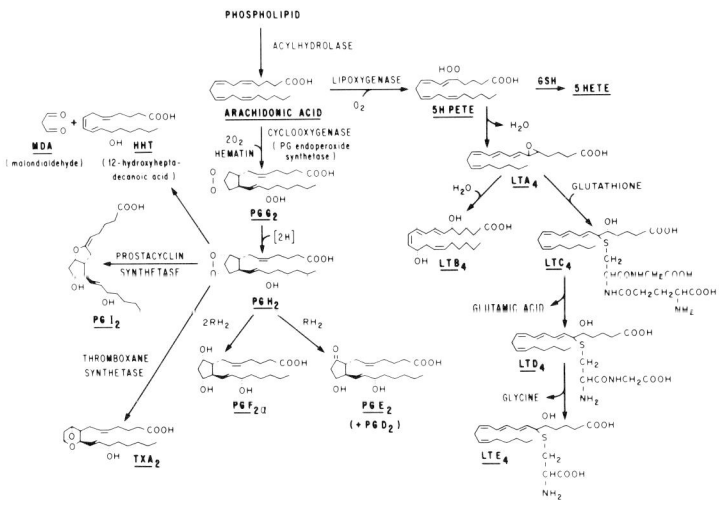

FIG. 10. SIMPLIFIED SCHEME OF MAJOR PATHWAYS OF ARACHIDONIC ACID OXIDATION VIA THE CYCLOOXYGENASE AND LIPOXYGENASE PATHWAYS.

of PGG$_2$ via the cyclooxygenase pathway involving the reaction of a fatty acid hydroperoxide with the iron-containing enzyme giving an enzyme FeII-peroxy radical complex. This complex then abstracts the 13-s-hydrogen from arachidonate and the resulting radical adds two molecules of oxygen and simultaneously cyclizes to yield PGG$_2$ which then acts as the intermediate in the formation of the cyclic prostanoids (see Fig. 11).

The "lipoxygenase" pathway, leading first to fatty acid hydroperoxides and then to the leukotrienes, also consists of enzymatically controlled oxidations leading, in each case to specific products and thus differing from the random attack seen in autoxidation. In Fig. 12 is shown a possible scheme for the formation of 5s-hydroperoxy-6-trans-8,11,14-cis-eicosatetraenoic acid (5-HPETE) and its conversion to 5s,6s,5(6)-oxide-7,9-trans,11,14-cis-eicosatetraenoic acid (LTA$_4$), the precursor of the other leukotrienes (2,3,10). This reaction is reminiscent of the proposed scheme for formation of epoxides on silica gel monolayers (22).

Thus we have seen that peroxidation products of the unsaturated fatty acids of the membrane lipids can occur as a result of the failure of the many protective systems in the face of attack by external oxidants or the continuous operation of certain oxidation enzymes and that they can also occur as necessary intermediates in the formation of certain hormones. It can also be shown that these products can attack the membrane constituents and damage the membranes in at least two ways. There is a good deal of evidence that such attacks occur and the products of the resulting reactions are found in the tissues and increase as a function of time and exposure to oxidants. What has not been shown is that these reactions lead to the observed phenomena of aging and that by slowing them longevity can be increased. This would seem to be a worthy goal for scientists in this field for some time to come.

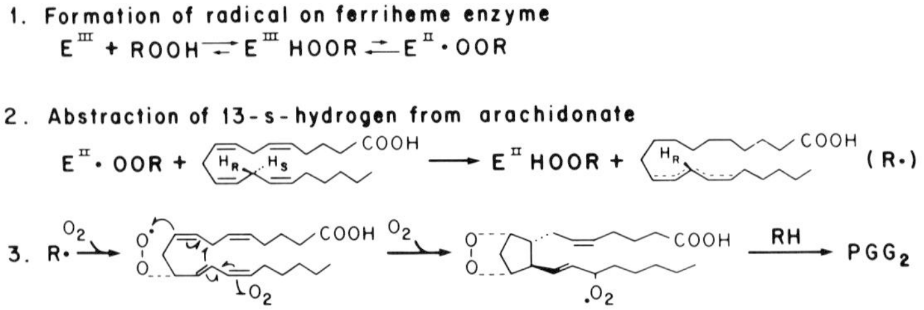

1. Formation of radical on ferriheme enzyme
   $E^{III} + ROOH \rightleftarrows E^{III} HOOR \rightleftarrows E^{II} \cdot OOR$

2. Abstraction of 13-s-hydrogen from arachidonate
   $E^{II} \cdot OOR +$ (H$_R$, H$_S$)~~~COOH $\longrightarrow E^{II}$ HOOR + (H$_R$)~~~COOH (R·)

3. R· $\xrightarrow{O_2}$ ... $\xrightarrow{O_2}$ ... $\xrightarrow{RH}$ PGG$_2$

FIG. 11. PROPOSED MECHANISM FOR FORMATION OF PGG$_2$ FROM ARACHIDONIC ACID (FROM HEMLER AND LANDS, 1980).

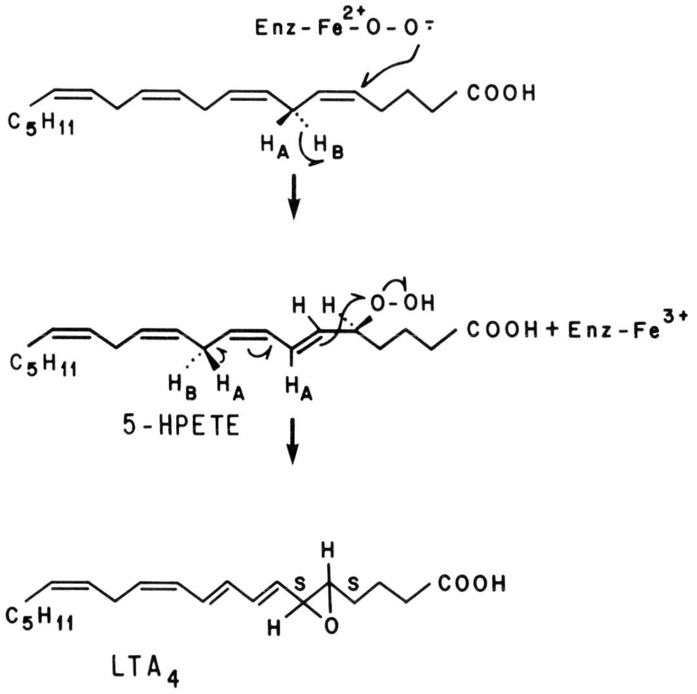

FIG. 12. PROPOSED MECHANISM FOR FORMATION OF HPETE AND LTA$_4$ FROM ARACHIDONIC ACID.

REFERENCES

1. Bielski, B.H.J., Arudi, R.L., and Sutherland, M.W. (1983): J. Biol. Chem., 258:4759-4761.
2. Borgeat, P., and Samuelsson, B. (1979). Proc. Natl. Acad. Sci. USA, 76:3213-3217.
3. Corey, E.J., Arai, Y., and Mioskowski, C. (1979): J. Am. Chem. Soc., 101:6748-6749.
4. Dillard, C.J., Dumelin, E.E., and Tappel, A.L. (1977): Lipids, 12:109-114.
5. Farmer, E.H. (1942): Trans. Faraday Soc., 38:340-348; (1946): 42:228-236.
6. Frankel, E.N. (1982): In: Progress in Lipid Research, Vol. 19, edited by R.T. Holman, pp. 1-22. Pergamon Press, Oxford.
7. Gardner, H.W., Kleiman, R., and Weisleder, D. (1974): Lipids, 9:696-706.
8. Hemler, M.E., and Lands, W.E.M. (1980): J. Biol. Chem., 255: 6253-6261.

9. Mead, J.F., and Elepano, M.G., Unpublished observations.
10. Popjak, G., Personal communication.
11. Porter, N.A., Weber, G.A., Weenen, H., and Kahn, J.A. (1980): J. Am. Chem. Soc., 102:5597.
12. Porter, N.A., Wolf, R.A., and Weenen, H. (1980): Lipids, 15: 163-167.
13. Porter, N.A., Lehman, L.S., Weber, B.A., and Smith, K.J. (1981): J. Am. Chem. Soc., 103:6447-6455.
14. Pryor, W.A., and Stanley, J.P. (1975): J. Org. Chem. 40: 3615-3617.
15. Riely, C.A., Cohen, G., and Lieberman, M. (1974): Fed. Proc., 33:233.
16. Sevanian, A., Mead, J.F., and Stein, R.A. (1979): Lipids, 14: 634-643.
17. Sevanian, A., Stein, R.A., and Mead, J.F. (1981): Lipids, 16: 781-789.
18. Shier, W.T., and Du Bourdieu, D.J. (1982): Biochem. Biophys. Res. Comm., 109:106-112.
19. Shimasaki, H., Ueta, N., and Privett, O.S. (1982): Lipids, 17: 878-883.
20. Tappel, A.L., and Dillard, G.J. (1981): Fed. Proc., 40:174-178.
21. Uri, N. (1961): In: Autoxidation and Antioxidants, edited by W.O. Lundberg, pp. 55-106. Interscience Publishers, New York.
22. Wu, G.-S., Stein, R.A., and Mead, J.F. (1977): Lipids, 12: 971-978.
23. Wu, G.-S., Stein, R.A., and Mead, J.F. (1978): Lipids, 13: 517-524.
24. Wu, G.-S. Sohlberg, E., Stein, R.A., Mead, J.F., and McElhaney, R.N. (1981): J. Am. Oil Chem. Soc., 58:591A.
25. Wu, G.-S., Stein, R.A., and Mead, J.F. (1982): Lipids, 17: 403-413.

Free Radicals in Molecular Biology, Aging, and Disease, edited by D. Armstrong et al. Raven Press, New York © 1984.

Free Radicals in Melanin Formation, Structure, and Reactions

Roger C. Sealy

National Biomedical ESR Center, Department of Radiology, Medical College of Wisconsin, Milwaukee, Wisconsin 53226

Free radicals are involved in many aspects of melanin chemistry. Potentially toxic oxygen radicals are produced during melanogenesis and melanin destruction during irradiation. However, many melanin reactions appear to be protective since they lead to the production of apparently non-damaging melanin radicals. Examples of such reactions include dissipation of visible and ultraviolet light and the scavenging of reactive radicals (e.g. O_2^-), excited states of sensitizer molecules, and singlet oxygen.

Melanins (26) are heterogeneous amorphous polymers from the oxidation of tyrosine, 3,4-dihydroxyphenylalanine (dopa) and related phenols and catecholamines. They are mostly found in skin, hair and eyes, where they act as photoprotective agents (29). These melanins have as major precursors dopa and cysteinyldopa (32) and are classified either as eumelanins or pheomelanins; eumelanins are largely derived from dopa, pheomelanins from cysteinyldopa (32). Melanins also are found in small amounts in the brain (this melanin is often termed "neuromelanin") and in the inner ear. In these locations the function of the melanin is not clear. Neuromelanin is thought to be derived from neurotransmitter substances, for example dopamine, norepinephrine and 5-hydroxytryptophan (1,27,42). Naturally occurring melanins are generally associated with protein (26).

Free radicals play a role in many reactions involving melanins (36), from their formation through to their destruction. Radical species are produced during melanogenesis; melanin polymers contain free radicals under all known experimental conditions; melanin reactions (including photoreactions) typically involve a change in free radical content, a change that may be reversible or irreversible. We shall consider these three aspects of melanin chemistry separately in the sections that follow. We will also consider the subject of melanin-related free radical toxicity. Here it is necessary to keep in mind that although many free radicals are extremely reactive and potentially toxic, this is not the case for all free radicals. (For example, the effectiveness of antioxidants is believed to reflect their ability to effect the conversion of a reactive and potentially damaging radical to a non-damaging radical derived from the antioxidant.) The range of reactivities of free radicals found in melanin systems is very wide.

RADICALS IN MELANOGENESIS

Melanins can be produced from their phenolic precursors by either enzymatic or chemical oxidation (26). Enzymatic oxidation may be with tyrosinase/O_2, peroxidase/H_2O_2 or other systems (e.g. mitochondrial incubations (44)). Autoxidation is perhaps the most common form of chemical oxidation of melanin precursors and is fast in aerated solutions. The melanin polymers that are the end products of these various oxidative procedures have similar chemical properties (26).

It is felt that the major enzymatic contribution to melanogenesis in vivo involves tyrosinase, an enzyme found in pigment cells (melanocytes) (19). The molecular intermediates formed in the conversion of dopa to melanin have been identified in the work of Raper (30) and Mason (24). Major intermediates are dopaquinone, dopachrome and 5,6-dihydroxyindole, as shown in Scheme 1.

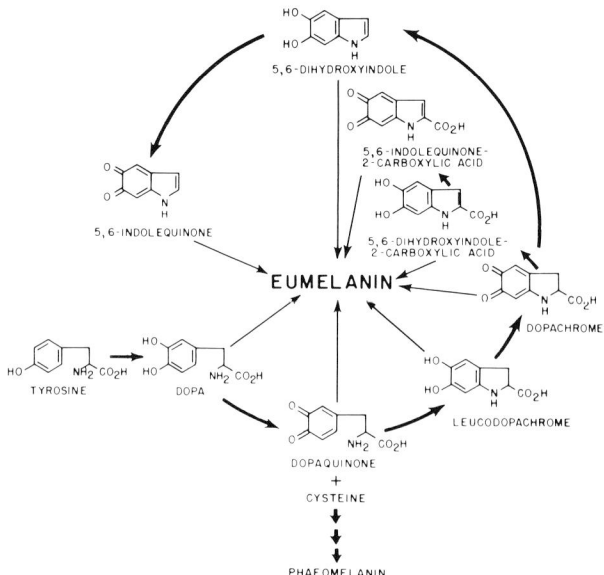

SCHEME 1. Melanogenesis of eumelanin from tyrosine and dopa. From (36), with permission.

It has generally been considered that only the oxidation of tyrosine through to dopaquinone is enzymatic, and that subsequent transformations are independent of enzyme. However, evidence has recently been presented (23) for factors that accelerate the breakdown of dopachrome and inhibit the oxidation of 5,6-dihydroxyindole.

Much of the work that has been carried out on melanogenesis has focussed on the above molecular intermediates. However, it seems clear (3,45) that free radicals also are generated during the oxidation and polymerization. Free radicals can be produced from melanin precursors either <u>directly</u>, via a one-electron oxidation, or <u>indirectly,</u> via an initial two-electron oxidation and a subsequent secondary reaction. Examples of direct and indirect radical production have been described in both enzymatic and chemical systems (e.g. reactions (1)-(3)).

$$2 \text{ catechol} + O_2 \xrightarrow{\text{tyrosinase}} 2 \underline{o}\text{-quinone} + 2H_2O \tag{1a}$$

$$\text{catechol} + \underline{o}\text{-quinone} \rightleftharpoons 2 \text{ semiquinone} + 2H^+ \tag{1b}$$

$$2 \text{ catechol} + H_2O_2 \xrightarrow{\text{peroxidase}} 2 \text{ semiquinone} + 2H_2O + 2H^+ \tag{2}$$

$$\text{catechol} + O_2 \xrightarrow{\text{autoxidation}} \text{semiquinone} + O_2^- + 2H^+ \tag{3}$$

Evidence for reactions 1a,b has come from studies of the oxidation of catechol with mushroom tyrosinase (25). Production of semiquinone was inferred to occur via the sequence shown. Production of semiquinones from dopa also has been demonstrated in a tyrosinase system (13). In the latter study it was found that radical concentrations could be greatly enhanced through spin stabilization, i.e. stabilization of radical products by complexation with metal ions. However, because of the short lifetime of dopaquinone, the mechanism of radical production from dopa may be more complex than is the case with catechol. Another oxidant (e.g. dopachrome) may be involved in the secondary step in which radicals are generated.

One-electron oxidations of melanin precursors by peroxidase/ H_2O_2 and by autoxidation also have been studied, in several instances using the spin stabilization approach. Semiquinone free radicals have been detected and identified from horseradish peroxidase-catalyzed oxidation of catechol (25), dopa (20), epinephrine and other catecholamines (22). The semiquinone products show little tendency to react with oxygen to form superoxide (21). In some cases radicals derived from products formed later in the melanogenesis pathway were identified, both in peroxidase and autoxidation systems (23,37). These secondary radicals are derived from either $\underline{o}$-quinones, aminochromes or dihydroxyindoles (e.g. Scheme 2). It appears (see below) that secondary radicals derived from oxidized products predominate in the structure of the final melanin polymers.

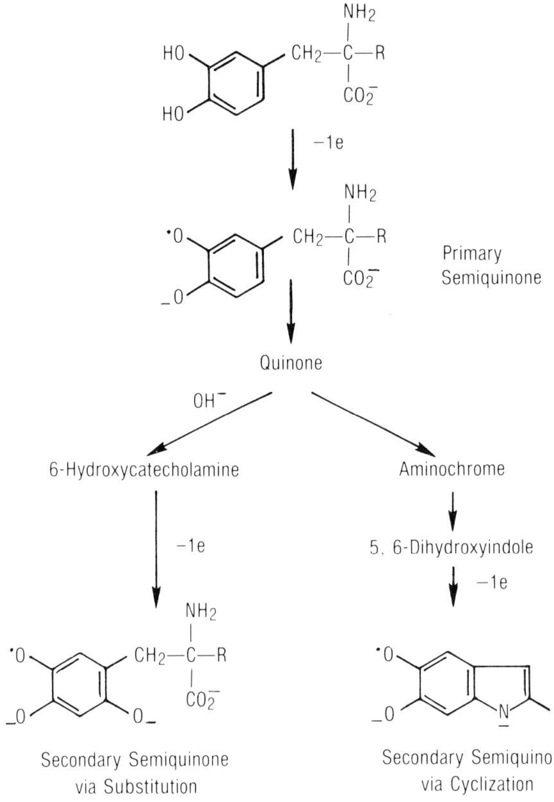

SCHEME 2. Primary and secondary free radicals formed during melanogenesis (via autoxidation).

There is evidence (9,17,31) that radicals (semiquinones, $O_2^{\overline{\cdot}}$) and simple molecular products (o-quinones, H_2O_2) generated during melanogenesis are toxic species. Thus tissues in which melanogenesis is occurring are subject to oxidative stress (18). Binding to protein has been shown to occur during metabolism of various catechols (e.g.(31)), and it has been argued that the active species are the semiquinone and/or quinone. Cells with tyrosinase activity may be more susceptible to damage by melanin precursors than cells lacking this activity (46).

RADICALS IN MELANIN STRUCTURE

Early ESR work by Commoner and his coworkers (7) showed that melanin in tissue contained free radicals. Subsequent extended studies of melanins by Blois et al. (2) and by Grady and Borg (16) identified one main type of ESR spectrum, which at X-band showed a single line of width 4 G with a g-value close to 2.004. More recent work on a variety of natural and synthetic

eumelanins and pheomelanins has revealed two distinct types of
ESR spectra (38), consistent with the presence of two kinds of
free radical. The first type of spectrum is found to the
largest degree in eumelanins and is that described above. The
second type is found to the largest degree in pheomelanins (39)
and shows three spectral features, an overall width of about 30
G, and g = 2.005. It was shown (38) that the two kinds of
spectra are associated with polymer units derived from dopa and
cysteinyldopa respectively and that several natural melanins
have ESR spectra that are composites of the two main types.
Parameterization of the composite spectra allows an estimate to
be made of the relative amounts of dopa and cysteinyldopa incorporated into the melanin polymer (38).

Information on the chemical structures of these polymer
radicals has been obtained from experiments with di- and tri-
positive metal ions. Changes in ESR spectra have been shown to
be consistent with complex formation between the metal ion and
a polymer radical that has a chelating structure (11). (This
complexation is analogous to that observed between metal ions and
semiquinone radicals from melanin precursors (11)). The structure
of the chelating radical can be further defined from linewidth
changes that reflect hyperfine splittings to metal ions; the
magnitude of the hyperfine splitting to a particular metal is
sensitive to the detailed radical structure, as has been demonstrated in experiments with complexes of radicals from melanin
precursors (14,41). Thus metal-complexed radicals in melanins
derived from dopa, catechol and cysteinyldopa have splittings to
cadmium isotopes of about 3.5, 7 and 15 G, consistent with the
complexed radicals being o-indolesemiquinones, o-semiquinones and
o-semiquinonimines respectively (35), as shown in Scheme 3.

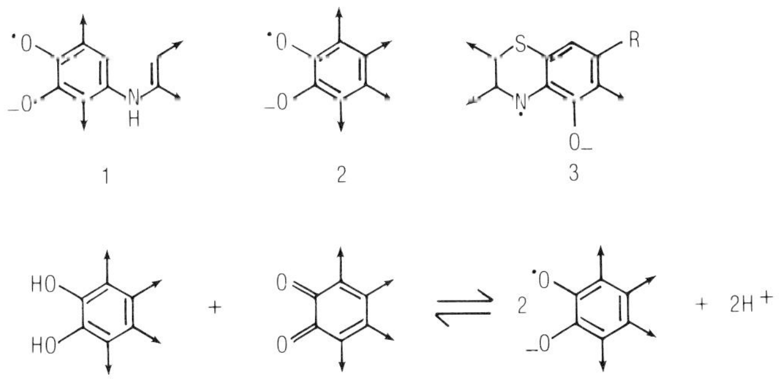

SCHEME 3. (Upper) Partial structures (1-3) of semiquinone
free radicals found in melanins from dopa, catechol and
cysteinyldopa respectively. (Lower) Example of a redox
equilibrium involving semiquinone free radicals.

The free radical content of melanins is typically around 2×10^{18} spins/g for an aqueous suspension at neutral pH. This value corresponds to about one free radical per 2000 polymer units (36). Values up to about ten times higher have been reported in some cases, particularly for melanins that have been dried (36). Although some reactivity of melanin free radicals is apparent, as for example with metal ions (see also below), it seems unlikely that these melanin radicals are toxic; they are clearly of limited mobility (being part of the structure of a macromolecule) and thermodynamically rather stable. They show no tendency to react with oxygen or other electron acceptors (36).

RADICALS IN MELANIN REACTIONS

Changes in melanin radical concentration can be induced by diamagnetic metal ions (11) and other reagents which do not cause a net oxidation or reduction of the polymer. For example, increases in temperature (5) and pH (6,16) both lead to increases in radical content of melanin suspensions. On the basis of such observations it has been proposed (36) that a majority (although not necessarily all) of the melanin free radicals exist in equilibrium with non-radical polymer units, for example the corresponding quinones and catechols (Scheme 3). This behavior again is analogous to that found for model radicals (reaction (1b)). In general, reagents that change the net redox level of the polymer also affect the radical content (33,36), in ways that are qualitatively consistent with the presence of equilibria such as that shown in the Scheme. Changes in radical content following modification of phenolic hydroxyl groups in the polymer also have been observed (33).

In view of the role of melanins in photoprotection, it is not surprising that their reactions that have received the most attention are those involving light. Most studies have been carried out on isolated pigment systems, in which two main kinds of reaction have been observed (36) during irradiation: increased production of melanin radicals and increased consumption of oxygen in the medium.

During irradiation of melanins with either visible or ultraviolet light the melanin free radical content is increased (8,12). Provided that light of > 300 nm is used the light-induced free radicals decay reversibly when irradiation is terminated, unless the irradiation is carried out at cryogenic temperatures in which case the light-induced radicals are partially stabilized. Both fast and slow components to the radical decay have been observed, and magnetic phenomena detected at short times after photolysis have been attributed to the free radicals having a triplet state precursor (12).

Irradiation of melanins in the presence of air promotes oxygen consumption (43). The oxygen is reduced to superoxide and hydrogen peroxide (4,10,34). The reaction is probably quite general: light-promoted oxygen consumption also has been observed in mammalian cells containing phagocytized melanin (28). Reduction of oxygen to $O_2^{\overline{\cdot}}$ and H_2O_2 is potentially a reaction of some consequence in view of the damaging nature of the products, which can generate hydroxyl radicals via a metal-ion catalyzed Haber-Weiss reaction. However, it remains to be demonstrated that production of reduced species of oxygen during irradiation leads to significant damage. The quantum yield for oxygen consumption is low for visible and near UV light (34). In addition, melanin is a scavenger for superoxide radicals (15) and, at least on the basis of its chemical structure, should also be an excellent scavenger for any hydroxyl radicals produced. It may even be possible that the net effect of oxygen consumption in vivo is protective (28), since anaerobic cells are less susceptible to light-induced damage.

Melanins quench excited states of other molecules with the production of melanin free radicals. This has been demonstrated (40) for synthetic melanin (from dopa) in the presence of the sensitizer Rose Bengal. Under anaerobic conditions the rate of melanin radical production was increased up to 20-fold by irradiation at the absorption maximum of the sensitizer. The reaction is reversible; no net degradation of either sensitizer or melanin occurs. In the presence of air the rate of oxygen consumption can be stimulated by two orders of magnitude above the unsensitized rate. The enhancement was greater in D_2O and quenched by azide, providing evidence for a reaction involving singlet oxygen. Oxygen consumption was correlated with polymer bleaching and irreversible production of melanin free radicals.

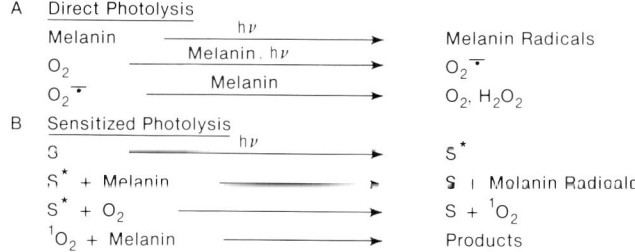

SCHEME 4. Mechanisms of free radical generation during photoreactions of melanins.

Known photoreactions of melanin are summarized below and in Scheme 4. Melanins (i) degrade light to heat, at least in part via production of melanin radicals; (ii) during irradiation reduce oxygen to superoxide and hydrogen peroxide; (iii) scavenge reactive free radicals (e.g. $O_2^{\overline{\cdot}}$) and excited states (including $^1O_2$). In each case an increase in the free radical content of the polymer occurs. Most changes in free radical content are reversible; exceptions occur with short-wave UV and reaction with $^1O_2$.

Acknowledgment. Much of the work described in this chapter was supported by grants from the National Institutes of Health (AM-26950, GM-29035 and RR-01008) and the National Science Foundation (PCM-7823206).

REFERENCES

1. Ambani, L.M. and Van Woert, M.H. (1973): Trans. Am. Neurol. Assoc. 7-10.
2. Blois, M.S., Zahlan, A.B., and Maling, J.E. (1964): Biophys. J. 4:471-490.
3. Blois, M.S. (1978): In: Photochemical and Photobiological Reviews, Vol. 3, edited by K.C. Smith, pp. 115-134. Plenum, New York.
4. Chedekel, M.R., Smith, S.K., Post, P.W., Pokora, A., and Vessell, D.J. (1978): Proc. Natl. Acad. Sci. (U.S.A.), 75:5395-5399.
5. Chio, S.-S., Hyde, J.S., and Sealy, R.C. (1980): Arch. Biochem. Biophys. 199:133-139.
6. Chio, S.-S., Hyde, J.S., and Sealy, R.C. (1982): Arch. Biochem. Biophys. 215:100-106.
7. Commoner, B., Townsend, J., and Pake, G.E. (1954): Nature 174:429-433.
8. Cope, F.W., Sever, R.J., and Polis, B.D. (1963): Arch. Biochem. Biophys. 100:171-177.
9. Dybing, E., Nelson, S.D., Mitchell, J. R., Sasame, H.A., and Gillette, J.R. (1976): Mol. Pharmacol. 12:911-920.
10. Felix, C.C., Hyde, J.S., Sarna, T., and Sealy, R.C. (1978): Biochem. Biophys. Res. Commun. 84:335-341.
11. Felix, C.C., Hyde, J.S., Sarna, T., and Sealy, R.C. (1978): J. Am. Chem. Soc. 100:3922-3926.
12. Felix, C.C., Hyde, J.S., and Sealy, R.C. (1979): Biochem. Biophys. Res. Commun. 88:456-461.
13. Felix, C.C. and Sealy, R.C. (1981): J. Am. Chem. Soc. 103:2831-2836.
14. Felix, C.C. and Sealy, R.C. (1982): J. Am. Chem. Soc. 104:1555-1560.
15. Goodchild, N.T., Kwock, L., and Lin, P.-S. (1981): In: Oxygen and Oxy-Radicals in Chemistry and Biology, edited by M.A.J. Rodgers and E. L. Powers, pp. 645-648. Academic Press, New York.
16. Grady, F.J. and Borg, D.C. (1968): J. Am. Chem. Soc. 90:2949-2952.
17. Graham, D.G., Tiffany, S.M., and Vogel, F.S. (1978): J. Invest. Dermatol. 70:113-116.
18. Hochstein, P. and Cohen, G. (1963): Ann. N.Y. Acad. Sci. U.S. 100:876-886.
19. Jimbow, K., Quevedo, W.C., Jr., Fitzpatrick, T.B., and Szabo, G. (1976): J. Invest. Dermatol. 67:72-89.

20. Kalyanaraman, B. and Sealy, R.C. (1982): Biochem. Biophys. Res. Commun. 106:1119-1125.
21. Kalyanaraman, B., Felix, C.C., and Sealy, R.C. (1982): Photochem. Photobiol. 36:5-12.
22. Kalyanaraman, B., Felix, C.C., and Sealy, R.C. (1983): J. Biol. Chem. in press.
23. Korner, A. and Pawalek, J. (1982): Science 217:1163-1165.
24. Mason, H.S. (1955): Adv. Enzymol. 16:105-184.
25. Mason, H.S., Spencer, E., and Yamazaki, I. (1961): Biochem. Biophys. Res. Commun. 4:236-238.
26. Nicolaus, R.A. (1968) Melanins, Hermann, Paris.
27. Okun, M., Edelstein, L., Or, N., Hamada, G., and Donnellan, B. (1970): Life Sci. 9:491-505.
28. Pajak, S., Hopwood, L.E., Hyde, J.S., Felix, C.C., Sealy, R.C., Kushnaryov, V.M., and Hatchell, M.C. (1983): Exp. Cell Res., in press.
29. Pathak, M. and Fitzpatrick, T.B. (1974): In: Sunlight and Man, edited by T.B. Fitzpatrick, M.A. Pathak, L.C. Harber, M. Seiji, and A. Kukita, p. 725. University of Tokyo Press, Tokyo.
30. Raper, H.S. (1972): Physiol. Rev. 8:245-284.
31. Rotman, A., Daly, J.W., and Creveling, C.R. (1976): Mol. Pharmacol. 12:887-899.
32. Prota, G. (1980): J. Invest. Dermatol. 75:122-127.
33. Sarna, T., Duleba, A., Korytowski, W., and Swartz, H.M. (1980): Arch. Biochem. Biophys. 200:140-148.
34. Sarna, T. and Sealy, R.C. (1984): Photochem. Photobiol. 39:69-74.
35. Sealy, R.C. (1984): Methods Enzymology 105:479-483.
36. Sealy, R.C., Felix, C.C., Hyde, J.S., and Swartz, H.M. (1980): In: Free Radicals in Biology, Vol. 4, edited by W.A. Pryor, pp. 209-259. Academic Press, New York.
37. Sealy, R.C., Puzyna, W., Felix, C.C., and Kalyanaraman, B., submitted for publication.
38. Sealy, R.C., Hyde, J.S., Felix, C.C., Menon, I.A., and Prota, G. (1982): Science 217:545-547.
39. Sealy, R.C., Hyde, J.S., Felix, C.C., Menon, I.A., Prota, G., Swartz, H.M., Persad, S., and Haberman, H.F. (1982): Proc. Natl. Acad. Sci. (U.S.A.) 79:2885-2889.
40. Sealy, R.C., Sarna, T., Wanner, L.J., and Reszka, K., submitted for publication.
41. Stegmann, H.B., Ulmschneider, R.B., Hieke, K., and Scheffler, K. (1976): J. Organometal. Chem. 118:259-287.
42. Uemura, T., Shimazu, T., Miura, R., and Yamano, T. (1980): Biochem. Biophys. Res. Commun. 93:1074-1081.
43. Van Woert, M.H. (1968): Proc. Soc. Exp. Biol. Med. 129:165-171.
44. Van Woert, M.H., Prasad, K.N., and Borg, D.C. (1967): J. Neurochem. 14:707-716.

45. Wertz, J.E., Reitz, D.C., and Dravnieks, F. (1961): In: *Free Radicals in Biological Systems*, edited by M.S. Blois, pp. 183-193. Academic Press, New York.
46. Wick, M.M., Byers, L., and Ratliff, J. (1979): *J. Invest. Dermatol.* 72:67-79.

Superoxide Dismutase: An Antioxidant Defense Enzyme

Hosni M. Hassan

Departments of Food Science and Microbiology, North Carolina State University, Raleigh, North Carolina 27695

Aging and oxygen-toxicity correlate in that both are mediated by oxygen free radicals. The accumulation of "age pigments", which are lipid peroxides, supports this notion. This presentation is focused on the toxic effects of oxy-radicals in eukaryotes and in prokaryotes. By using such a simple model as Escherichia coli a great deal of knowledge is learned in a short time period and the information gained is generally helpful in studying the more complex eukaryotic organisms, including man.

MOLECULAR OXYGEN AND OXY-RADICALS

Dioxygen, in the ground state, is paramagnetic containing two unpaired electrons with parallel spins. This leads to a spin restriction which hinders the insertion of pairs of electrons and favors a univalent pathway of reduction (65). The sequential univalent reduction of dioxygen leads to the production of several partially reduced intermediates (9,65): superoxide anion radicals (O_2^-), hydrogen peroxide (H_2O_2), and hydroxyl radicals ($OH\cdot$). Due to their extreme reactivity, these intermediates may perturb a wide spectrum of important biological macromolecules such as nucleic acids, proteins, lipids and polysaccharides. Nevertheless, the univalent pathway of oxygen reduction does occur in respiring cells (13,26) and the damaging effect of these reactive species must be accommodated. Indeed, to circumvent the destructive damage caused by these intermediates, multiple defenses arose and have persisted (13,14,26).

The primary defense is to prevent the generation of these reactive forms of partially reduced oxygen. This is accomplished by the cytochrome oxidase system which brings about the tetravalent reduction of oxygen without a significant release of any of these reactive intermediates. The second line of defense is provided by enzymes that catalytically scavenge the intermediates of oxygen reduction. The O_2^- is eliminated by superoxide dismutases

(SOD), which dismutase $2O_2^-$ to H_2O_2 and O_2. The H_2O_2 is then removed by catalases to form H_2O. The removal of both O_2^- and H_2O_2 will prevent the formation of $OH\cdot$, which is fortunate since the enzymatic scavenging of $OH\cdot$ is impossible due to its extreme reactivity. These enzymatic defenses are also supplemented by a third level of biochemical defenses (i.e. vitamin E, ascorbic acid, and reduced glutathione).

Biological Sources

O_2^-, H_2O_2 and $OH\cdot$ are generated in all respiring cells. Attempts to measure the rates of O_2^-, H_2O_2 and $OH\cdot$ generation in intact cells are usually hampered by the ubiquity of superoxide dismutases and hydroperoxidases (i.e. catalase and peroxidases). Recently, the steady state levels of O_2^- and H_2O_2 in intact mitochondria were estimated to be about 10^{-11} M and 10^{-7} M, respectively (6). Hydrogen peroxide may be produced directly by the two-electron reduction of oxygen or indirectly by the dismutation of O_2^-. The rates of O_2^- and H_2O_2 generated in washed mitochondrial membranes, where SOD and hydroperoxidases are removed, are usually much higher than in intact mitochondria and account for 1 to 5% of the total oxygen consumed (6).

In living cells, there are many biological reactions that are known to produce substantial amounts of O_2^- and H_2O_2. Thus, the autoxidations of hydroquinones (39,44), leucoflavins (3,44), catecholamines (45), thiols (47), tetrahydropterins (51), and reduced ferredoxins (43,50) have all been shown to generate O_2^-. Hemoglobin and myoglobin in their oxygenated forms, have been shown to liberate O_2^- as they are converted to methemoglobin and metmyoglobin, respectively (32,46). Several enzymes, including xanthine oxidase (34), aldehyde oxidase (56,57), dihydroorotic dehydrogenase (1), and several of the flavin dehydrogenases (41) produce O_2^-. Superoxide radical has also been demonstrated to be an intermediate formed during the catalytic action of galactose oxidase (18), indoleamine dioxygenase (28), 2-nitropropane dioxygenase (29), and several other enzymes. Most of the reactions that generate O_2^- will also generate H_2O_2. For example, xanthine oxidase can simultaneously produce both O_2^- and H_2O_2 at a ratio that is dependent upon the pH, pO_2 and substrate concentration (12).

The generation of $OH\cdot$ in living cells during aerobic metabolism has been demonstrated by using electron paramagnetic resonance spectrometry (EPR) and spin traps (11,58). However, exact quantitative measurements of $OH\cdot$ made in biological systems are much more difficult than that of O_2^- and H_2O_2 because of its highly non-selective reactivity with the many organic molecules present in the cells.

Cytotoxicity

Of the partially reduced oxygen intermediates, H_2O_2 is the least reactive while O_2^- is much less reactive than $OH\cdot$ (5,9). It should be realized, however, that because of its slow reactivity O_2^- may be more specific in its damaging reactions than $OH\cdot$ which reacts indiscriminately with most organic and inorganic molecules. In support of this rational, recent studies have shown that lactate dehydrogenase is inactivated to a greater extent by O_2^- than by $OH\cdot$ (2); and that O_2^- is more toxic in Stretpococcus sanguis (10). Furthermore, exposure of human erythrocyte ghosts to superoxide radicals causes disruption of the membranes, while exposure to hydroxyl radicals causes a decrease in membrane fluidity as a result of the peroxidation of the membranal lipids (60).

It has been demonstrated that enzymically, photochemically, or electrochemically generated superoxide radicals are toxic and destructive to living cells. For example: fluxes of O_2^- kill bacteria, inactivate viruses, lyse erythrocytes, destroy granulocytes, damage myoblasts in culture, depolymerize hyaluronate, inactivate enzymes, damage DNA, decompose methional to ethylene, and initiate lipid peroxidation (for complete reference listing see 20 and 26). Furthermore, paraquat, which is known to generate O_2^-, is mutagenic in Salmonella typhimurium TA98 and TA100 (48). There is also mounting evidence that the cytotoxicity of many antitumor and xenobiotic compounds is due to the partially reduced oxygen species produced during the redox cycling of these compounds (7,24,30,31). Recent studies have also shown that hydroxyl radicals induce main-chain scission of poly-ribonucleic acids (68). In several of the studies listed above, superoxide dismutase, catalase, or compounds known to scavenge $OH\cdot$, were effective in protecting against the toxicity of O_2^-. This observation led to the proposal that O_2^- and H_2O_2 can react and generate $OH\cdot$ (4), which is the same damaging agent generated during ionizing radiation (61,66). The interaction between O_2^- and H_2O_2 was later shown to require chelated iron as a catalyst (33). This reaction is known to many scientists as the "iron catalyzed Haber-weiss reaction" and to others (10a) as "the superoxide-driven Fenton chemistry". Regardless of the name given to this reaction, the fact remains that O_2^- and H_2O_2 can interact to generate $OH\cdot$, which can damage the delicate structure of the cell.

OXY-RADICALS DEFENSE SYSTEM

It is clear that efficient removal of O_2^- and H_2O_2 will minimize the chances for $OH\cdot$ formation. Also, the presence of chemical antioxidants and of hydroxyl radical scavengers will "mop up" any dangerous oxy-radicals tht may have escaped the first line of defense. Now, we will consider the properties and the regulation of superoxide dismutase, that constitute the first line of defense.

Superoxide Dismutases (EC 1.15.1.1.)

Chemical and physical properties.
Superoxide dismutase was first isolated in 1938 as a green copper protein (38) whose biological function was thought to be copper storage. Its catalytic function, however, was later discovered by McCord and Fridovich (35) while examining the reduction of cytochrome c by xanthine oxidase. The enzyme is ubiquitous. It is found in virtually all oxygen consuming organisms (36), in some aerotolerant anaerobes (64), and in some obligate anaerobes (27). All superoxide dismutases are metalloproteins which catalyze the following reaction

$$O_2^- + O_2^- + 2H^+ \longrightarrow H_2O_2 + O_2 \quad (1)$$

at a rate constant equal to 2×10^9 M^{-1} S^{-1}, which is close to the diffusion limit. Essentially, three distinct types of superoxide dismutases, based on the metal ion in their active sites, have been observed from a wide range of organisms. Thus, there are superoxide dismutases that contain copper and zinc (CuZnSOD), or iron (FeSOD), or manganese (MnSOD). The FeSOD and MnSOD are characteristic of prokaryotes and are related since they share extensive sequence homologies (19,62); on the other hand the CuZnSODs, which are characteristic of eukaryotic cytosols, show no sequence hemology with the Mn/FeSODs and probably represent an independent line of evolution (19,63). The mitochondria contain a MnSOD which has a high degree of sequence homology with the prokaryotic enzyme (19,62). The finding of a CuZnSOD in the symbiotic prokaryote, Photobacterium leiognathi (55), represents the only exception to the rule that this class of enzyme is characteristic of cytosols, however, recent studies (40) have concluded that gene transfer from the host (the pony fish) to the bacterium may have taken place during the course of evolution.

The mechanism of dismutation by all superoxide dismutases is the same where the active site metal undergoes a cycle of reduction and reoxidation. In the first step, the metal is reduced by O_2^- and one molecule of oxygen is released. In the second step, another molecule of O_2^- oxidizes the reduced metal at the active site and results in the formation of H_2O_2 (14). A general mechanism for catalysis is presented by Eqs. 2 and 3; where E is SOD and M is the metal in the active site.

$$E \cdot M^n + O_2^- \longrightarrow E \cdot M^{n-1} + O_2 \quad (2)$$

$$E \cdot M^{n-1} + O_2^- + 2H^+ \longrightarrow E \cdot M^n + H_2O_2 \quad (3)$$

Regulation of SOD biosynthesis.
The biosynthesis of superoxide dismutases, in most biological systems, seems to be under rigorous controls. Exposure to high concentrations of oxygen have been shown to increase SOD biosynthesis

in E. coli (16), Streptococcus faecalis (15), Saccharomyces cerevisiae (17), adult rats (8), and many other organisms. It has been shown that when E. coli was maintained in a glucose-limited chemostat, under constant and abundant aeration, the activity of superoxide dismutase increases in proportion to the increase in specific growth rate and to the rate of respiration (21). These results provided the first clue that the inducer for superoxide dismutase is not molecular oxygen itself but rather a product of its metabolism. It also raised the hypothesis that the level of superoxide dismutase within the cell correlates with the intracellular level of superoxide radical generated during growth in the presence of oxygen. It was also noted that in glucose limiting trypticase-soy/yeast extract medium the content of superoxide dismutase in E. coli remains low, however, after glucose is exhausted, the cells shift to an aerobic metabolism and produce more superoxide dismutase (22). These results indicate that the rate of superoxide radical production in these cells is low during glucose fermentation, but increases during the oxidative phase of metabolism and that cells modulate the level of superoxide dismutase to meet the increasing need for the scavenging of superoxide radical. The effect of glucose on superoxide dismutase in not due to a classic catabolite repression since adding cyclic AMP has no effect on its synthesis (22). Moreover, it was observed that increasing the intracellular production of superoxide radical, via the cyclic oxidation-reduction of paraquat at a constant pO_2, causes increased synthesis of MnSOD (22-24). Several redox-active compounds were shown to act like paraquat in inducing MnSOD (24), via their ability to increase the intracellular flux of O_2^-. The induction of MnSOD by paraquat is prevented by inhibitors of transcription or of translation, but not by inhibitors or replication (24). The nature of the inducer and the repressor molecules is under intensive studies.

Physiological function. In view of the data available in the literature up to this date, it seems logical to conclude that superoxide dismutases scavenge O_2^- *in vivo* as they do *in vitro* and that they protect against many of the ill-effects of oxygen. If SOD has another function (10a), that will remain to be proven.

The superoxide radical, the substrate for SOD, can be made only in the presence of oxygen. Several studies *in vitro* and *in vivo* have demonstrated that O_2^- is cytotoxic, and that superoxide dismutases protect against such toxicity. A positive correlation is found between aerotolerance and the concentration of superoxide dismutase in anaerobic microorganisms (64). Loss of superoxide dismutase activity results in oxygen intolerance (25). Increased levels of MnSOD in E. coli achieved by a variety of growth and physiological conditions (i.e. oxygenation, increasing the specific growth rate and growth in the presence of redox-active compounds) imparted an increase resistance against oxygen toxicity (21-24). Similar findings were reported in eukaryotes. High levels of

SOD in rat lungs, induced by exposure to 85% oxygen, positively correlated with their tolerance to 100% oxygen (8). Superoxide dismutase also protects against radiation damage (42,54,67), and against paraquat-induced mutations in S. typhimurium (48). Recent studies have also shown that the capacity of the antioxidant enzymes, in the bovine eye lens, decline with age and that the level of SOD is significantly lower in cataractous human lenses than in normal clear lenses (53).

It is clear that the cytotoxicity of molecular oxygen is held in check by the delicate balance between the rates of generation of the partially reduced oxygen species and the rates of their removal by the different defense mechanisms (4,5); any shift in this delicate balance can lead to cellular damage. In general, the uncontrolled generation of the partially reduced oxygen species has been associated with cancer (52), aging (49,59) and inflammation (37).

ACKNOWLEDGEMENT

This work was supported in part by grant PCM-8213853 from the National Science Foundation.

REFERENCES

1. Aleman, V., and Handler, P. (1967): J. Biol. Chem., 242:4087-4096.
2. Armstrong, D.A., and Buchanan, J. D. (1978): Photochem. Photobiol., 28:743-755.
3. Ballou, D., Palmer, G., Massay, V. (1969): Biochem. Biophys. Res. Commun., 36:898-904.
4. Beauchamp, C. O., and Fridovich, I. (1970): J. Biol. Chem., 245:4641-4646.
5. Bielski, B. H. J., and Richter, H. W. (1977): J. Am. Chem. Soc., 99:3019-3023.
6. Boveris, A., and Cadenas, E. (1983): In: Superoxide dismutase, edited by L. W. Oberley, Vol. 2, pp. 15-30. CRC Press, Boca Raton, Florida.
7. Cone, R., Hasan, S. K., Lown, J. W., and Morgan, A. R. (1976): Can. J. Biochem., 54:219-223.
8. Crapo, J. D., and Tinerney, D. L. (1974): Am. J. Physiol., 226:1401-1407.
9. Czapski, G. (1971): Annu. Rev. Phys. Chem., 22:171-208.
10. DiGuiseppi, J., and Fridovich, I. (1982): J. Biol. Chem., 257:4046-4051.
10a.Fee, J. A. (1982): Trends Biochem. Sci., 7:84-86.
11. Finkelstein, E., Rosen, G. M., and Rauckman, E. J. (1980): Arch. Biochem. Biophys., 200:1-16.
12. Fridovich, I. (1970): J. Biol. Chem., 245:4053-4057.
13. Fridovich, I. (1975): Annu. Rev. Biochem., 44:147-159.

14. Fridovich, I. (1978): *Science*, 201:875-880.
15. Gregory, E. M., and Fridovich, I. (1973): *J. Bacteriol.*, 114:543-548.
16. Gregory, E. M., Yost, F. J.,Jr., and Fridovich, I. (1973): *J. Bacteriol.*, 115:897-991.
17. Gregory, E. M., Goscin, S. A., and Fridovich, I. (1974): *J. Bacteriol.*, 117:456-460.
18. Hamilton, G. A., Adolf, P. K., deJersey, J., Dubois, G. C., Dyrkacz, G. R., and Libby (1978): *J. Am. Chem. Soc.*, 100:1899-1912.
19. Harris, J. I., and Steinman, H. M. (1977): In *Superoxide Dismutases*. Edited by A. M. Michelson, J. M. McCord, and I. Fridovich. pp. 225-230. Academic Press, New York.
20. Hassan, H. M. (1980): In *Biological roles of copper*, Ciba Foundation Symposium 79, pp. 125-142. Excerpta Medica.
21. Hassan, H. M., and Fridovich, I. (1977): *J. Bacteriol.*, 130:805-811.
22. Hassan, H. M., and Fridovich, I. (1977): *J. Bacteriol.*, 132:505-510.
23. Hassan, H. M., and Fridovich, I. (1977): *J. Biol. Chem.*, 252:7667-7672.
24. Hassan, H. M., and Fridovich, I. (1979): *Arch. Biochem Biophys.*, 196:385-395.
25. Hassan, H. M., and Fridovich, I. (1979): *Rev. Infect. Dis.*, 1:357-367.
26. Hassan, H. M., and Fridovich, I. (1980): In: *Enzymatic Basis of Detoxication*, edited by W. B. Jakoby, Vol. 1, pp. 311-332. Academic Press, New York.
27. Hewitt, J., and Morris, J. G. (1975): *FEBS Lett.*, 50:315-318.
28. Hirata, F., Ohnishi, T., and Hayaishi, O. (1977): *J. Biol. Chem.*, 252:4637-4642.
29. Kido, T., Soda, K., and Asada, K. (1978): *J. Biol. Chem.*, 253:226-232.
30. Lorentzen, R., and Ts'o, P.O.P. (1977): *Biochemistry* 16:1467-1473.
31. Lorentzen, R., Leska, S., McDonald, K., and Ts'o, P.O.P (1979):*Cancer Res.*, 39:3194-3198.
32. Lynch, R. E., Thomas, J. E., and Lee, G. R. (1977): *Biochemistry*, 16:4563-4567.
33. McCord, J. M., and Day, E. D., Jr. (1978): *FEBS Lett.*, 86:139-142.
34. McCord, J. M., and Fridovich, I. (1968): *J. Biol. Chem.*, 243:5753-5760.
35. McCord, J. M., and Fridovich, I. (1969): *J. Biol. Chem.*, 244:6049-6055.
36. McCord, J. M., Keele, B. B., Jr., and Fridovich, I. (1971): *Proc. Natl. Acad. Sci. U.S.A.*, 68:1024-1027.
37. McCord, J. M., Stokes, S. H., and Wong, K. (1979): In *Advances in Inflammation Res.* Edited by G. Weissman, B. Samuelsson, and R. Paoletti, Vol. 1, pp. 273-280. Raven Press, New York.

38. Mann, T., and Keilin, D. (1938): Proc. Roy. Soc. (London)., B126:303-315.
39. Marklund, S., and Marklund, G. (1974): Eur. J. Biochem., 47:469-474.
40. Martin, J., and Fridovich, I. (1981): J. Biol. Chem. 256:6080-6089.
41. Massey, V., Strickland, S., Mayhew, S. G., Howell, L. G., Engel, P. C., Mathews, R. G., Schuman, M., and Sullivan, P. A. (1969): Biochem. Biophys. Res. Commun., 36:891-897.
42. Michelson, A. M., and Buckingham, M. E. (1974): Biochem. Biophys. Res. Commun., 58:1079-1086.
43. Misra, H. P., and Fridovich, I. (1971): J. Biol. Chem., 246:6886-6890.
44. Misra, H. P., and Fridovich, I. (1972): J. Biol. Chem., 247:188-192.
45. Misra, H. P., and Fridovich, I. (1972): J. Biol. Chem., 247:3170-3175.
46. Misra, H. P., and Fridovich, I. (1972): J. Biol. Chem., 247:6960-6962.
47. Misra, H. P. (1974): J. Biol. Chem., 249:2151-2155.
48. Moody, C. S., and Hassan, H. M. (1982): Proc. Natl. Acad. Sci. U.S.A., 79:2855-2859.
49. Munkers, K. D. (1979): Mech. Age. Develop., 10:249-260.
50. Nakamura, S., and Kimura, T. (1972): J. Biol. Chem., 247:6462-6468.
51. Nishikimi, M. (1975): Arch. Biochem. Biophys., 166:273-279.
52. Oberley, L. W. (1983): In Superoxide dismutase, edited by L. W. Oberley, Vol. 2, pp. 127-165. CRC Press, Boca Raton, Florida.
53. Ohrloff, C., Hockwin, O., Olson, R., and Dickman, S. (1984): Current Eye Res., 3:109-115.
54. Petkau, A., Chelack, W. S., and Plaskash, S. D. (1976): Int. J. Radiat. Biol. Relat. Stud. Phys. Chem. Med., 29:297-299.
55. Puget, K., and Michelson, A. M. (1974): Biochem. Biophys. Res. Commun., 58:830-838.
56. Rajagopalan, K. V., Fridovich, I., and Handler, P. (1962): J. Biol. Chem., 237:922-928.
57. Rajagopalan, K. V., and Handler, P. (1964): J. Biol. Chem. 239: 2022-2026.
58. Rauckman, E. J., Rosen, G. M., and Kitchell (1979): Mol. Pharmocol., 15:131-137.
59. Reiss, U., and Gershon, D. (1976): Eur. J. Biochem., 63:617-623.
60. Rosen, G. M., Barber, M. J., and Rauckman, E. J. (1983): J. Biol. Chem., 258:2225-2228.
61. Smith, K. C., and Heys, J. E. (1968): Radiat. Res., 33:129-141.
62. Steinman, H. M., and Hill, R. L. (1973): Proc. Natl. Acad. Sci. U.S.A., 70:3725-3729.
63. Steinman, H. M. (1978): J. Biol. Chem., 253:8708-8720.
64. Tally, F. P., Goldin, H. R., Jacobus, N. V., and Gorbach, S. L. (1977): Infect. Immun., 16:20-25.

65. Taube, H. (1965): J. Gen. Physiol., 49:Suppl. 29-52.
66. Van Hemmen, J. J. (1971): Nature (London), 231:79-80.
67. Van Hemmen, J. J., and Meuling, W. J. A. (1975): Biochem. Biophys. Acta., 402:133-141.
68. Washino, K., Denk, O., and Schnabel, W. (1983): Z. Naturforsch. 38c:100-106.

Free Radical Injury During Inflammation

R. F. Del Maestro

Brain Research Laboratory, Department of Clinical Neurological Sciences, Victoria Hospital, University of Western Ontario, London, Ontario, Canada N6A 4G5

In an oxygen environment, the survival of aerobic organisms depends on a complex interaction between the free radical byproducts of oxidative metabolism and the ability of each organism to control these compounds (13,21,23). In some organisms, the controlled release of oxygen derived free radicals by specialized inflammatory cells has been harnessed to perform a bactericidal role (5). These same free radical species have also been implicated in a host of disease processes (13,21,23,47). Indeed a classification of disease states in which free radical generation may play a role has been proposed (13). A precarious balance would appear to exist within an individual cell's internal and external microenvironment between the generation and control of these species. An appreciation of the chemical origins, the reactions and the control of free radicals is necessary to adequately characterize their role in any disease state. The aim of this review will be to outline some current free radical concepts which are influencing research in the field of free radical induced injury during inflammation.

Concept Definition

In this article, some essential biochemical concepts and reactions are discussed and it seems appropriate to clearly define some of these before proceeding. A "free radical" is any atom, group of atoms or molecules in a particular state with one unpaired electron occupying an outer orbital. A "biradical" is a molecule containing two unpaired electrons in outer orbitals and molecular oxygen (O_2) is an example. Molecular oxygen in its ground state is a triplet ($^3\Sigma g^-$) because the two unpaired electrons have parallel or unpaired spins. Table I demonstrates other possible electron states of the oxygen molecule. Singlet oxygen, $O_2(^1\Delta g)$, is by definition not a free radical since both electrons occupy the same orbital. Another excited singlet state ($^1\Sigma g^+$) exists with paired electron spins; however, its extremely short lifetime suggests that it may not be an important species (9).

TABLE I

Excited States of Molecular Oxygen

State	Configuration of highest occupied orbitals
$^1\Sigma g^+$	↑ ↓
$^1\Delta g$	↓↑
$^3\Sigma g^-$	↑ ↑

The parallel electron spin configuration of O_2 hinders the direct addition of a pair of electrons (↓↑ configuration) to the molecule because a bond would only form after an electron spin inversion. The one electron reduction of O_2 or univalent pathway in which no electron spin inversion occurs would predominate over the two electron reduction (divalent pathway). The generation of the superoxide anion radical, O_2^-, hydrogen peroxide, H_2O_2, hydroxyl radical, $OH\cdot$ and H_2O result from the complete reduction of O_2 by the univalent pathway (Figure 1).

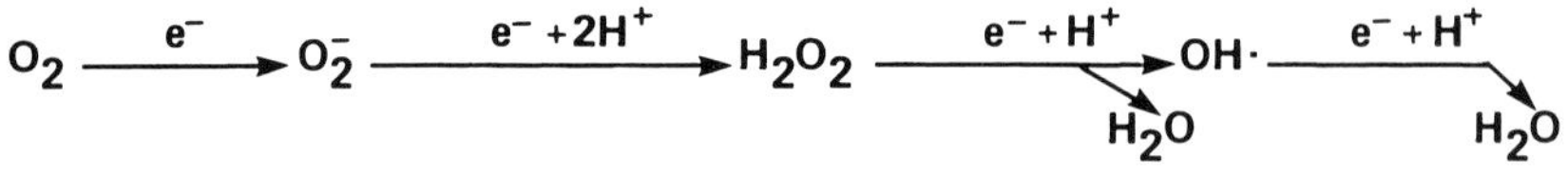

$$O_2 \xrightarrow{e^-} O_2^- \xrightarrow{e^- + 2H^+} H_2O_2 \xrightarrow{e^- + H^+} OH\cdot \xrightarrow{e^- + H^+} H_2O$$
$$H_2O \qquad\qquad H_2O$$

Fig. 1.
Molecular pathway for the univalent reduction of oxygen.

The ability of aerobic organisms to survive in an O_2 microenvironment depends on the removal and control of these reactive intermediates from the univalent pathway (22,23). A hierarchy of control or scavenging mechanisms have evolved to deal with these intermediates and may be divided into enzymatic, hydrophobic, hydrophilic and structural groups (13).

Scavenging Mechanisms
Enzymatic

The electron spin restriction of O_2 has been circumvented by the evolution of a variety of enzymes capable of the divalent and tetravalent reduction of O_2. The cytochrome oxidase system localized on the inner mitochondrial membrane tetravalently reduces the major portion of O_2 reduced by aerobic cells (3). A small but significant proportion of O_2^- continues to occur by the univalent pathway and this appears to have resulted in the evolution of a family of enzymes collectively known as superoxide dismutases (SOD) which catalytically scavenge O_2^- (reaction 1)(Fig. 2).

$$O_2^- + O_2^- + 2H^+ \longrightarrow H_2O_2 + O_2 \qquad (1)$$

The rate of reaction 1 is fast spontaneously; however, at intracellular concentrations of SOD, this reaction may be increased by a factor of 10^9 (23). Hydrogen peroxide, the divalent reduction product, is controlled intracellularly by catalase (CAT)(reaction 2).

$$2H_2O_2 \longrightarrow 2H_2O + O_2 \qquad (2)$$

and by peroxidases (reaction 3)

$$H_2O_2 + RH_2 \longrightarrow 2H_2O + R \qquad (3)$$

It has been suggested that glutathione peroxidase (GSHPx) may be the most important intracellular mechanism for the decomposition of H_2O_2 (29).

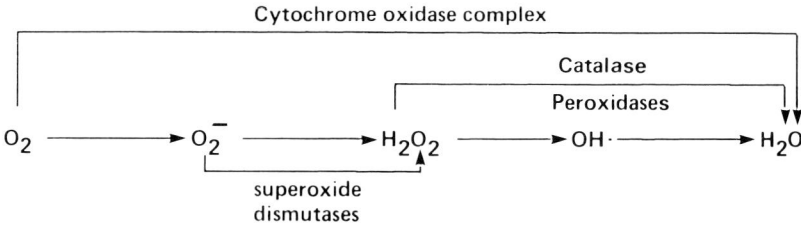

FIG. 2.
Enzymatic mechanisms to bypass and prevent the accumulation of reactive intermediates. Slightly modified from Del Maestro (13).

Hydrophobic

The only sustained hydrophobic regions associated with cells are the various lipid membranes. These are composed of a variety of polyunsaturated and saturated fatty acids (50). Pryor (47) has suggested that low rates of lipid peroxidation occur in all aerobic cell membranes and these may be modulated by various hydrophobic scavengers including vitamin E (a - tocopherol) and vitamin A (B - carotenes). Lipid peroxidation reactions occurring in hydrophobic microenvironments may be terminated by the presence of hydrophobic scavengers. Lipid hydroperoxides are excellent substrates for GSHPx but this soluble enzyme may not reduce lipid hydroperoxides in membranes (41). The coupling of the action of various phospholipases which would remove lipid hydroperoxides from membranes and subsequent reduction by GSHPx may be a possible important mechanism by which GSHPx may act.

Hydrophilic

Mechanisms which control free radical reactions in ionic or H_2O compartments of both extracellular and intracellular microenvironments may be especially important during inflammation. Compounds such as ascorbic acid, cysteine and reduced glutathione are just some of the many compounds which may be involved (13,21). Plasma components such as ceruloplasmin and transferrin may also play a role (1,27).

Structural

The cells of aerobic organisms possess balanced combinations of the scavenging mechanisms discussed evolved to deal with the free radicals generated in their individual microenvironments. The reactivity of the free radicals generated along with the biochemical environment determine the extent of biomolecular and subsequent cellular injury sustained by a given tissue. The extracellular space contains low concentrations of SOD and CAT (30,40,51). Therefore interstitial space macromolecules and the plasmalemma may be preferentially vulnerable to an extracellular generation of O_2^-. Except for degenerative cells like erythrocytes, CAT and GSHPx tend to be localized to distinct compartments. CAT is localized predominately to peroxisomes while GSHPx to the cytosol and mitochondrial matrix space (52).

The Activated Inflammatory Cell

Metchnikoff's (43) discovery of the role of phagocytes in the inflammatory process by their release of cytases (lysosomal products) has heralded almost a century of continuing research into the role of phagocytes and their products on the inflammatory response. Inflammatory cells like polymorphonuclear leukocytes possess an impressive bactericidal arsenal consisting of at least three major components. These are the release of enzymes and other reactive components from lysosomes and specific granules, the generation of O_2^- and the resultant family of reactive species, and the formation of oxidized arachidonic acid products. Although this review will focus on free radical induced alterations, it is clear that a host of mechanisms are utilized by inflammatory cells to kill bacteria and that the tissue injury which may result will be the cumulative result of all components of this arsenal.

Prerequisites for Free Radical Induced Injury During Inflammation

Four prerequisites seem to be necessary before suggesting a role for O_2 derived free radical species during inflammation. These are (1) the presence of inflammatory cells such as polymorphonuclear leukocytes, macrophages and monocytes; (2) appropriate stimuli which result in the induction of a respiratory burst with the generation of O_2^- and H_2O_2; (3) low concentrations of scavenging enzymes in the extracellular space and (4) the presence of metal complexes chelated such that OH· and possibly other oxidizing species can be generated. In any given tissue if these prerequisites are fulfilled, then free radical induced injury may be hypothesized as a possible cause of tissue injury. However, it should be also considered that in some conditions such as irradiation induced injury or that resulting from some toxins and drugs (48) prerequisited 1 and 2, that is a source of radicals, may be provided by the irradiation, toxins and drugs themselves.

Free Radical Induced Injury

Babior et al. (6) were the first to report that the activation of polymorphonuclear leukocytes resulted in the reduction of O_2 to O_2^-. The activation of polymorphonuclear leukocytes, macrophages and monocytes on exposure to a whole host of stimuli such as bacteria, fungi immune complexes, etc. resulted in a co-ordinated series of biochemical events called the "respiratory burst" (5). Products such as H_2O_2 and OH· which were derived from O_2^- were soon shown to play active bacterial roles (5,7). The respiratory burst may be considered a collection of metabolic events which together allow a short but sustained release of O_2^- derived activated products.

Polymorphonuclear leukocytes contain two distinct types of granules: primary granules or lysosomes containing myeloperoxidase, acid hydrolases, glycosidases and cationic proteins and specific granules containing alkaline phosphatase, lactoferrin and lysozyme (8). Myeloperoxidase catalyzes the oxidation of a variety of substrates by H_2O_2 (reaction 4) and products such as hypochlorite (33) and $O_2(^1\Delta g)$ (49) have been suggested as the bactericidal component of this reaction system.

$$Cl^- + H_2O_2 \longrightarrow OCl^- + H_2O_2 \longrightarrow O_2 (^1\Delta g) + Cl^- + H_2O \quad (4)$$

Foote et al. (20) using specific methods to assess O_2 ($^1\Delta g$) generation concluded that O_2 ($^1\Delta g$) does not play a major role in the bactericidal activity of polymorphonuclear leukocytes and other inflammatory cells. It seems reasonable to conclude that the myeloperoxidase system plays an active role in bactericidal killing but that the major portion of this role does not depend on O_2 ($^1\Delta g$).

The low levels of SOD and CAT present extracellularly and in such specialized liquids as plasma, cerebrospinal fluid, synovial fluid and lymph (39,40,51) result in the accumulation of H_2O_2 via the spontaneous dismutation of O_2^- and O_2^- mediated reduction of metal chelates may result in the generation of $OH\cdot$ and other active radical species (schemes I, II, III)(10,24).

Scheme I

$$Me^n + O_2^- \longrightarrow Me^{n-1} + O_2$$
$$Me^{n-1} + H_2O_2 \longrightarrow Me^n + OH^- + OH\cdot$$

Scheme II

$$Me^n + O_2^- \longrightarrow (Me - O_2)^{n-1}$$
$$(Me - O_2)^{n-1} + H_2O_2 \longrightarrow Me^n + O_2 + OH^- + OH\cdot$$

Scheme III

$$Me^n + H_2O_2 \longrightarrow (Me - OOH)^{n-1} + H^+$$
$$(Me - OOH)^{n-1} + O_2^- \longrightarrow (Me - O)^{n-1} + O_2 + OH^-$$
$$(Me - O)^{n-1} + H^+ \longrightarrow (Me - OH)^n \longrightarrow Me^n + OH\cdot$$

The generation of the oxidizing species appears to be crucially dependent on the concentration and reactivity of the metal chelates present (13). Ambruso and Johnston (2) have shown that iron complexed to lactoferrin results in increased $OH\cdot$ generation and these authors speculate that bacteria themselves may provide iron to transferrin which may transfer it to lactoferrin as the pH drops in the phagosome.

Since the prerequisites for free radical induced injury appear to be fulfilled during inflammatory processes, free radical induced injury can be invoked as a possible cause of damage to the macromolecules of the extracellular space, the membranes and protein component of cells and the microvascular environment.

Extracellular Space

The term connective tissue is applied to a variety of tissues characterized by a low cell content and a large content of high-molecular weight intercellular materials. Subcutaneous tissue, vascular wall cartilage, bone and synovial fluid are some examples. The main macromolecular components of these tissues may be divided into fibrous proteins (collagen and elastin) and connective tissue polysaccharides; glycosaminoglycans such as hyaluronic acid and mucopolysaccharides such as keratin and chondrotin sulfate. Proteoglycan aggregates are the building blocks of the extracellular space and consist of mucopolysaccharides attached to protein cores strung like branches along a hyaluronic acid core (11).

McCord (42) was the first to suggest that the release of free radical species from inflammatory cells and subsequent degradation of hyaluronic acid by O_2^- derived activated species may be important elements in inflammation induced tissue injury. Substrate-xanthine oxidase radical generating systems have been used as models for activated polymorphonuclear leukocytes and employed by a number of groups to assess the mechanism of degradation (13,15,25,42). The hypothesis diagramatically illustrated in Figure 3 was consistent with the observed results.

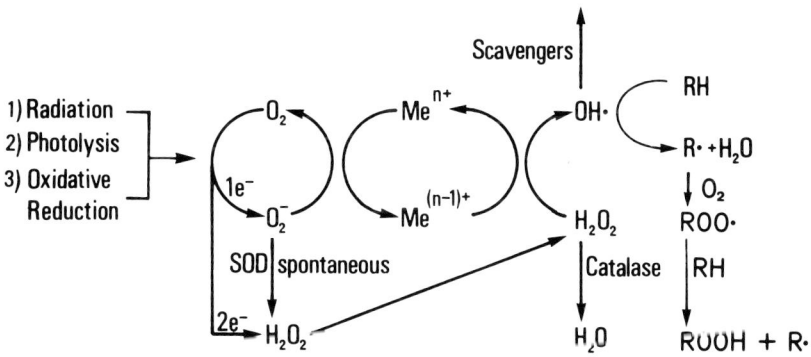

FIG. 3.
Schematic illustration of the concept in which the electron flux from a variety of initiation reactions results in O_2 reduction to O_2^-. Metal chelates (Me^{n+}) may be reduced by O_2^- and these ($Me^{(n-1)+}$), then react with H_2O_2 generated via O_2^- dismutation or by the divalent pathway ($2e^-$) to generate $OH\cdot$ which may react with a variety of compounds including lipids to generate other radicals ($R\cdot$). In the case of lipids, O_2 may react with $R\cdot$ to form lipid peroxide radicals ($ROO\cdot$) which by abstracting a hydrogen can initiate chain reactions and form lipid hydroperoxides (ROOH). Reproduced with permission from Del Maestro (13).

Greenwald (25) demonstrated that polymorphonuclear leukocytes in vitro activated by phorbol myristate acetate in the presence of ferrous iron resulted in hyaluronic acid degradation suggesting that Figure 3 may be a possible mechanism of hyaluronic acid degradation in vivo. Fibrous protein collagen exposed in vitro to an enzyme radical generating system has also been shown to induce structural modifications (26).

It would seem reasonable to suggest that during inflammation in conditions in which the prerequisites for the generation of free radicals are fulfilled that free radical interactions may result in significant alteration in both the permeability and structural characteristics of inflamed tissue (Fig. 4 and 5). Caution must be exercised in projecting a role for radical species in extracellular injury. Since the data presently available is derived from in vitro systems and extrapolation to human disease states is fraught with difficulty.

Cell Membranes and Proteins

Major alterations in cellular function may result from the initiation of propagating lipid free radical chain reactions in cellular membranes. The susceptibility of plasmalemmal membranes to extracellular free radical generation has been stressed (13,21). Lipid peroxidative injury is sustained by liposomal and erythrocyte membranes on exposure to extracellular enzyme generated radical species (31,32). Myoblast growth and differentiation are altered by enzymatic O_2^- fluxes (44). Erythrocytes (53) and tumour cells (28) lyse on exposure to activated polymorphonuclear leukocytes and phagocytosing cells are damaged by the radical they release.

The role played by the extracellular generation of radical species on subsequent intracellular alterations is not well understood. It has been demonstrated that O_2^- can cross the erythrocyte membrane (37). Increased lysosomal activity was seen in glial cells exposed to an extracellular free radical flux (13) and lysosomal membranes have been damaged by an in vitro radical generating system (19). Weiss (53) has suggested that radical induced injury to erythrocytes may be depending on an interaction of H_2O_2 with haemoglobin forming a cytotoxic complex. Although the intracellular environment may be less susceptible to free radical injury due to its protective scavenging mechanisms under certain conditions, extracellular free radical generation may initiate the formation of specific intracellular metabolic consequences which contribute to cellular injury.

Enzymatic proteins may also be inactivated by free radical species although the mechanisms of injury remain unclear (4,34,36,45). The biomembrane and protein constituents of cells may also sustain free radical induced injury during inflammation.

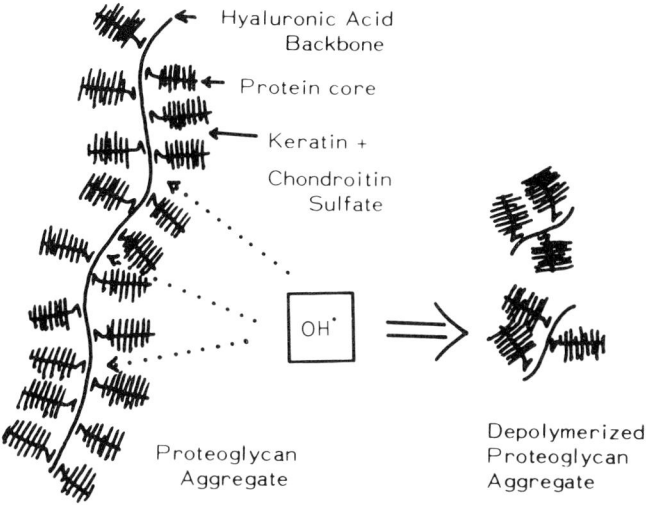

FIG. 4
Suggested mechanism of proteoglycan aggregate degradation by OH•. Reproduced with permission from Del Maestro and Alexander (15).

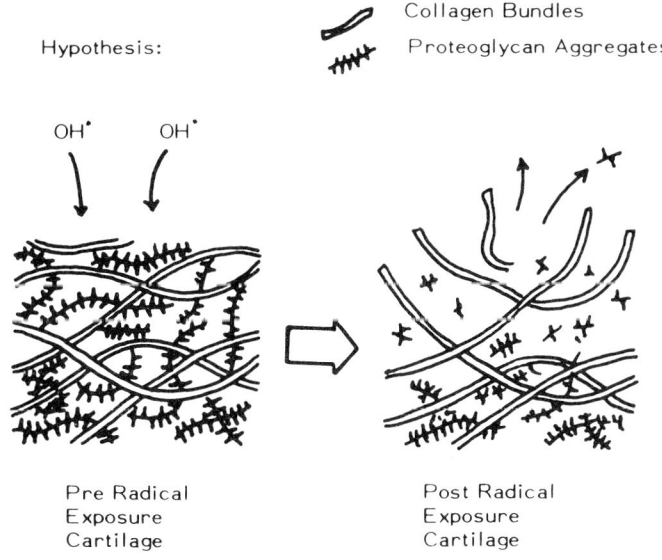

FIG. 5
Suggested hypothesis for the radical induced injury to articular cartilage. Reproduced with permission from Del Maestro and Alexander (15).

Microvascular Environment

The application of the hypoxanthine-xanthine oxidase free radical generating system to the cheek pouch of the hamster resulted in increased microvascular permeability (16,17)(Fig. 6) and altered granulocytes adhesion (18). A significant decrease was seen in granulocyte velocity; however, granulocyte rolling frequency increased and adherent granulocytes were observed.

The macromolecular permeability increase was inhibited by SOD, CAT and OH· scavengers such as DMSO (Fig. 7) while only SOD prevented the abnormal granulocyte endothelial interactions seen suggesting that O_2^- plays a role in this process. The effectiveness of SOD, CAT and OH· scavengers in modifying macromolecular leakage suggested that OH· or more likely an OH·-derived product may be principally involved in the increased microvascular permeability seen. Ley et al. (35) demonstrated that the intravascular microinjection of active xanthine oxidase into 40 um arterioles of the hamster cheek pouch was associated with increased macromolecular permeability and that SOD, chemically modified to increase its half-life prior to intraarteriolar microinjection of active enzyme resulted in the prevention of macromolecular leakage.

Fridovich and Porter (24) have shown that arachidonic acid can be co-oxidized by a substrate - xanthine oxidase system and that the major products are 5- and 15-hydroperoxyeicosatetraenoic acid (5- and 15-HPETE). Leukotrienes, LTA_4, LTB_4, LTC_4 and LTE_4 can result from 5-HPETE (30,38). LTB_4 causes reversible leukocyte adhesion and the other compounds increased macromolecular extravasation when applied to the cheek pouch (12). The in vitro incubation of purified arachidonic acid with an O_2^- generating system has been shown to result in the generation of chemotactic lipids which have been related to OH generation (46).

The hypothesis that the enzymatic generation of O_2^- on the cheek pouch generates chemotactic compounds (LTB_4 may be one of these) which causes granulocyte adhesion and possibly LTC_4, LTD_4 and LTE_4 which results in macromolecular extravastion although intriguing is only speculative (14).

SUMMARY

The role played by activated O_2 species is beginning to be unravelled through the use of both in vitro and in vivo models. The study and interpretation of free radical interaction is difficult. However, only through an understanding of these reactive species will we be able to assess their contribution to the inflammatory process.

FREE RADICAL INJURY DURING INFLAMMATION

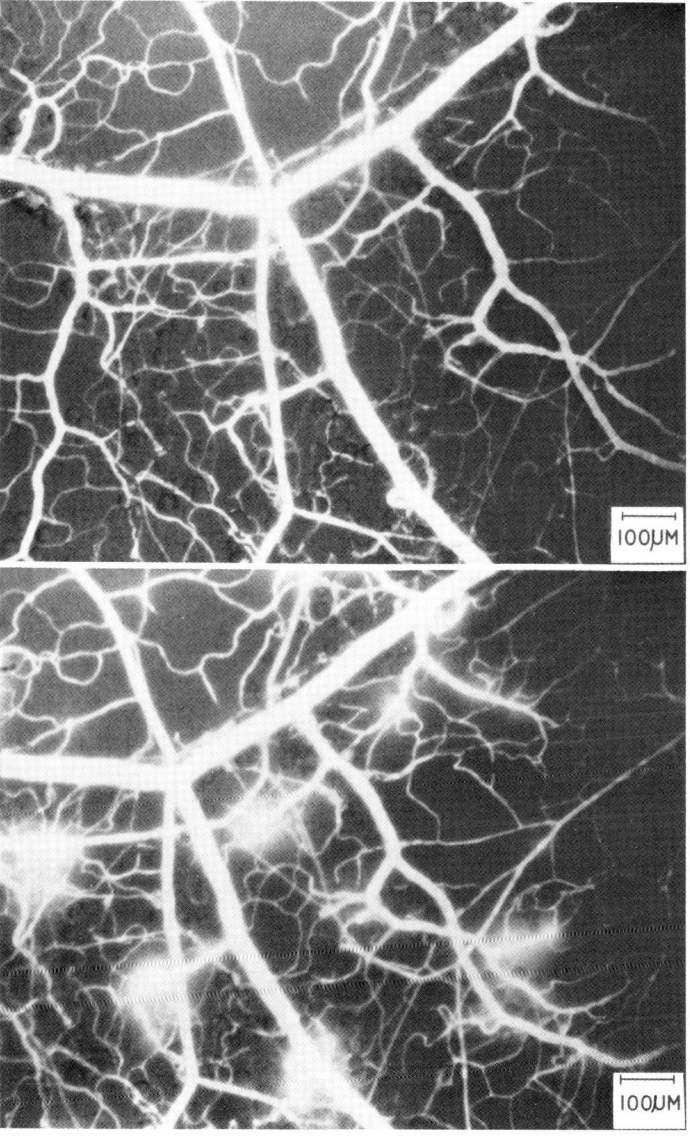

FIG. 6
Micrographs of cheek pouch microvasculature taken in fluorescent light at x 35 magnification. Upper panel is prior to application of 0.96 mM hypoxanthine and 0.05 units/ml of xanthine oxidase while lower panel is the same region 5 min following the application of substrate and enzyme demonstrating FITC dextran extravasation from postcapillary venules.

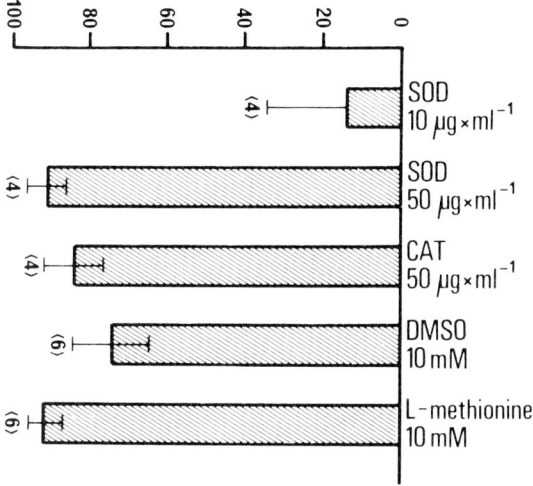

FIG. 7
Percent decrease in the mean leakage sites/cm^2 ($\pm$SE) at 10 min following application of 0.05 unit/ml of xanthine oxidase in the presence of the substances tested as compared to the mean leakage site/cm^2 in their absence. Number in brackets represents number of animals. Reproduced with permission from Del Maestro et al. (17).

ACKNOWLEDGEMENTS

The author wishes to thank Dr. Karl-E. Arfors, Jakob Bjork and Manfred Planker, collaborators in these studies and Jo-Ann Dunn for excellent secretarial assistance.

These studies were supported by the Medical Research Council of Canada. Dr. Del Maestro is a recipient of a Canadian Life Insurance Medical Scholarship.

REFERENCES

1. Al-Timimi, D.J. and Dormandy, T.L. (1977): The Inhibition of Lipid Autoxidation by Human Caeruloplasmin, Biochem. J. 168:283-288.

2. Ambruso, D.R. and Johnston, Jr., R.B. (1981): Lactoferrin Enhances Hydroxyl Radical Production by Human Neurtrophils, Neutrophil Particulate Fractions and an Enzymatic Generating System, J. Clin. Invest. 67:352-360.
3. Antonini, E., Brunori, M., Greenwood, C. and Malmstrom, B.G. (1970): Catalytic Mechanisms of Cytochrome Oxidase, Nature 228:936-937.
4. Armstrong, D.A. and Buchanan, J.B. (1978): Reactions of O_2 H_2O_2 and Other Oxidants with Sulfhydryl Enzymes, Photochem. Photobiol. 28:743-455.
5. Babior, B.M. (1978): Oxygen-Dependent Microbial Killing by Phagocytes. Part I, N. Engl. J. Med. 298:659-668.
6. Babior, B.M., Kipnes, R.S. and Curnette, J.T. (1973): Biological Defense Mechanisms. The Production by Leukocytes of Superoxide, a Potential Bactericidal Agent, J. Clin. Inves. 52:741-744.
7. Babior, B.M., Curnette, J.T. and Kipnes, R.S. (1975): Biological Defense Mechanisms. Evidence for the Participation of Superoxide in Bacterial Killing by Xanthine Oxidase, J. Lab. Clin. Med. 85:235-244.
8. Bainton, D.F. (1972): In: Phagocytic Mechanisms in Health and Disease, edited by R.C. William and H.H. Fudenberg, pp. 123-133. Intercontinental Medical Book Corp., New York.
9. Bellus, D. (1978): In: Singlet Oxygen. Reactions with Organic Compounds and Polymers, edited by B. Ranby and J.F. Rabek, pp. 86-87. John Wiley and Sons, New York.
10. Brawn, K. and Fridovich, I. (1980): In: Free Radicals in Medicine and Biology, edited by D.H. Lewis and R.F. Del Maestro, pp. 9-18. Acta. Physiol. Scand. Suppl.,
11. Comper, W.D. and Laurent, T.C. (1978): Physiological Function of Connective Tissue Polysaccharides, Physiol. Rev. 58:255-315.
12. Dahlen, S.-E., Björk, J. Hedquist, P., Arfors, K.-E., Hammarstrom, S., Lindgren, J.-A. and Samuelsson, B. (1981). Leukotrienes Promote Plasma Leakage and Leukocyte Adhesion in Postcapillary Venules: In Vivo Effects with Relevance to the Acute Inflammatory Response, Proc. Natl., Acad. Sci. U.S.A. 78:3887-3891.
13. Del Maestro, R.F. (1980): An Approach to Free Radicals in Medicine and Biology, Acta. Physiol. Scand. Suppl. 492:153-168.
14. Del Maestro, R.F. (1982): Role of Superoxide Anion Radicals in Microvascular Permeability and Leukocyte Behaviour, Can. J. Physiol. Pharmacol. 60:1406-1414.

15. Del Maestro, R.F. and Alexander, I. (1981): In: The Inflammatory Process, an Introduction to the Study of Cellular and Humoral Mechanisms, edited by P. Venge and A. Lindbom, pp. 113-143. Almquist and Wiksell, Stockholm.
16. Del Maestro, R.F., Björk, J. and Arfors, K.-E. (1981a): Increase in Microvascular Permeability Induced by Enzymatically Generated free Radicals, I, In Vivo Study, Microvasc. Res. 22:239-254.
17. Del Maestro, R.F., Björk, J. and Arfors, K.-E. (1981b): Increase in Microvascular Permeability Induced by Enzymatically Generated free Radicals, II, Role of Superoxide Anion Radical, Hydrogen Peroxide and Hydroxyl Radicals, Microvasc. Res. 22:255-270.
18. Del Maestro, R.F., Planker, M. and Arfors, K.-E. (1982): Evidence for the Participation of Superoxide Anion Radical in Altering the Adhesive Interaction Between Granulocytes and Endothelium, In Vivo, Int. J. Microcirc.: Clin. Exp. I:105-120.
19. Fong, K.L., McCay, P.B., Poyer, J.L., Keele, B.B. and Misra, H. (1973): Evidence that Peroxidation of Lysosomal Membranes is Initiated by Hydroxyl Free Radicals Produced During Flavin Enzyme Activity, J. Biol. Chem. 248:7792-7797.
20. Foote, C.S., Abakerli, R.B., Clough, R.L. and Shook, F.C. (1980): In: Biological and Clinical Aspects of Superoxide and Superoxide Dismutase, edited by W.H. Bannister and J.V. Bannister, pp. 222-230. Elsevier/North Holland, New York.
21. Freeman, B.A., and Crapo, J.D. (1982): Biology of Disease, Free Radicals and Tissue Injury, Lab. Invest. 47:412-426.
22. Fridovich, I. (1975): In: Ann. Rev. Biochem., edited by E.E. Snell, pp. 147-159.
23. Fridovich, I. (1978): The Biology of Oxygen Radicals, Science 201:875-880.
24. Fridovich, S.E. and Porter, N.A. (1981): Oxidation of Arachidonic Acid in Micelles by Superoxide and Hydrogen Peroxide, J. Biol. Chem. 256:260-265.
25. Greenwald, R.A. (1980): In: Biological and Clinical Aspects of Superoxide and Superoxide Dismutase, edited by W.H. Bannister and J.V. Bannister, pp. 160-171. Elsevier/North Holland, New York.
26. Greenwald, R.A. and Moy, W.W. (1979): Inhibition of Collagen Gelation by Action of the Superoxide Radical, Arthritis Rheum. 22:251-259.
27. Gutteridge, J.M.C. (1977): The Protective Action of Superoxide Dismutase on Metal-Ion Catalysed Peroxidation of Phospholipids, Biochem. Biophys. Res. Comm. 77:379-386.

28. Hafeman, D.G. and Lucas, Z.J. (1979): Polymorphonuclear Leukocyte-Mediated, Antibody-Dependent, Cellular Cytotoxicity against Tumour Cells: Dependence on Oxygen and the Respiratory Burst, J. Immunol. 123:55-62.
29. Halliwell, B. (1978): Biochemical Mechanisms Accounting for the Toxic Action of Oxygen on Living Organisms: The Key Role of Superoxide dismutase, Cell Biol. Int. Rep. 2:113-128.
30. Higgs, G.A., Flower, R.J. and Vane, J.R. (1979): A new Approach to Anti-inflammatory Drugs, Biochem. Pharmacol. 28:1959-1961.
31. Kellogg, E.W. and Fridovich, I. (1975): Superoxide, Hydrogen Peroxide and Singlet Oxygen in Lipid Peroxidation by an Xanthine Oxidase System, J. Biol. Chem. 250:8812-8817.
32. Kellogg, E.W. and Fridovich, I. (1977): Liposome Oxidation and Erythrocyte Lysis by Enzymatically Generated Superoxide and Hydrogen Peroxide, J. Biol. Chem. 252:6721-6728.
33. Klebanoff, S.J. (1975): In: The Phagocytic Cell in Host Resistance, edited by J.A. Bellanti and D.H. Dayton, Raven Press, New York.
34. Lavelle, F., Michelson, A.M. and dimitrijevic, L. (1973): Biological Protection by Superoxide Dismutase, Biochem. Biophys. Res. Commun. 55:350-357.
35. Ley, K. and Arfors, K.-E. (1982): Changes in Macromolecular Permeability by Intravascular Generation of Oxygen Derived Free Radicals, Microvas. Res. 24:25-33.
36. Lin, W.S., Armstrong, D.A. and Lai, M. (1978): Effects of Superoxide Dismutase, dithiothreitol and Formate Ion on the Inactivation of Papain by Hydroxyl and Superoxide Radicals in Aerated Solutions, Int. J. Radiat. Biol. 33:231-243.
37. Lynch, R.E. and Fridovich, I. (1978): Effects of Superoxide on the Erythrocyte Membrane, J. Biol. Chem. 253:1838-1845.
38. Malmsten, C.L. (1981): In: The Inflammatory Process, an Introduction to the Study of Cellular and Humoral Mechanisms, edited by P. Venge and A. Lindbom, pp. 73-102. Almquist and Wiksell, Stockholm.
39. Marklund, S. (1980): In: Free Radicals in Medicine and Biology, edited by D.H. Lewis and R.F. Del Maestro, pp. 19-23.
40. Marklund, S., Holme, E. and hellner, L. (1982): Superoxide Dismutase in Extracellular Fluids, Clinic. Chimica. Acta. 126:41-51.
41. McCay, P.B., Gibson, D.D., Fong, K. and Hornbrook, K.R. (1976): Effect of Glutathione Peroxidase Activity on Lipid Peroxidation in Biological Membranes, Biochim. Biophys. Acta. 431:459-468.

42. McCord, J.M. (1974): Free Radicals and Inflammation: Protection of Synovial Fluid by superoxide dismutase, Science 185:529-531.
43. Metchnikoff, E. (1905): Immunity in Infective Diseases, Johnson Reprint Corp., New York and London.
44. Michelson, A.M. and Buckingham, M.E. (1974): Effects of Superoxide Radicals on Myoblast Growth and Differentiation, Biochem. Biophys. Res. Commun. 58:1079-1086.
45. Nohl, H. Breuninger, V. and Hegner, O. (1978): Influence of Mitochondrial Radical Formation of Energy-Linked Respiration, Eur. J. Biochem. 90: 385-390.
46. Perez, H.D., Weksler, B.B. and Goldstein, I.M. (1980): Generation of a Chemotactic Lipid From Arachidonic Acid by Exposure to a Superoxide-Generating System, Inflammation 4:313-328.
47. Pryor, W.A. (1976): In: Free Radicals in Biology, edited by W. Pryor, pp. 1-49. Academic Press, New York.
48. Pryor, W.A. (1980): In: Molecular Basis of Environmental Toxicity, edited by R.S. Bhatnager, pp. 3-36. Ann Arbor Science Publishers Inc., Ann Arbor.
49. Rosen, H. and Klebanoff, S.J. (1977): Formation of Singlet Oxygen by the Myeloperoxidase Mediated Antimicrobial System, J. Biol. Chem. 252:48-3-4810.
50. Rouser, G., Nelson, G.J., Fleicher, S. and Simon, G. (1968): In: Biological Membranes, Physical Fact and Function, edited by D. Chapman, pp. 5-69. Academic Press, New York.
51. Salin, M.I. and McCord, J. (1975): Free Radicals and Inflammation. Protection of Phagocytosing Leukocytes by Superoxide Dismutase, J. Clin. Invest. 56:1319-1323.
52. Sunde, R.A. and Hoekstra (1980): Structure, Synthesis and Function of Glutathione Peroxidase, Nutrition Reviews 38:265-273.
53. Weiss, S.J. (1980): The Role of Superoxide in the Destruction of Erythrocyte Targets by Human Neutrophils, J. Biol. Chem. 255:9912-9917.

Free Radicals in Molecular Biology, Aging, and Disease, edited by D. Armstrong et al. Raven Press, New York © 1984.

Free Radical-Induced Changes in the Surface Morphology of Isolated Hepatocytes

Martyn T. Smith, *Hjördis Thor, *Sarah A. Jewell, *Giorgio Bellomo, Martha S. Sandy, and *Sten Orrenius

*Department of Biomedical and Environmental Health Sciences, School of Public Health, University of California, Berkeley, California 94720; *Department of Forensic Medicine, Karolinska Institute, Stockholm 104 01, Sweden*

Freshly isolated rat hepatocytes prepared by the collagenase perfusion technique described in (20) appear perfectly spherical when incubated in suspension and are covered with surface microvilli (FIG. 1A). Such hepatocytes contain high levels (35 - 50 nmol/10^6 cells) of reduced glutathione (GSH), which plays a protective role against both oxidative stress and free radical injury by acting as a readily available nucleophile and as a substrate for the enzyme glutathione peroxidase (33). Other essential cofactors such as ATP and pyridine nucleotides are also present at physiological levels. Toxic agents which are known to induce oxidative stress via the formation of free radicals, such as menadione (2-methyl-1,4-naphthoquinone) and t-butylhydroperoxide (t-BH), cause a rapid depletion of GSH via its oxidation to oxidized glutathione (GSSG) and subsequent release from the hepatocytes (1,35). Scanning electron microscopy reveals that one of the most immediate consequences of this GSH depletion is a marked change in the surface morphology of the hepatocytes, known as "plasma membrane blebbing" (FIG. 1B and FIG. 2A). Membrane blebbing only seems to occur when GSH has been depleted below a certain level, indicating the critical importance of GSH in protecting the hepatocyte against this type of oxidative injury. Moreover, the blebbing induced by t-BH is made worse and induced at lower t-BH concentrations by the presence of N,N-bis-(2-chloroethyl)-N-nitrosourea, which inhibits glutathione reductase and thereby prevents the reduc-

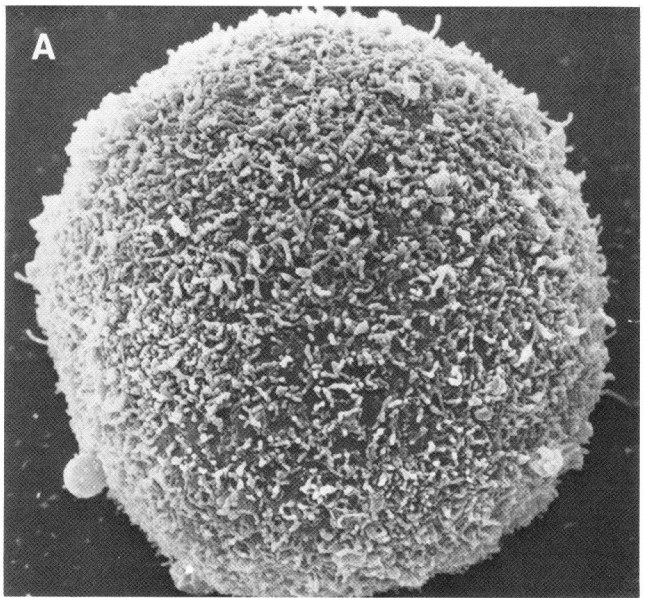

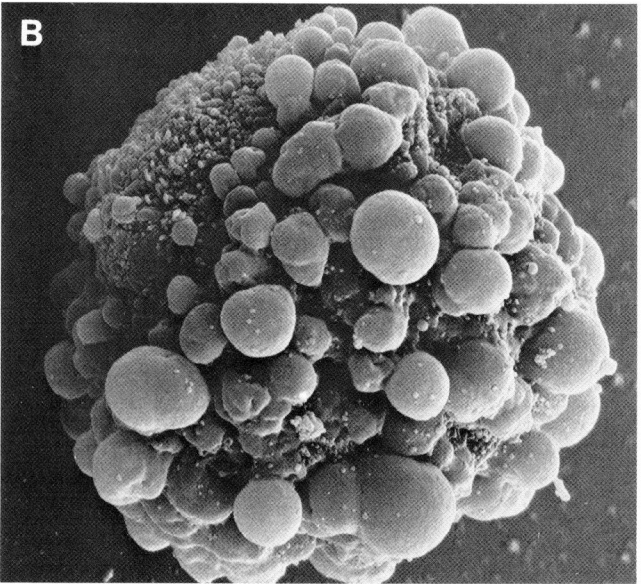

FIG. 1. Scanning electron micrographs of typical hepatocytes incubated for 30 min. in the absence (A) or presence (B) of 200 µM menadione. Magnification: x 2400-3200.

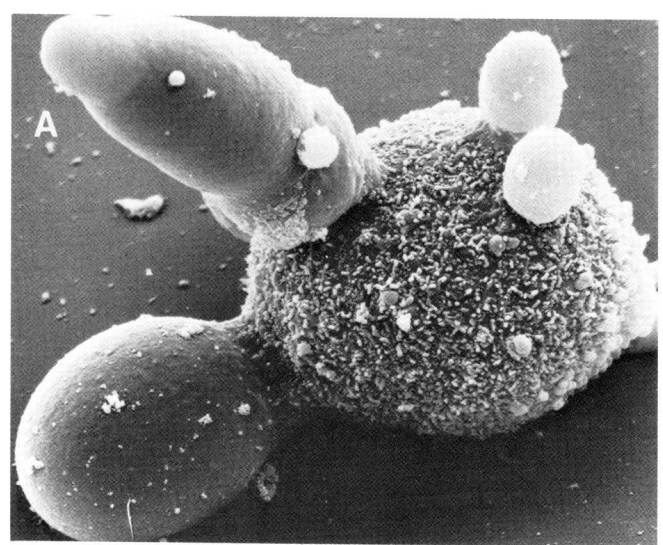

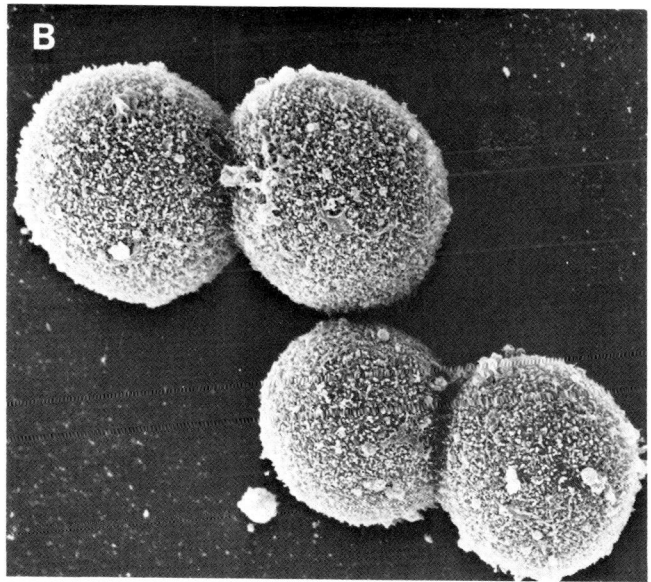

FIG. 2. Scanning electron micrographs of typical hepatocytes incubated for 30 min with 2mM t-BH in the absence (A) or presence (B) of 2mM dithiothreitol. Magnification: × 2400-3200.

tion of GSSG back to GSH causing a more rapid lowering of cellular GSH levels (1). The critical role of thiol oxidation in the blebbing process is further illustrated by the fact that dithiothreitol (a free radical scavenging thiol) completely protects against the blebbing induced by t-butylhydroperoxide (FIG. 2B).

Free radical-induced oxidative stress therefore leads to surface blebbing in isolated hepatocytes following the depletion of GSH. The formation of blebs is an early sign of toxic injury to the hepatocytes and precedes other indications of damage and cell death. We have been investigating the mechanism(s) involved in bleb formation in an attempt to identify a common sequence of events underlying irreversible cellular injury. A major goal of our investigations has been to identify the elusive "point of no return" at which the potential for recovery is lost. Our studies to date indicate that this "point of no return" may be an uncontrollable rise in the cytosolic concentration of free Ca^{2+}, $[Ca^{2+}]_i$, as a result of its release from intracellular stores. We therefore suggest that free radical-induced alterations in Ca^{2+} homeostasis could play a critical role in the irreversible toxic injury and terminal differentiation of living cells.

POSSIBLE MECHANISM(S) OF BLEB FORMATION

It is widely considered that cell surface morphology is determined by the organization of cortical microfilaments associated with the plasma membrane (19,38,40). This notion is supported by the fact that two classes of compounds, the cytochalasins and phalloidins, which are known to disrupt cortical microfilament structure, cause the formation of blebs on the surface of hepatocytes similar to those shown in Figures 1B and 2A (19,25,33,41). Thus, bleb formation appears to occur as a result of a disruption of cortical microfilament structure. We can find no evidence, however, that compounds which produce reactive intermediates and deplete cellular GSH interact directly with microfilament cytoskeletal components. It seems more likely that the key alterations in microfilament structure, which produce membrane blebbing, are brought about indirectly via alterations in levels of regulatory cofactors or ions. For example, alterations in the cytosolic concentrations of ATP (7), Ca^{2+} (32), Mg^{2+} (16) and/or H^+ (8) would significantly affect the structure of the hepatocyte cytoskeleton. Preliminary studies from our laboratories suggest that alterations in intracellular free Mg^{2+} and H^+ concentration are of little importance in bleb formation. For example, the transient application of 15 mM NH_4Cl, which loads cells with an excess of intracellular H^+ (30), does not cause blebbing in isolated hepatocytes (M.T. Smith and M.S. Sandy, unpublished observations), and variation of extracellular Mg^{2+} concentration between 0 and 1 mM has no effect on bleb formation or toxicity induced by t-BH (31). Further studies on the role, if any, of alterations in intracellular free Mg^{2+} and H^+ concentrations in bleb formation are presently being performed in our laboratory using intracellularly-trapped fluorescent indicators and other techniques (29).

MEMBRANE BLEBBING AS A RESULT OF ATP DEPLETION - STUDIES WITH ANTIMYCIN A

The polymerization of G-actin (monomeric form) to F-actin (filamentous form) is dependent upon ATP, one mole of bound ATP being converted to ADP for every monomeric actin subunit polymerized (9). Actin-myosin interactions in the cytoskeleton of eukaryotic cells are also dependent upon the activity of F-actin-activated myosin ATPase (10). Thus, the microfilaments of the hepatocyte cytoskeleton require ATP for the maintenance of normal cell shape and a lack of ATP could result in actin depolymerization, a breakdown of the actomyosin network and hence plasma membrane blebbing.

Figure 3A shows that antimycin A (25 μM) lowers the level of ATP in isolated hepatocytes, presumably by inhibiting the flow of electrons in the mitochondrial transport chain. There is a simultaneous appearance of blebs on the surface of the hepatocytes (FIG. 3B), which increase in incidence with time. However, there is no significant alteration in cell viability (FIG. 3C) at these early stages. Although antimycin A is very effective at producing membrane blebbing it does not cause a complete loss of cellular ATP. At least 30% of the cellular ATP remains (FIG. 3A) and the majority of this ATP is located in the cytosolic compartment (S. Orrenius and H. Thor, unpublished observations). Thus, there should be sufficient ATP available to keep the cytoskeletal architecture intact. An alternative explanation of antimycin A's efficacy in producing membrane blebbing may be that its inhibition of mitochondrial electron transport causes the release of Ca^{2+} from this compartment. Figure 4 shows that 20 μM antimycin A is capable of releasing all of the Ca^{2+} from the same pool in hepatocytes as the uncoupler FCCP (carbonyl cyanide p-trifluoromethoxyphenylhydrazone) i.e., the mitochondrial pool (1,14). Antimycin A could therefore cause the release of mitochondrial Ca^{2+} into the cytosol, resulting in a sustained rise in $[Ca^{2+}]_i$ as mitochondrial regulation of cytosolic Ca^{2+} is impaired by the cessation of mitochondrial Ca^{2+} uptake. Let us now consider how such a rise in $[Ca^{2+}]_i$ could cause plasma membrane blebbing, altered cellular function and eventually cell death.

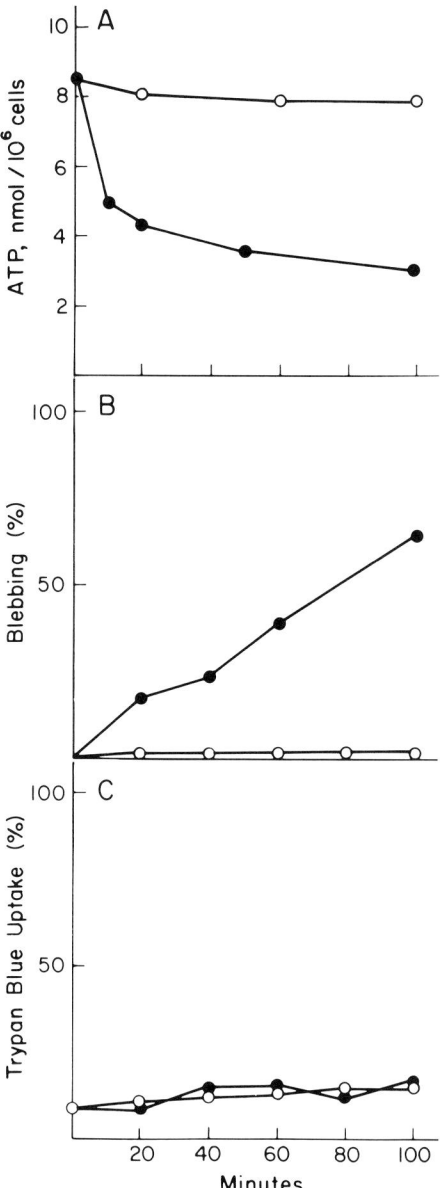

FIG. 3. Effect of antimycin A (25 µM) on ATP levels (A); bleb formation (B); and viability (C) of isolated hepatocytes. Data from typical incubations in the presence (●) and absence (○) of antimycin A are shown.

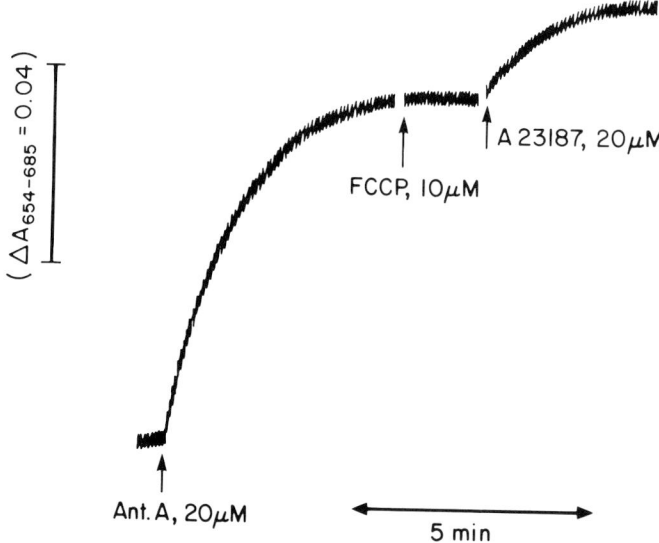

FIG. 4. Calcium release from hepatocytes induced by antimycin A, FCCP and A23187. Cells were incubated and prepared for spectrophotometry as described in (14). Extracellular arsenazo III was used as the indicator of Ca^{2+} release (14).

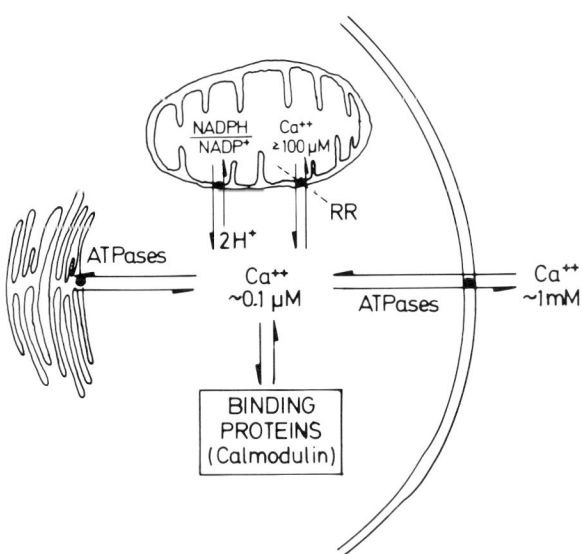

FIG. 5. Overview of Ca^{2+} compartmentation and regulation in hepatocytes (RR=ruthenium red).

BLEBBING AS A RESULT OF ALTERED INTRACELLULAR Ca^{2+} HOMEOSTASIS

Ca^{2+} is a very biologically active ion and its concentration in the cytosol of hepatocytes is maintained at a very low level (0.1 µM) (5,24) through the concerted action of specific translocases in the plasma membrane, endoplasmic reticulum and inner mitochondrial membrane (43), as well as binding proteins, such as calmodulin (6) (FIG. 5). A change in $[Ca^{2+}]_i$ would alter the structure of the hepatocyte cytoskeleton because Ca^{2+} and its associated binding proteins play such a pivotal role in regulating cytoskeletal structure (39). One manner in which $[Ca^{2+}]_i$ might be altered is by the influx of extracellular Ca^{2+} into the cell, since there is a 10,000-fold difference in $[Ca^{2+}]_i$ and the plasma free Ca^{2+} concentration (FIG. 5). We recently showed, however, that several differently acting cytotoxins, including carbon tetrachloride and t-BH cause more surface blebbing and are more toxic to isolated hepatocytes in the absence of extracellular Ca^{2+} than in its presence (31,34). Furthermore, in the isolated perfused rat liver the level of extracellular Ca^{2+} had no significant effect on t-BH toxicity, as determined by lactate dehydrogenase release (FIG. 6). We can

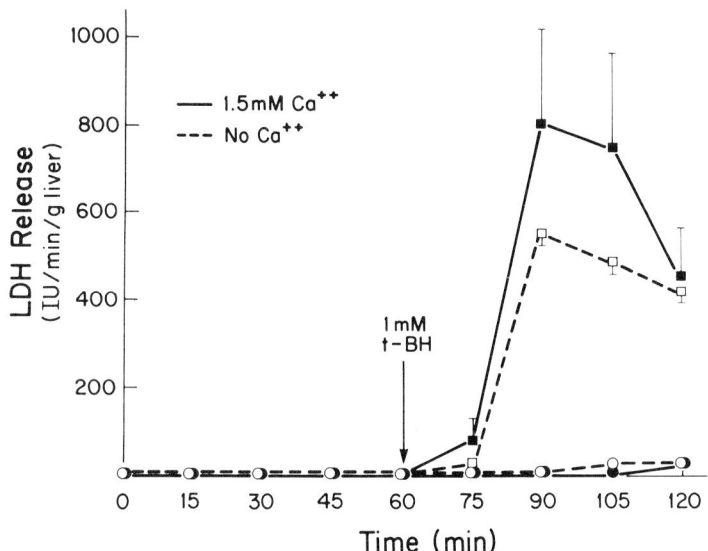

FIG. 6. t-BH induced toxicity to the isolated perfused rat liver determined as the release of lactate dehydrogenase (LDH) into the perfusate. Data represents the mean ± S.D. of 4 perfusions in the presence (■,□) and absence (●,o) of 1 mM t-BH.

therefore find no essential role for an influx of extracellular Ca^{2+} in the initiation of toxic injury to liver cells.

A redistribution of intracellular Ca^{2+} could also cause a significant rise in $[Ca^{2+}]_i$, because the mitochondria and endoplasmic reticulum actively accumulate large amounts of sequestered Ca^{2+}. The fact that the Ca^{2+} ionophore A23187 is able to produce membrane blebbing and cell death in the absence of extracellular Ca^{2+} (14) further supports the idea that a redistribution of intracellular Ca^{2+} is responsible for surface blebbing and that an influx of extracellular Ca^{2+} is not involved. A disturbance of intracellular Ca^{2+} homeostasis is therefore the most likely cause of membrane blebbing in hepatocytes.

Figure 7 shows that the incubation of hepatocytes with the free radical-generating toxins t-BH and menadione causes a rapid depletion of GSH and loss of cell Ca^{2+}. Our studies with a non-disruptive technique capable of determining the Ca^{2+} content of both the mitochondrial and extramitochondrial compartments of intact hepatocytes have shown that this loss of cell Ca^{2+} reflects the mobilization of Ca^{2+} from both intracellular compartments (1,14,35). The mobilization and depletion of Ca^{2+} from the mitochondrial compartment seems to be dependent on the oxi-

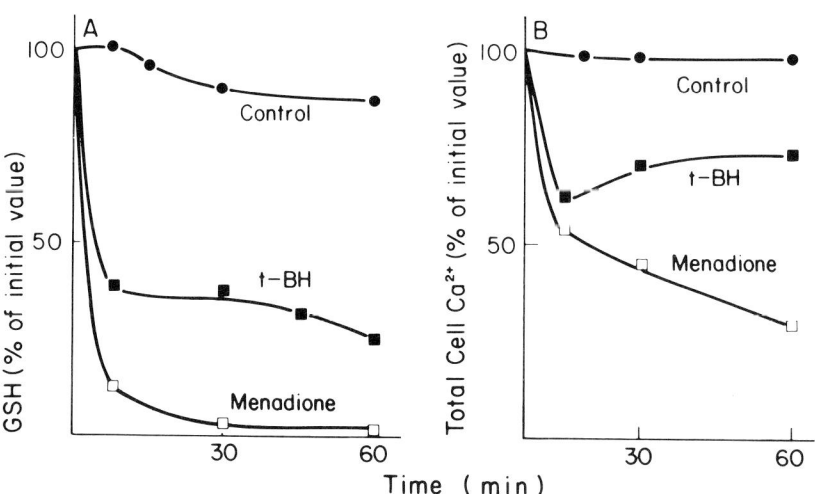

FIG. 7. Levels of cellular GSH (A) and Ca^{2+} (B) during incubation of hepatocytes with 2mM t-BH (■); 200 µM menadione (□) and no addition (●).

dation of intramitochondrial GSH and pyridine nucleotides (2,18) and is not due to a non-specific increase in membrane permeability (21). Thus, Ca^{2+} release occurs via a specific efflux process which is sensitive to the mitochondrial NAD(P)H and thiol redox states.

The majority of the Ca^{2+} present in the extramitochondrial compartment consists of that sequestered by the endoplasmic reticular Ca^{2+}-ATPase (14,43). Since the depletion of this Ca^{2+} pool seemed to be closely associated with the loss of GSH and surface blebbing (14), we developed a spectrophotometric assay for microsomal Ca^{2+} sequestration (15). Studies utilizing this method have shown that the liver endoplasmic reticular Ca^{2+}-ATPase is highly susceptible to oxidative damage (15). For example, both t-BH and menadione inhibit Ca^{2+} sequestration in rat liver microsomes (Table 1). This inhibitory effect is, however, completely prevented by the presence of GSH (Table 1). Thus, it is highly probable that the primary site of damage is the thiol group(s) of the Ca^{2+}-ATPase, a conclusion supported by the findings of Moore and co-workers (21,22).

Menadione and t-BH are therefore able to inhibit Ca^{2+} sequestration processes in both the mitochondria and endoplasmic reticulum through their ability to oxidize pyridine nucleotides and intracellular thiols. Incubation of hepatocytes with these compounds causes the release of Ca^{2+} into the cytosol which can-

TABLE 1. Effect of preincubation with menadione or t-butylhydroperoxide on Ca^{2+} sequestration by rat liver microsomes

Preincubation	Ca^{2+} sequestered nmol/mg protein
without substrate	7.5
t-butylhydroperoxide, 200 μM	1.4
t-butylhydroperoxide, 200 μM plus GSH, 2mM	7.2
menadione, 0.6 mM + NADPH, 1mM	2.3
menadione, 0.6 mM + NADPH, 1mM plus GSH, 2mM	6.9

All incubations were performed at 25°C.
Preincubations were for 25 min with t-BH and for 10 min with menadione.

not be resequestered. Under normal circumstances this would cause a transient rise in $[Ca^{2+}]_i$ followed by a return to normal levels as the plasma membrane Ca^{2+}-translocase removed Ca^{2+} from the cell. If, however, this removal of Ca^{2+} were inhibited there could be a more sustained rise in $[Ca^{2+}]_i$ which could have toxic consequences. Table 2 shows that t-BH also inhibits the activity of the hepatic plasma membrane Ca^{2+}-translocase in isolated vesicle preparations and that GSH protects against this effect. More detailed studies (3) have shown that this inhibitory activity is related to the oxidation of membrane thiol groups. Thus, oxidative stress brought about by free radical-generating toxins is capable of disrupting the normal sequestration and removal processes which regulate Ca^{2+} homeostasis in liver cells (FIG. 8). This may in turn lead to a sustained rise in $[Ca^{2+}]_i$ which could cause plasma membrane blebbing by altering the organization of the hepatocyte microfilament system. Obviously, measurements of $[Ca^{2+}]_i$ under various conditions and at multiple time points are required to substantiate this hypothesis and a cause and effect relationship between raised $[Ca^{2+}]_i$ and blebbing has yet to be shown. Such studies are, however, technically quite difficult. We have attempted to measure $[Ca^{2+}]_i$ during blebbing using several procedures with little success. We attempted to use the null-point titration method described in (24) and although a general upward trend in the measured $[Ca^{2+}]_i$ was observed during toxic injury, we could not obtain consistent results. Moreover, this method requires the destruction of the plasma membrane with digitonin so that simultaneous measurements of $[Ca^{2+}]_i$ and blebbing cannot be made. The use of the fluorescent indicator of $[Ca^{2+}]_i$, 'quin 2'

TABLE 2. Effect of preincubation with t-butylhydroperoxide on Ca^{2+} sequestration by plasma membrane vesicles

Preincubation	Ca^{2+} sequestered nmol/mg protein
without substrate	9.3
GSH, 1 mM	14.2
t-butylhydroperoxide, 1 mM	4.0
t-butylhydroperoxide, 1 mM plus GSH, 1 mM	8.4

All incubations were performed at 37°C.
Preincubations were for 5 min.

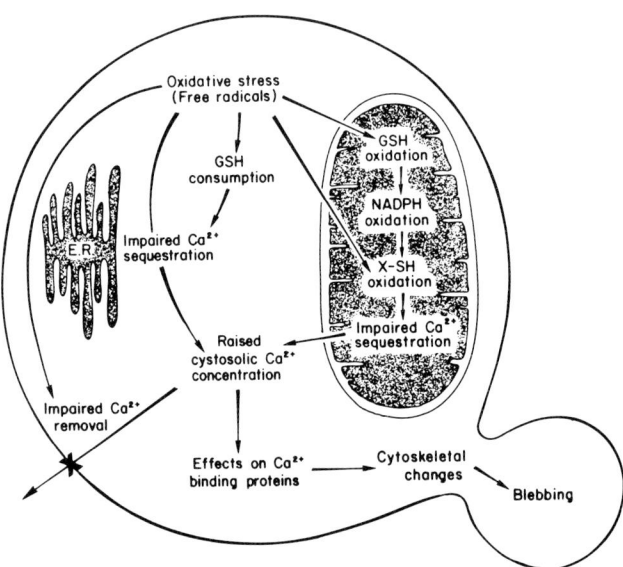

FIG. 8. Schematic illustration of possible mechanisms by which free radicals may alter intracellular Ca^{2+} homeostasis and cytoskeletal structure resulting in bleb formation.

(36), which can be trapped in the cytoplasm of intact viable cells (28,36,37) has also been advocated as being suitable for toxicity studies with isolated hepatocytes (26). Unfortunately, the excitation-emission wavelength characteristics of 'quin 2' are very similar to those of the reduced pyridine nucleotides NADH and NADPH, which are present at high concentrations in viable hepatocytes. Thus, any changes in $[Ca^{2+}]_i$ and hence 'quin 2' fluorescence brought about by t-BH, menadione and a variety of other toxins tend to be masked by changes in pyridine nucleotide autofluorescence (M.T. Smith and R.Y. Tsien, unpublished observations). 'quin 2' can be used, however, with isolated hepatocytes to study hormone-induced changes in $[Ca^{2+}]_i$ because the changes in 'quin 2' fluorescence occur several seconds prior to changes in autofluorescence (5), presumably because hormones act rapidly at specific receptors on the cell surface and do not require metabolism to elicit their effects. Unfortunately this is not the case with hepatotoxins requiring metabolic activation to exert their toxic effects. We are therefore concentrating our efforts on developing new Ca^{2+}-sensitive dyes having emission-excitation maxima at longer wavelengths for use with isolated hepatocytes.

CYTOSKELETAL CHANGES DURING BLEBBING

We have also investigated the functional organization of the hepatocyte microfilament system during toxic injury to elucidate the connection between altered cytoskeletal structure and blebbing. By using the method of Blikstad et al. (4) to determine the polymerization state of the cytoskeletal actin, we found that the monomeric:filamentous (G:F) actin ratio does not change during blebbing induced by a variety of compounds, including t-BH and menadione. Thus, plasma membrane blebbing need not be the result of F-actin depolymerization to G-actin in cortical microfilaments, but could reflect other kinds of structural reorganization, such as actin cross-linking of the type induced by α-actinin or vinculin (39) or actin filament severing and bundling as mediated by gelsolin, villin, and other actin binding proteins (39). Villin is of special interest because at low Ca^{2+} concentrations (≤ 0.1 μM) it cross-links actin filaments into bundles, but when the Ca^{2+} concentration is raised above 1 μM the bundles are dispersed and the filaments fragmented (10,11,39). This type of dispersion and fragmentation of actin microfilaments as a result of a rise in free $[Ca^{2+}]_i$ may well be the mechanism involved in bleb formation, but any firm conclusions can only result from detailed electron microscopic studies of changes in cortical microfilament structure along with simultaneous measurement of $[Ca^{2+}]_i$.

CONCLUSIONS

Substantial evidence has been acquired which links the oxidative stress and toxicity of free radical-generating chemicals to disturbances in intracellular Ca^{2+} homeostasis. One of the earliest manifestations of this disturbed Ca^{2+} homeostasis appears to be an alteration in surface morphology known as plasma membrane blebbing. Such blebbing occurs well before cell death and seems to result from changes in hepatocyte cytoskeletal organization. Soluble thiols, notably GSH, are important in protecting cells from this type of injury. Measurements of the cytosolic free Ca^{2+} concentration in hepatocytes during blebbing and toxicity will be an important next step in our efforts to gain an understanding of the mechanisms involved in free radical-mediated tissue injury. Whether or not a sustained rise in cytosolic free Ca^{2+} is the common "point of no return" in toxic cell death remains an open question worthy of future investigations.

IMPLICATIONS FOR THE MECHANISM OF CELLULAR AGING

According to the free radical theory of aging proposed by Harman (12), aging results from the deleterious effects of free radicals produced in the normal course of cellular metabolism. One important deleterious effect of free radicals may be the progressive inhibition of the activity and synthesis of enzyme translocases and binding proteins involved in maintaining intracellular Ca^{2+} homeostasis. We have shown here that the Ca^{2+}-translocase enzymes of cells are particularly susceptible to free radical damage. Their progressive inhibition would alter the ionic environment of the cell, including an adjustment to $[Ca^{2+}]_i$, and may lead to a build-up of cellular Ca^{2+}. Such an increase in cell Ca^{2+} could be recognized by Ca^{2+}-dependent enzymes, such as transglutaminase (27), as a signal for terminal differentiation and may initiate new programs of gene expression (17). Recent work has shown that calcium is a key element in controlling the proliferation and differentiation of both mesenchymal (42) and epidermal (13) cells in culture. Thus, free radical-induced alterations in intracellular Ca^{2+} homeostasis could lead to cellular aging by initiating terminal differentiation. This hypothesis obviously requires rigorous testing, which is a goal of future experiments in our laboratory.

ACKNOWLDGEMENTS

Supported by the Swedish Medical Research Council and the Swedish Council for Planning and Coordination of Research to S.O. and by the National Foundation for Cancer Research and Northern California Occupational Health Center to M.T.S. We are grateful to Professors R.Y. Tsien, D.P. Jones and U. Lindberg for much helpful discussion, to Mr. S. Thorold for preparing the electron micrographs and E. Symanski for preparing this manuscript.

REFERENCES

1. Bellomo, G., Jewell, S.A., Thor, H. and Orrenius, S. (1982): Proc. Natl. Acad. Sci. USA, 79:6842-6846.

2. Bellomo, G., Jewell, S.A. and Orrenius, S. (1982): J. Biol. Chem., 257:11558-11562.
3. Bellomo, G., Mirabelli, F., Richelni, P. and Orrenius, S. (1983): F.E.B.S. Letts., in press.
4. Blikstad, I., Markey, F., Carlsson, L., Persson, T. and Lindberg, U. (1978): Cell, 15:935-943.
5. Charest, R., Blackmore, P.F., Berthon, B. and Exton, J.H. (1983): J. Biol. Chem., 258:8769-8773.
6. Cheung, W.Y. (1980): Science, 207:19-27.
7. Clarke, M. and Spudich, J.A. (1977): Ann. Rev. Biochem. 46:797-822.
8. Condeelis, J. and Vahey, M. (1982): J. Cell Biol., 94:466-474.
9. Fisher, M.M. and Phillips, M.J. (1979): In: Progress in Liver Diseases, edited by H. Popper and F. Schattner, Vol. 6, pp. 105-121, Grune and Stratton, N.Y.
10. Glenney, J.R., Bretscher, A. and Weber, K. (1980): Proc. Natl. Acad. Sci. USA, 77:6458-6462.
11. Glenney, J.R., Kaulfus, P. and Weber, K. (1981): Cell, 24:471-480.
12. Harman, D. (1956): J. Gerontol., 11:298-300.
13. Hennings, H., Michael, D., Cheng, C., Steinert, P., Holbrook, K. and Yuspa, S.H. (1980): Cell, 19:245-254.
14. Jewell, S.A., Bellomo, G., Thor, H., Orrenius, S. and Smith, M.T. (1982) Science, 217:1257-1259.
15. Jones, D.P., Thor, H., Smith, M.T., Jewell, S.A. and Orrenius, S. (1983): J. Biol. Chem., 258:6390-6393.
16. Korn, E.D. (1978): Proc. Natl. Acad. Sci. USA, 75:588-599.
17. Levenson, R. and Housman, D. (1981): Cell, 25:5-6.
18. Lötscher, H.R., Winterhalter, K.H., Carafoli, E. and Richter, C. (1979): Proc. Natl. Acad. Sci. USA, 76:4340-4344.
19. Mesland, D.A.M., Los, G. and Spiele, H. (1981): Exp. Cell Res., 135:431-435.
20. Moldéus, P., Högberg, J. and Orrenius, S. (1978): Methods Enzymol., 52, 60-71.
21. Moore, G., Jewell, S.A., Bellomo, G. and Orrenius, S. (1983): F.E.B.S. Letts., 153:289-293.
22. Moore, L. (1982): Biochem. Pharmacol., 31:1465-1467.
23. Moore, L., Davenport, G.R. and Landon, E.J. (1976): J. Biol. Chem., 251:1197-1201.
24. Murphy, E., Coll, K., Rich, T.L. and Williamson, J.R. (1980): J. Biol. Chem., 255:6600-6608.
25. Prentki, M., Chaponnier, C., Jeanrenaud, B., and Gabbiani, G. (1979): J. Cell Biol. 81:592-607.
26. Recknagel, R.O. (1983): Life Sciences, 33:401-408.

27. Rice, R.H. and Green, H. (1978): J. Cell Biol., 76:705-711.
28. Rink, T.J., Smith, S.W. and Tsien, R.Y. (1982): F.E.B.S. Letts., 148:21-26.
29. Rink, T.J., Tsien, R.Y. and Pozzan, T. (1982): J. Cell Biol., 95:189-196.
30. Roos, A. and Boron, W.F. (1981): Physiol. Rev., 61:296-434.
31. Sandy, M.S. and Smith, M.T. (1983): Submitted for publication.
32. Schliwa, M. (1981): Cell, 25:587-590.
33. Smith, M.T. and Orrenius, S. (1984): In: Drug Metabolism and Drug Toxicity, edited by J.R. Mitchell and M.G. Horning, pp. 71-95, Raven Press, N.Y.
34. Smith, M.T., Thor, H. and Orrenius, S. (1981): Science, 213:1257-1259.
35. Thor, H., Smith, M.T., Hartzell, P., Bellomo, G., Jewell, S.A. and Orrenius, S. (1982): J. Biol. Chem., 257:12419-12425.
36. Tsien, R.Y. (1981): Nature, 290:527-528.
37. Tsien, R.Y., T. Pozzan and T.J. Rink, (1982): J. Cell Biol., 94:325-334.
38. Weatherbee, J.A. (1981): Int. Rev. Cytol., 12 (suppl.): 113-176.
39. Weeds, A. (1982): Nature, 296:811-816.
40. Weihing, R.R. (1979): Methods Achiev. Exp. Pathol., 8:42-109.
41. Weiss, E., Sterz, I., Frimmer, M. and Kroker, B. (1973): Beitr. Pathol., 150:345-356.
42. Whitfield, J.F., MacManus, J.P., Rixon, R.H., Boynton, A.L., Youdale, T. and Swierenga, S. (1976): In Vitro, 12:1-18.
43. Williamson, J.R., Cooper, R.H. and Hoek, J.B. (1981): Biochim. Biophys. Acta, 639:243-295.

Metabolic Rate, Free Radicals, and Aging

R. S. Sohal

Department of Biology, Southern Methodist University, Dallas, Texas 75275

Two main lines of investigation have been followed in the literature to potentially gain an insight into the mechanisms underlying the aging process. One involves the identification of regimens which extend the maximum, rather than the average, life span of experimental populations. For example, low caloric intake and reduced metabolic rate have been shown to influence the rate of aging (1). The second is concerned with the identification of constitutional traits which are associated with species specific longevity. For example, ratio of brain weight to body weight (2) and rate of energy dissipation (3-4) have been correlated with species-specific longevity in mammals. The objective of this article is to discuss the possible relationship between metabolic rate, free radicals and aging.

It was first recognized by Rubner (4) that the total amount of energy dissipated by mammalian species, with several fold differences in longevity, tends to be relatively constant, around 200 kcal/g. More recently, calculations by Cutler have confirmed this inference in non-primate mammals (see 3). Primates apparently consume 3 to 4 times more energy /g body weight per life span than the other mammals (3). It must, however, be emphasized that such correlations indicate statistical tendencies only and many deviations from the predicted relationship have been found. Furthermore, no firm conclusions regarding cause and effect can ever be drawn from such correlations. Thus, it is certainly preferable to experimentally investigate the relationship between metabolic rate and aging in the same species. For instance, metabolic rate of poikilotherms and mammalian hibernators can be manipulated experimentally by variations in the ambient temperature. Average and maximum life spans of a variety of insects and other species have been shown to be longer, at lower temperatures than at higher temperatures (for review, see 5). Similarly, Turkish hamsters which undergo hibernation have been found to live relatively longer in cold, in proportion to the length of time spent under hibernation (6).

Thermal variations, however, cause changes in rates of almost all reactions besides metabolic rate. Furthermore, the rate of oxygen consumption cannot be assumed to be strictly temperature-dependent. For example, rate of oxygen consumption in milkweed bugs kept at 25°C and 30°C throughout life, was higher at 30°C than at 25°C only during the first two-thirds of the life span (7). In the last trimester of life, bugs kept at 25°C exhibited a higher rate of oxygen consumption than those at 30°C. Rate of oxygen consumption in insects is also modulated by the previous thermal conditions under which they were kept (8). Another important factor influencing the metabolic rate of flying insects in the laboratory is the size of the housing containers (9). Within the viable range, insects are more active physically at higher temperatures than at lower temperatures (10). If flying is possible, rate of oxygen utilization will be much higher than that attributable to the temperature-dependent increase in basal metabolic rate. Flight in insects is strongly influenced by ambient temperature and has distinct thermal thresholds. For example, using radar-Doppler instrumentation, it was found that houseflies cannot fly below 16°C (10). With a gradual increase in temperature, a 10-fold increase in flying and a 15-fold increase in walking activity occurred between 15 and 24.5°C, followed by a small decline until 29°C.

In view of the sources of complications mentioned above, it was inevitable that studies conducted in different laboratories, on the relationship between aging and temperature-induced metabolic rate, would give variable and often contradictory results (see 11). The main point of controversy arising from numerous studies conducted on this subject concerned the validity of the rate of living theory of Pearl (12) and the threshold theory of Maynard Smith (13). The former theory was based on studies on the mortality of a variety of organisms under different environmental conditions. In general, it was observed that the shapes of the survivorship curves of well-fed and starved populations were essentially similar, i.e., rectangular. These and other results led to the formulation of the rate of living theory which postulates that duration of life depends on the rate at which a fixed sum of energy is expended. Maynard Smith tested the rate of living theory using protocols which involved transference of adult Drosophila from one temperature to another. Results of these studies were found to be discordant with the predictions of the rate of living theory. The sources of discrepancy between the two points of view have been discussed previously (see 5,14) and will not be detailed here. In summary, it was pointed out that the underlying assumption by Maynard Smith and others that ambient temperature is strictly proportional to metabolic rate at all ages is wrong.

A different regime that depended on physical activity was, consequently, developed in this laboratory to alter the metabolic rate of insects (5,15). The level of physical activity was altered by a variety of means including surgical removal of wings, variation in the volume of housing containers and alterations of sex ratios of populations (16). Physical activity was measured, separately, as walking and flying activity using radar-Doppler. Results of these studies have been reviewed elsewhere (14). In general, it was found that life spans of flies are longer under regimens which reduce the level of physical activity. For example, the average and the maximum life spans of flies confined within small vials, where they were unable to fly, were about twice those kept in one-cubic foot cages, where flying was possible. The total average amount of oxygen consumed during life was found to be statistically similar in flies kept under conditions of high and low physical activity (17). Age-specific death rates (slopes of Gompertz plots), which are believed to be indicative of aging rates were significantly higher in flies kept under conditions of relatively high physical activity (14).

In order to determine whether or not the differences in the life spans of flies that were maintained under identical conditions corresponded to the differences in the levels of their spontaneous physical activity, walking and flying activity of individual flies was compared in relation to life span (16). It was found that those flies which exhibited relatively greater tendency for spontaneous flying tended to have shorter life spans than the relatively more sluggish flies. In toto, results of the studies mentioned above were believed to be in accord with the rate of living theory rather than the threshold theory.

Next, an effort was made to determine if levels of physical activity and the accompanying rates of oxygen consumption also affect age-associated changes besides mortality. It should be mentioned here that at present there is a great dearth of cellular and biochemical markers of aging, which can be reliably employed, to compare rates of aging under different physiological conditions. According to Strehler (18), lipofuscin accumulation comes closest to fulfilling the requirements of the criterion for aging-specific changes.

Currently, lipofuscin is measured by two different methods. The classical method is based on morphometric measurement of lipofuscin granules in histological sections. The more recent method, originally developed by Tappel and coworkers (19,20), employs the measurement of blue-emitting fluorescent material in organic solvent extracts of tissues. Although a quantitative correlation between the two methods has never been established, many workers have enthusiastically embraced the fluorometric technique due to its relative convenience and sensitivity. However, measurements of lipofuscin in the same tissue by the two methods have been found to be in total disagreement in some cases (21,22). Therefore, in this discussion, the term lipofuscin will be used to refer specifically to the granular structures detectable in tissue sections by microscopy, whereas, the blue-emitting fluorescent material in chloroform extracts of tissue homogenates will be called "soluble fluorescent material" (SFM).

Rate of lipofuscin accumulation was compared in three

different cell types of houseflies kept under varied conditions of physical activity. Lipofuscin accumulation was found to be faster in the short-lived flies, kept under high activity conditions, as compared to the long-lived, low activity flies (23,24). However, the maximal level of lipofuscin reached during life was similar in the two groups, but was attained at about twice the age in the low activity flies as compared to the high activity flies. Similarly, the rate of accumulation of soluble fluorescent material (SFM) was relatively faster under conditions of high physical activity (23,25). The maximum concentration of SFM reached during life in the high and the low activity groups was similar, but was achieved at a later age in the low activity flies. Individual flies, kept under identical housing conditions (each fly confined in a 3-liter container), showed differences in SFM concentration at three weeks of age. These differences were correlated to the levels of spontaneous physical activity in individual flies (26). Flies exhibiting relatively higher spontaneous physical activity tended to also contain higher concentrations of SFM (26). Altogether, results of these studies indicated that flies which are physically more active tend to die relatively earlier and accumulate lipofuscin granules and SFM at a faster rate.

A similar relationship, between metabolic rate and lipofuscin, has been found in Turkish hamsters by Lyman et al. (6,27). They found that the hamsters which hibernated lived longer than the non-hibernating controls (6). The rate of lipofuscin accumulation in the brain and the heart of hibernators was slower than in the controls (27).

Since rates of lipofuscin and SFM accumulations in the housefly are correlated with life expectancy rather than chronological age, it can be suggested that accrual of lipofuscin and SFM is associated with physiological rather than chronological aging.

Cumulatively, results of the studies cited above suggest that a relationship may exist between metabolic rate, aging and formation of lipofuscin and SFM. The fundamental question arising from these findings however concerns the mechanism by which physical activity modulates life span and cellular changes associated with aging. Obviously, an increase in physical activity also increases the rates of many intracellular processes including the rate of oxygen consumption. It is possible that the effects of physical activity on aging may be due to the involvement of processes other than those directly related to oxygen utilization. However, it can be reasonably hypothesized that oxygen metabolism may play a role in aging. There is a large body of evidence indicating that oxygen utilization, even under normal physiological conditions, entails the production of potentially deleterious free radicals and hydroperoxides (28). A small proportion of oxygen utilized by cells is apparently reduced by single electron additions producing, first, superoxide radical (O_2^-) which is converted to H_2O_2 by the activity of superoxide dismutase (SOD). In turn, H_2O_2 is eliminated by catalase and glutathione peroxidase (29); however, the latter enzyme is absent in insects including the housefly. Although both O_2^- and H_2O_2 are cytotoxic, the main agent of oxidative damage is believed to be the hydroxyl radical (OH·) which is generated by the reaction of O_2^- and H_2O_2 in the presence of transition metal ions (30). In addition to the enzymes SOD, catalase and peroxidases, whose combined functions tend to decrease the probability of OH· generation, cells also possess non-enzymatic defenses against free radicals. Glutathione, β-carotene, α-tocopherol and ascorbic acid are the most well known endogenous antioxidants (29).

There is some evidence to suggest that, despite the existence of enzymatic and non-enzymatic antioxidant defenses, a small proportion of free radicals escape elimination (28). It has been amply demonstrated that the intermediates of oxygen reduction can cause damage to cellular constituents. For example, fluxes of experimentally-generated O_2^- have been shown to cause deleterious changes including lipid peroxidation, inactivation of enzymes, lysis of membranes and nicking of DNA (for references, see 31). It is possible, indeed probable, that a small proportion of free radicals that escape elimination by antioxidant defenses cause damage to cellular constituents. Such damage would be slow but constant and, if unrepaired, will accumulate with age.

Prolonged, exhaustive, exercise in rats has been shown to cause a 2-3 fold increase in the concentration of an unidentified free radical species in muscle and liver (32). Prolonged exercise in the rat also resulted in increased levels of lipid peroxidation products, loss of sarcoplasmic reticulum and decrease in mitochondrial respiratory control values (33).

Accumulation of lipofuscin and SFM during aging has been suggested to result from free radical-induced lipid peroxidation damage to cellular organelles (19,20). It has been postulated by Tappel and coworkers (19,20) that malondialdehyde, a product of lipid peroxidation, causes cross-linking of amine-containing molecules to produce Schiff base substances which are responsible for the characteristic fluorescence of lipofuscin and lipid-soluble fluorescent pigment. They have demonstrated that substances with Schiff base characteristics can be produced in vitro by an interaction between the products of lipid perox- idation and amine-containing molecules such as proteins, nucleic acids and phospholipids. Gutteridge (34) has reported that in addition to lipid peroxidation, malondialdehyde-like substances are also produced as a result of free radical damage to organic molecules such as amino acids, DNA and carbohydrates. Deficiency of antioxidants in the diet often shortens life span (35) and causes an increase in the rate of ceroid (structures resembling lipofuscin) and SFM accumulation (36). Basson and coworkers (37) have reported that the rate of SFM accumulation is faster in rats undergoing treadmill physical training as compared to the sedentary controls. Results of these studies are consistent with the view that lipofuscin and SFM may be products of free radical reactions. There is however no direct in vivo evidence demonstrating a link between free radicals and accumulation of lipofuscin and SFM.

The hypothesis that free radicals may be the cause of molecular damage underlying the aging process was originally proposed by Harman (see 35), but the view that oxygen is potentially toxic at all concentrations and that its deleterious effects are due to the production of free radicals was first enunciated clearly by Gerschman (38). Initial efforts, to investigate the role of free radicals in aging, were made by determining the effects of antioxidant intake on life spans of organisms (for review, see 35). In general, results of these studies indicated that supplemental feeding of antioxidants caused only a slight increase in the average life span of some species, whereas, maximum life spans remained relatively unaffected. Analyses of mortality rates indicated that antioxidant supplementation tends to prevent only the premature deaths in a population and has no effect on age-dependent mortality rates (1). Prolongation of average life span is apparently due to the removal of antioxidant deficiency in some organisms rather than slowing of the aging process.

The logic behind the experiments involving antioxidant intake was that if free radicals play a causal role in aging, antioxidant administration should quench a certain proportion of these radicals, thus sparing some biological molecules from damage. Apparently, serious consideration was not given to the intracellular concentrations of antioxidants and rate constants of free radical reactions for assessing the feasibility of antioxidant-dependent amelioration of aging process. For example, the concentration of antioxidants in cells required to quench even a small fraction of hydroxyl radicals is theoretically beyond dietary means.

The idea that innate levels of antioxidant defenses may be associated with species-specific longevity was first advanced by Cutler and coworkers (see 3). They reported a correlation between maximum life span of several mammalian species and the ratio between SOD activity and total energy dissipation during life. Species expending greater amounts of energy per gram body weight during life tended to have relatively higher SOD activity.

The role of different antioxidant defenses in the aging process was first experimentally examined in this laboratory using adult housefly as a model system. Initially, studies were conducted to determine if aging in the housefly was associated with a general decline in the efficiency of the mechanisms protective against the intermediates of oxygen metabolism (39). The rate of oxygen consumption, activities of SOD and catalase, and levels of inorganic peroxides, glutathione (GSH and GSSG) and chloroform-soluble antioxidants were measured in adult male

houseflies of different ages. Rate of oxygen consumption
declined in old flies. Activities of total- and cyanide-
insensitive-SOD and catalase decreased during the second half
of life span whereas concentration of inorganic peroxides
increased in older flies. Levels of total glutathione and GSH
decreased during later half of life whereas GSSG increased
during this period. The concentration of chloroform-soluble
antioxidants sharply declined during the first half of life.
These findings indicated that aging in the housefly is correlated
with an attrition in the antioxidant defense mechanisms.

It is now well known that antioxidant defense mechanisms tend
to have an overlapping function (40). For example, O_2^- is
dismutated by SOD and H_2O_2 is eliminated by catalase. However,
GSH can also eliminate O_2^- and H_2O_2 (40). Therefore, to ascertain
if aging is associated with a decline in the efficiency of the
mechanisms protective against free radicals, it is imperative to
obtain information about the various overlapping antioxidant
defenses.

In an effort to learn if SOD activity was a determinant of
longevity, houseflies were administered diethyldithiocarbamate
(DDC) in their drinking water (41). It should be pointed out
that DDC, a copper chelator, is an effective inhibitor of
cupro-zinc SOD activity, it also tends to inactivate other
copper-containing proteins. DDC, however, can also act as an
antioxidant. Nonetheless, at present DDC is the most widely
used SOD inhibitor and a more specific one is presently
unavailable. Despite these limitations, DDC offers a useful
experimental approach to evaluate the involvement of SOD in
aging. In the housefly, DDC administration caused a decrease
in activities of both cyanide-sensitive (presumably cupro-zinc-
SOD) as well as cyanide-insensitive (manganese-SOD) SOD (41).
Unexpectedly, DDC-administered flies had a longer average life
span than the controls. However, DDC caused a reduction in the
metabolic rate of flies and an elevation in GSH levels. Reduced
metabolic rate would tend to increase life span whereas higher
GSH levels can partially replace SOD activity, as also stated
above. It was postulated that reduction in metabolic rate and
increase in GSH can together compensate for the loss of SOD
activity, whereby, the average life span remains unaltered.

Since houseflies lack glutathione peroxidase, catalase would
seem to play a crucial role in the elimination of H_2O_2. However,
total inhibition of catalase activity, induced by 3-amino-
triazole, had no effect on life span of flies (42). The rate of
oxygen consumption was greatly reduced and GSH levels were
considerably elevated in catalase-inhibited flies. These results
were interpreted to suggest that loss of catalase activity was
compensated by adaptive responses in metabolic rate and GSH
levels. Oxidative stress, induced by the administration of
diamide (43) and paraquat (44), also produced a decline in
metabolic rate with simultaneous increase in GSH levels.

In general, results of the studies on oxidative stress in

houseflies have indicated that a complex balance exists between products of oxygen metabolism and antioxidant defenses.

Studies on the correlation between oxidative stress, induced by chemical means, and age-associated changes in the housefly such as the accumulations of lipofuscin and soluble fluorescent material are currently in progress. Results of preliminary studies indicate that oxidative stress does not cause an increase in the concentration of thiobarbituric acid reacting substances which is unlike that occurring during normal aging. Further experimental work is needed to resolve whether or not the effect of metabolic rate on aging in the housefly is exerted through the generation of oxygen-centered free radicals.

Acknowledgements: Research of the author has been supported by grants from the National Institutes of Health (AG00171) and the Glenn Foundation for Medical Research.

REFERENCES

1. Sacher, G.S.(1977):In:The Biology of Aging, edited by C.E. Finch and L. Hayflick, pp. 582-638. Van Nostrand Reinhold, New York.
2. Sacher, G.A.(1978):BioScience, 28:497-501.
3. Tolmasoff, J.M., Ono, T. and Cutler, R.G.(1980):Proc. Nat. Acad. Sci. U.S.A., 27:2777-2781.
4. Rubner, M.(1908):Das Problem der Lavensdaur. Oldenbourg, Munich.
5. Sohal, R.S.(1976):In:Interdisciplinary Topics in Gerontology, vol. 9, pp. 25-40. S. Karger, Basel.
6. Lyman, C.P., O'Brien, R.G., Green, G.C. and Papafrangos, E.D.(1981):Science, 212:668-670.
7. McArthur, M.C. and Sohal, R.S.(1982):J. Geront., 37: 268-274.
8. Dehnel, P.A. and Segal, E.(1956):Biol. Bull., 111:53-61.
9. Ragland, S.S. and Sohal, R.S.(1975):Exp. Geront., 10: 279-289.
10. Buchan, P.B. and Sohal, R.S.(1981):Exp. Geront., 16:223-228.
11. Lints, F.A.(1971):Gerontologia, 17:33-51.
12. Pearl, R.(1928):The Rate of Living. University of London Press, London.
13. Maynard Smith, J.(1963):Nature, 199:400-402.
14. Sohal, R.S.(1981):In:Age Pigments, edited by R.S. Sohal, pp. 303-316. Elsevier/North Holland, Amsterdam.
15. Ragland, S.S. and Sohal, R.S.(1973):Exp. Geront., 8:135-145.
16. Sohal, R.S. and Buchan, P.B.(1981):Exp. Geront., 16: 157-162.
17. Sohal, R.S.(1982):Age, 5:21-24.
18. Strehler, B.L.(1977):Time, Cells and Aging. Academic Press, New York.
19. Tappel, A.L.(1975):In:Pathology of Cell Membranes, edited by B.F. Trump and A.V. Arstila, vol. 1, pp. 145-170.

Academic Press, New York.
20. Tappel, A.L. (1980):In:Free Radicals in Biology, edited by W.A. Pryor, vol. 4, pp. 1-27. Academic Press, New York.
21. Bieri, J.G., Tolliver, T.J. and Robinson, W.G. (1980): Lipids, 15:10-13.
22. Eldred, G., Miller, G.V., Stark, W. and Feeney Burns, L. (1982): Science, 216:757-759.
23. Sohal, R.S. and Donato, H. (1979): J. Geront., 34:489-496.
24. Sohal, R.S. (1981): Exp. Geront., 16:347-355.
25. Sohal, R.S. and Donato, H. (1978): Exp. Geront., 13:335-341.
26. Sohal, R.S. and Buchan, P.B. (1981): Mech. Ageing Dev., 15:243-249.
27. Papafrangos, E.D. and Lyman, C.P. (1982): J. Geront., 37:417-421.
28. Chance, B., Sies, H. and Boveris, A. (1979): Physiol. Rev., 59:527-603.
29. Fridovich, I. (1978): Science, 201:875-880.
30. Halliwell, B. (1981): In: Age Pigments, edited by R.S. Sohal, pp. 1-62. Elsevier/North Holland, Amsterdam.
31. Brawn, K. and Fridovich, I. (1980): Acta Physiol. Scand. Suppl., 492:9-18.
32. Davies, K.J.A., Quintanilha, A.T., Brooks, G.A. and Packer, L. (1982): Biochem. Biophys. Res. Comm., 107:1198-1205.
33. Quintanilha, A.T., Packer, L., Davies, J.M.S., Rancanelli, T.L. and Davies, K.J.A. (1982): Ann. N.Y. Acad. Sci., 393:32-37.
34. Gutteridge, J.M.C. (1982): Int. J. Biochem., 14:649-653.
35. Harman, D. (1982): In: Free Radicals in Biology, edited by W.A. Pryor, vol. 5, pp. 255-275. Academic Press, New York.
36. Wolman, M. (1981): In: Age Pigments, edited by R.S. Sohal, pp. 265-281. Elsevier/North Holland, Amsterdam.
37. Basson, A.B.K., Terblanche, S.E. and Oelofsen, W. (1982): Comp. Biochem. Physiol., 71A:369-374.
38. Gerschman, R. (1959): In: Symp. Spec. Lect. XXI Intern. Congress Physiol. Sciences, Buenos Aires, pp. 222-226.
39. Sohal, R.S., Farmer, K.J., Allen, R.G. and Cohen, N.R. (in press): Mech. Ageing Dev.
40. Forman, H.J. and Fisher, A.B. (1981): In: Oxygen and Living Processes, edited by D.L. Gilbert, pp. 235-249. Springer-Verlag, New York.
41. Sohal, R.S., Farmer, K.J., Allen, R.G. and Ragland, S.S. (in press): Mech. Ageing Dev.
42. Allen, R.G., Farmer, K.J., and Sohal, R.S. (1983): Biochem. J., 216:503-506.
43. Allen, R.G., Farmer, K.J. and Sohal, R.S. (in press): Comp. Biochem. Physiol.
44. Allen, R.G., Farmer, K.J., Newton, R.K. and Sohal, R.S. (in press): Comp. Biochem. Physiol.

Free Radical Involvement in the Formation of Lipopigments

Donald Armstrong

Department of Ophthalmology, University of Florida College of Medicine, J. Hillis Miller Health Center, Gainesville, Florida 32610

Autofluorescent lipopigments are generally referred to as either lipofuscin, an inert material that accumulates in all normally aging tissues (72), or ceroid (3). These two classes of lipopigments may be associated with various pathological or inherited conditions such as seen in certain of the pre-senile dementias (8,15,22,23), in metabolic disorders of children (57), in nutritional alteration (21,30,38,46,48,64,69,78), or following exposure to toxic chemicals (39). In two inherited disorders, namely ceroid-lipofuscinosis (3) and the Chediak-Higashi syndrome (58), ceroid type lipopigments (Batten's syndrome) are a prominent feature. These are summarized in the following table:

LIPOFUSCIN FORMATION	CEROID FORMATION
Aging (4,21,72)	Neurological disease (21)
Vitamin A deficiency (64)	Batten syndrome (3,4)
	Kufs pre-senile dementia (3,8,15,23)
	Huntington's disease (22)
	Chediak-Higashi syndrome (58)
	Mucopolysaccharidoses and the Sanfillipo's syndrome (57)
	Vitamin E deficiency (30,38,64,78)
	Protein restriction (46,69)
	Chloroquine toxicity (39)

In Figures 1 through 4, the ultrastructural appearance of these lipopigments is demonstrated. Lipofuscin in the brain of a 70 year old person is shown in Fig. 1 (mag. = 9,728X) and that from a patient with Batten's disease with ceroid deposition, is shown in Fig. 2 (mag. = 9,282X) for comparison. Depending upon the type of disease, inclusions may be in a curvilinear (Fig. 2a; mag. = 35,176X) or fingerprint (Fig. 2b; mag. = 74,000X) configuration. Lipofuscin in the aging dog brain (Fig. 3; mag. = 3,250X) is similar to human brain and the ceroid in dogs affected with canine ceroid-lipofuscinosis (Fig. 4; mag. = 10,928X) consists of membranous profiles with fingerprint patterns (Fig. 5; mag. = 96,743X) like that in figure 2b. Each of these lipopigments, fluorescence when excited by blue light (300-400 nm).

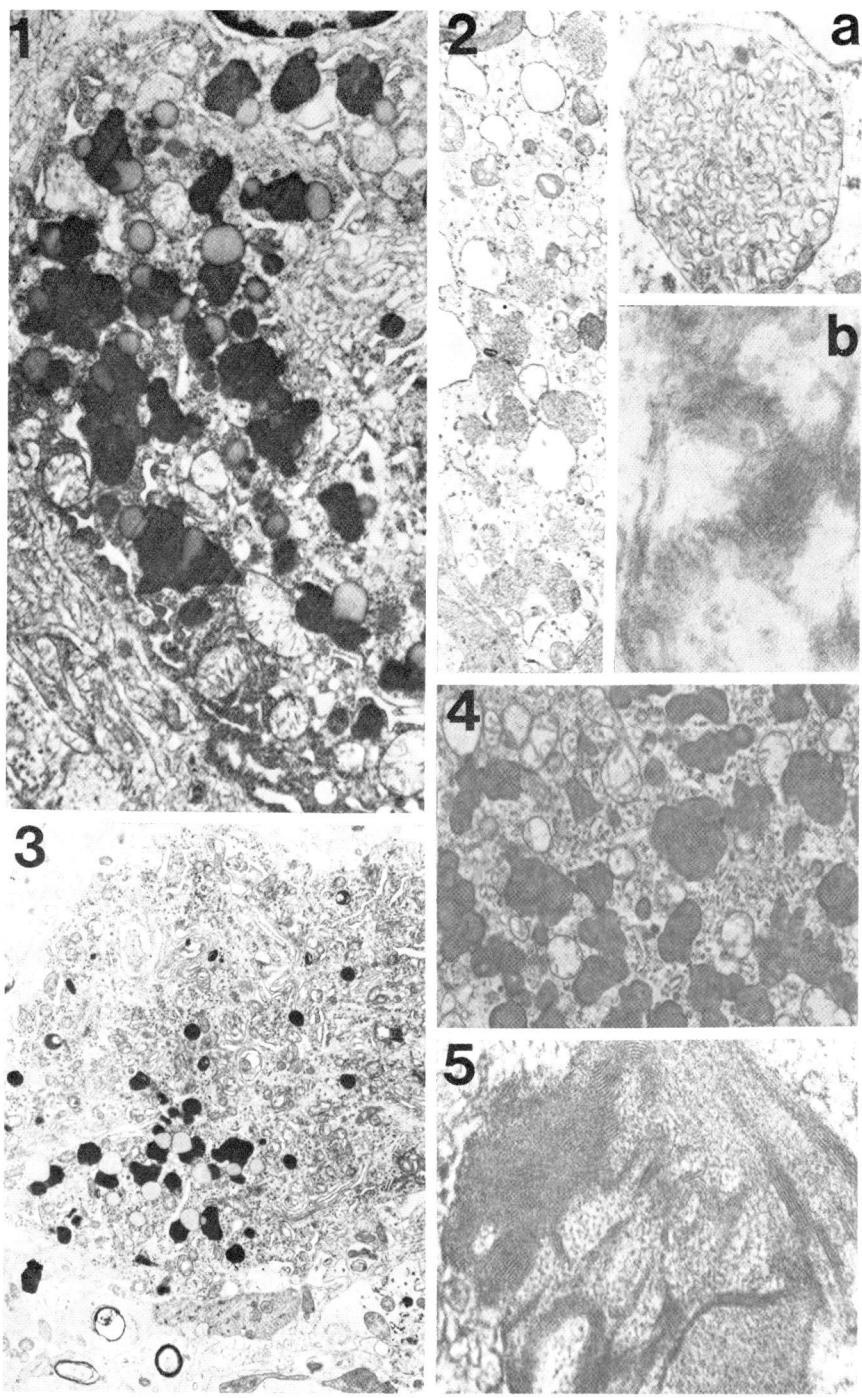

Figures 1 to 5.

Both lipopigments are thought to arise via reactions initiated by free radicals and subsequently formed by lipid peroxide metabolites (malonyldialdehyde or MDA) during the break-down of damaged biomolecules from various sub-cellular compartments. The former involves the initial binding of polymers to pre-formed lipid globules and transforming them into increasingly denser structures, whereas the latter begin with characteristic ultrastructure and increase in size and number:

Lipid globule $\xrightarrow{\text{peroxidation, metals, poly MDA}}$ Lipid complex $\xrightarrow{\text{fusion with lysosomes}}$ Lipofuscin

Focal lipid damage $\xrightarrow{\text{peroxidation, metals, MDA}}$ Membranous profiles $\xrightarrow{\text{fusion with lysosomes}}$ Ceroid

Histochemical and biochemical analyses have shown that lipopigments are rich in polyunsaturated fatty acids (PUFA), and metal ions (iron, copper) which are presumably involved in catalyzing the autoxidation of PUFA to lipid peroxides. At equal levels of oxidation, the longer-chain PUFA with more unsaturated sites i.e.; docosahexaneoic acid, will produce more radicals and peroxides (~ 8/molecule) than shorter-chain ones, i.e.; linolenic acid (~ 2/molecule). This chapter will discuss the available evidence to implicate such reactions in the genesis of lipopigment granules.

FREE RADICALS IN TISSUES

The occurrence of free radicals in biological material has been summarized by Slater (73). Employing electron spin resonance (ESR) techniques (40), normal frozen samples have broad and complex peaks which are attributed to the individual contributions from flavins, semiquinones, cytochrome P_{450}, prostaglandin intermediates (38), melanins (68), metallocomplexes (iron, copper) (34,85) and sub-cellular organelles (11,45,73), to name a few. Reactive, rather long-lived species such as superoxide and perhydroxyl radicals (16,63), organic peroxy radicals (6) and hydroxyl free radicals (55) are also present in cells (10). Free radicals (especially hydroxyl) are known to damage PUFA and proteins (19,65,67) through interactions between the various reactive species (42), metals and ascorbate (13,24,88), or other radicals (12,56). As a result, colored products with fluorescent characteristics of lipopigments are formed (65).

FREE RADICALS ASSOCIATED WITH LIPOPIGMENTS

The first report of a free radical component in lipofuscin was published by Van Woert et al. (84). These authors isolated particles from human aged cardiac muscle, extracted them with chloroform-methanol (2:1) and analyzed the resulting lipid and insoluble fractions by ESR. The microwave absorption for intact particles produced a single resonance peak at g = 2.005, which was also found in the insoluble fraction. No detectable signal was found in the lipid extract. Because of the similarity of this signal to that observed from the substantia nigra, they concluded that cardiac lipofuscin contained a melanin component. A similar finding was observed in human plasma where that lipofuscin had an ESR signal compatible with melanin (31). Indeed, electron microscopy of myocardium reveals dense (black) globular structures consistent with the appearance of melanin (70).

Recently, particulate fractions of lipofuscin isolated from childhood, adult and aged human brain were examined and compared to ceroid particles isolated from the brain of the canine ceroid-lipofuscinosis model (85). Resonance peaks were observed over a wide magnetic field region for all samples. The peaks were compatible with organic radicals ($g = 2.2$ to 1.9), as well as complexes of non-blue, type 2 copper ions and high-spin ferric iron which were seen at lower magnetic fields. Furthermore, the radical concentration in cerebral tissue increased with age. An in vitro system, which produced lipopigments rapidly, gave in addition to the metal-ion complexes, a signal compatible with an oxygen-centered radical.

FREE RADICALS, METALS AND LIPID PEROXIDATION

It is well known that oxygen is toxic to cells when greater than ambient pressures are encountered (28). Its toxicity is due to the formation of inorganic superoxide free radicals ($O_2^{-\cdot}$), which undergo a dismutation reaction catalyzed by superoxide dismutase (SOD) to produce hydrogen peroxide (H_2O_2). Superoxide can also react with ferric (Fe^{3+}) complexes (reactions 1 and 2), generating ferrous iron which then act as a catalyst to decompose H_2O_2 and generate hydroxyl free radicals ($OH^{\cdot}$) (reaction 3) (26,28,82,87):

$$2\,O_2^{-\cdot} + 2H^+ \xrightarrow{\text{SOD}} H_2O_2 + O_2 \quad (1)$$

$$O_2^{-\cdot} + Fe^{3+}\,\text{complex} \longrightarrow Fe^{2+}\,\text{complex} + O_2 \quad (2)$$

$$Fe^{2+}\,\text{complex} + H_2O_2 \longrightarrow OH^{\cdot} + OH^- + Fe^{3+}\,\text{complex} \quad (3)$$

Non-enzymatic reactions involving heme-containing compounds or iron complexes (9,13,43,77,82) also catalyze lipid peroxidation (32,36,80,86). Copper ions produce $O_2^{-\cdot}$ if glutathione (GSH) is available, and H_2O_2 and OH (25,32) are formed by the following reactions:

$$Cu^+ + O_2 \longrightarrow O_2^{-\cdot} + Cu^{2+} \quad (4)$$

$$2\,GSH + O_2 \xrightarrow{Cu^{2+}} H_2O_2 + GSSH \quad (5)$$

$$Cu^+ + H_2O_2 \longrightarrow OH + OH^- + Cu^{2+} \quad (6)$$

It has been stated (26) that copper catalyzed degradation of H_2O_2 (step 6) may be more active than iron (step 3).

Hydroxyl free radicals are extremely reactive (17) and cause damage to cellular membranes i.e.; increased permeability (42), because they attack PUFA (reaction 7) to form lipid free radicals (reaction 8). Due to the proximity of other PUFA, hydrogen abstraction (47) may take place to form lipid peroxides and new free radicals (reaction 9):

$$PUFA + OH^{\bullet} \longrightarrow L^{\bullet} \qquad (7)$$

$$L^{\bullet} + O_2 \longrightarrow LOO^{\bullet} \qquad (8)$$

$$LOO^{\bullet} + PUFA \longrightarrow LOOH + L^{\bullet} \qquad (9)$$

Unchecked, this process becomes a chain reaction with involvement of additional PUFA. Mono-substituted peroxides may in turn undergo reactions with ascorbate and ferrous complexes (44), or di-peroxides can react with cuprous ions (41,43) to form a variety of lipid free radicals (14,77,81). Lipid peroxides readily decompose to liberate highly reactive carbonyl fragments; the most prominent being malondialdehyde (MDA) and these either polymerize or cross-link with amine groups of proteins (65) and phospholipids (20) to form a Schiff's base (27). Both of these compounds exhibit autofluorescence and are found in lipopigments (71,72).

Recent experiments have demonstrated the reactivity of lipid peroxides and MDA with regard to metal ions. In Figure 6A, pure docosahexanoic acid peroxide (0.2 mM) was prepared (27) and a 200 μl aliquot analyzed by ESR, before and after addition of 10 mM EDTA.

The peroxide signal appears to be due almost entirely to copper complexed to the lipid. Binding was not as strong as to the EDTA and the complex was slightly different from the ESR spectrum for the lipid peroxide standard. In a separate experiment, manganese was also found in association with this sample.

In Figure 6B, MDA was analyzed under similar conditions. The region around $g = 2$ shows two peaks instead of the three noted for the lipid peroxide. However, following chelation, the spectra for the two compounds were identical. From these results, it is clear that the signal detected in each sample is almost entirely due to copper complexed to these molecules. This suggests that lipid peroxides and MDA may be important chelators of copper in living tissues.

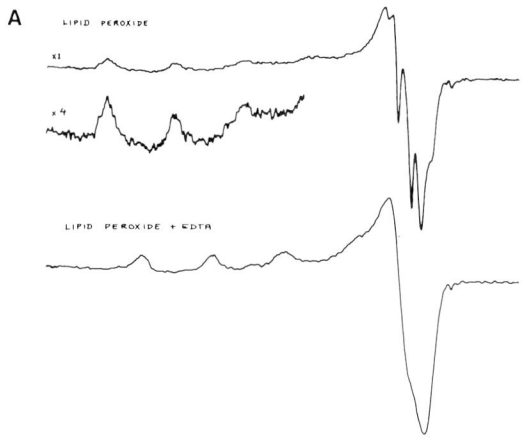
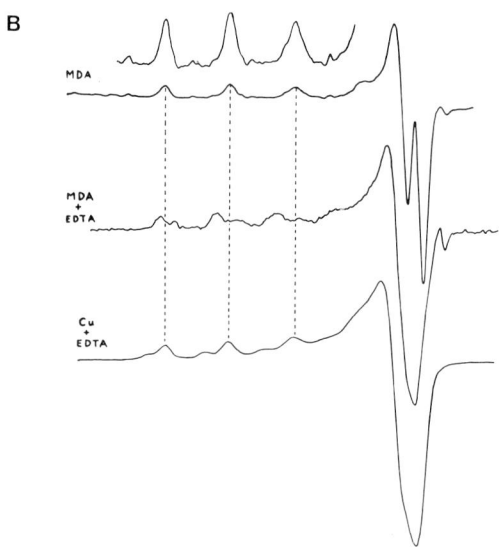

Figure 6A and B. Recordings were carried out at 77K, 4.2 mW microwave power, 4 gauss modulation and are the sum of 42 individual spectra. Sweep time was 100 sec. and microwave frequency was 9.493 GHz.

Noda et al. (51) have found that in young rat brain under hyperbaric conditions, increasing amounts of lipid peroxides and MDA are produced as a function of age. Nohl et al. (52) have also found that lipid peroxides, $O_2^{\cdot -}$ and H_2O_2 increase in mitochondria during the aging process. Membranes from young cells when exposed to peroxide, reflect differences in protein patterns seen in senescent membranes (61). Hochstein et al. have shown that this is due to Cu^{2+} induced lipid peroxidation (32) and MDA polymerization of membrane proteins and lipids (33), resulting in the production of fluorophors and membrane rigidity. Finally, lipid peroxidation is increased in antioxidant deficient diets (66) and lipopigment concentration is elevated (5,7,64,89).

INHIBITION OF LIPOPIGMENT FORMATION BY FREE RADICAL SCAVENGERS

Many years ago, Tappel (79) proposed that α-tocopherol acted as an antioxidant by scavenging free radicals in biological systems. More recently, it has been shown that 1 molecule of the vitamin protects 220 PUFA molecules, emphasizing just how effective this lipid soluble vitamin really is (18). In conjunction with ascorbate, the potency of α-tocopherol is increased (44) and this interaction may be due to a recycling of the α-tocopherol free radical through its reaction with ascorbate (60):

$$FR^{\cdot} + \text{Vit. E} \longrightarrow \text{Vit. E} \tag{10}$$

$$\text{Vit. E}^{\cdot} + \text{Vit. C} \longrightarrow \text{Vit. E} + \text{Vit. C} \tag{11}$$

Other compounds with antioxidant properties which alter lipopigment formation are butylated hydroxytoluene, nordihydroguaiaretic acid and N,N'-diphenyl p-phenylenediamine (79,89). In addition to these compounds, drugs such as dimethylaminoethyl p-chlorophenoxyacetate (Centrophenoxin[1]) and N-acetyl homocysteine thiolactone (Cithiolone) are known to decrease the rate of lipofuscin formation in vivo (51,88) as well as "dissolve" it in vitro (52,79). Recently, Cithiolone has been shown to dissolve ceroid in cultured cells taken from dogs with ceroid-lipofuscinosis (80). Although the exact mechanisms are unknown, their ability to regulate lipopigment turnover----a process linked to free radical interactions----suggests antioxidant capabilities.

PEROXIDE REGULATING SYSTEMS

There are many enzymes that regulate the intracellular level of H_2O_2 and lipid peroxide and thus control the potential availability of $OH^{\cdot}$ concentration within cells (52). The major enzymes are catalase (CAT), PPD-peroxidase (POD), GSH-peroxidase (GSH-Px), GSH-reductase (2,35), and/or NADPH-cytochrome c reductase (63). Furthermore, individual types of cells do not seem to be protected by a full complement of these enzymes and may be subject to greater damage and lipopigment formation. This is quite evident in the retina where cell types are well

---

[1] Promonta Labs., Hamburg

separated (2). Figure 7 shows that retinal pigment epithelium (RPE) and rod outer segments (ROS) only have 1 enzyme, whereas the retina is protected by several.

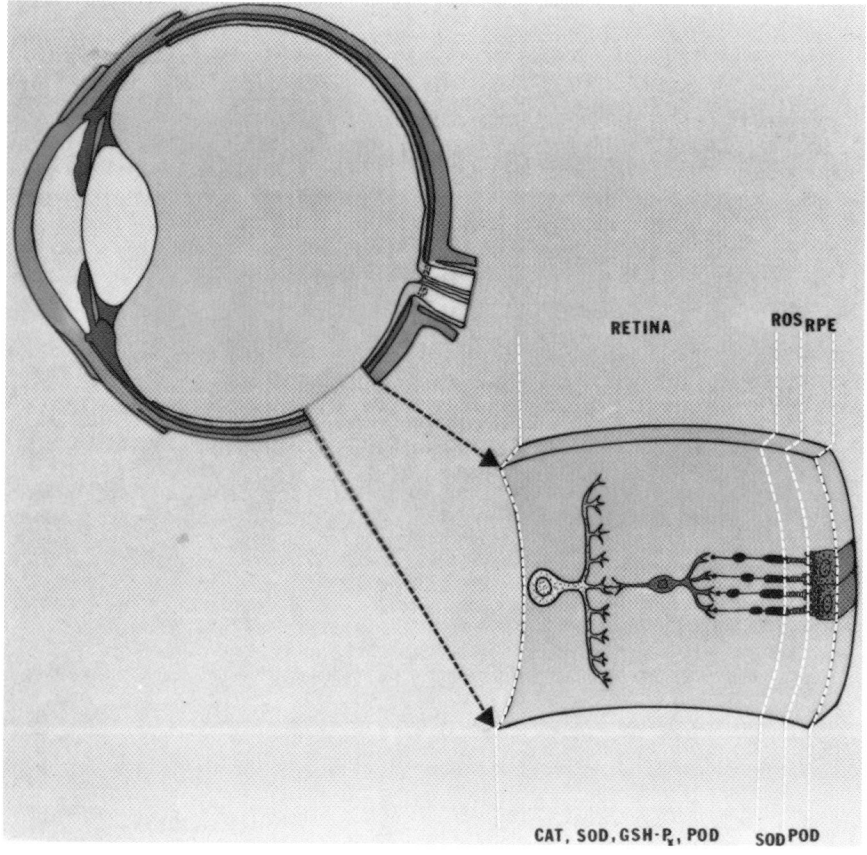

Figure 7. Schematic representation of enzyme activities across the canine retina, rod outer segments and retinal pigment epithelium. Acronyms are explained in the text.

During aging when free radicals are increasing and lipopigments are accumulating within cells, several papers have reported changes in antioxidant enzyme activity. For example, in whole tissue, PPD-peroxidase appears to decrease with age and could account for an increase in H_2O_2 (1,52) and lipid peroxides (4). In mitochondria, catalase and GSH-peroxidase increase with age (53), suggesting an inductive response to the elevated peroxides.

SUMMARY

Free-radicals are produced by many reactions and are present in all respiring cells (Fig. 8). Such reactions are regulated by a number of protective enzymes but during aging, changes in activity may occur, so that free radicals increase and damage cells. Due to their high degree of unsaturation, the biomolecules most susceptible must be the phospholipids which contain polyunsaturated fatty acids, and structural proteins within the cell membrane, or sub-cellular organelles. Once damaged, lipid free radicals react with molecular oxygen to form unstable lipid peroxides and these rapidly undergo metal catalyzed degeneration, liberating carbonyl fragments which cross-link and/or polymerize essential cellular molecules. These processes lead to the formation of autofluorescent compounds which are un-degradable and thus accumulate in a progressive manner during aging.

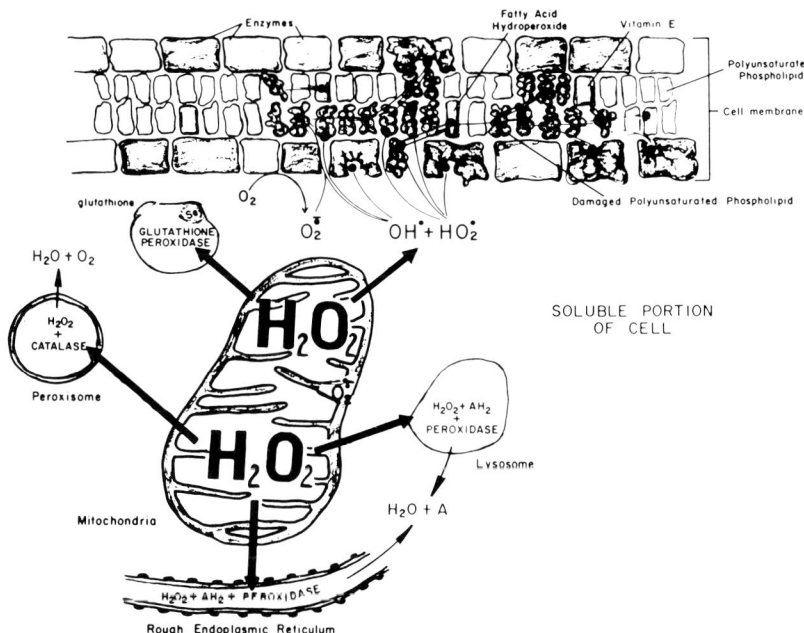

Figure 8. A schematic representation of free radical formation and the manner in which damaging reactions might occur.

Many years ago, Harman proposd an intriguing theory of free radical involvement in aging (29). Since then, much indirect, albeit strong, suggestive evidence has been reported to support that hypothesis. The difficulty in obtaining direct evidence is due to the fact that free radicals are extremely short-lived (59) and may not be measurable when analyzed by ESR techniques. However, now that in vitro systems are becoming more sophisticated, natural and induced autoxidation processes can be examined and the events leading to the formation of lipopigments determined. Geropharmacology will be aided greatly by continued improvement of this methodology and the production of compounds which regulate lipopigment deposition in aging cells.

ACKNOWLEDGEMENTS

The author expresses appreciation to Dr. A. Vistnes for performing ESR studies on the lipid peroxide and MDA samples. This work was supported by NIH grant EY 04083, the Children's Brain Disease Foundation, the Paul F. Glenn Foundation for Medical Research and an unrestricted departmental grant from Research to Prevent Blindness, Inc., New York. Dr. Armstrong is the recipient of a Research Career Development Award, K04 EY 00236.

REFERENCES

1. Armstrong, D., Rinehart, R., Dixon, L., and Reigh, D. (1978): AGE, 1:8-12.
2. Armstrong, D., Santangelo, H., and Connole, E. (1981): Curr. Eye Res., 1:225-242.
3. Armstrong, D., Koppang, N., and Rider, J. (1982): Ceroid-Lipofuscinosis (Batten's disease), pp. 1-421, Elsevier Biomedical Press, Amsterdam.
4. Armstrong, D. (1984): In: Relations Between Normal Aging and Disease, edited by H. Johnson, J. Ordy, and R. Kohn, (in press), Raven Press, New York.
5. Balin, A. (1982): In: Testing the Theories of Aging, edited by R. Adelman and G. Roth, pp. 137-182, CRC Press, Inc., Boca Raton.
6. Benedetto, C., Bocci, A., Dianzani, M., Ghiringhello, B., Slater, T., Tomasi, A., and Vannini, V. (1981): Cancer Res., 41:2936-2942.
7. Blockett, A. and Hall, D. (1981): J. Gerontol., 36:529-533.
8. Boehme, D., Cottrell, J., Leonberg, S., and Zeman, W. (1970): Brain, 94:745-760.
9. Bucher, J., Tien, M., and Aust, S. (1983): Biochem. Biophys. Res. Comm., 111:777-784.
10. Chance, B., Sies, H., and Boveris, A. (1979): Physiol. Rev., 59:527-605.
11. Conti, F., Segre, A., Eremenko, T., Zaniratti, S., Elia, G., Benadetto, A., and Volpe, P. (1981): Cancer Biochem. Biophys., 5:195-199.
12. Del Maestro, R., Thaw, H., Bjork, S., Planker, M., and Arfors, K. (1980): Acta Physiol. Scand., Suppl. 492:43-57.
13. Devasagayan, T., Pushpendran, C., and Eapen, J. (1981): Biochem. Biophys. Acta, 750:91-97.
14. Dirks, R. and Faiman, M. (1982): Brain Res., 248:355-360.
15. Dowson, J. (1982): Br. J. Psychiatry, 140:142-148.

16. Fee, J. and Valentine, J. (1977): In: Superoxide and Superoxide Dismutases, edited by A. Michelson, J. McCord, and I. Fridovich, pp. 19-60, Academic Press, New York.
17. Fong, K., McCay, P., Poyer J., Misra, H., and Keele, B. (1976): Chem. Biol. Interac., 15:77-89.
18. Fukazama, K., Tokumura, A., Ouchi, S., and Tsukatani, H. (1982): Lipids, 17:511-513.
19. Funes, J. and Karel, M. (1981): Lipids, 16:347-379.
20. Garg, H., Awasthi, Y., and Srivastana, S. (1981): J. Neurosci. Res., 6:771-783.
21. Glees, P. and Hasan, M. (1976): Lipofuscin in Neuronal Aging and Disease, pp. 1-68, Georg Thieme Publ., Stuttgart.
22. Goebel, H., Heipertz, R., Scholz, W., Iqbal, K., and Tellez-Nagel, I. (1978): Neurology, 28:23-31.
23. Goebel, H., Praak, H., Seidel, D., Doshi, R., Marsden, C., and Gullotta, F. (1982): Clin. Neuropathol., 1:151-162.
24. Grosch, W. (1976): Z. Lebensm. Unters.-Forsch., 160:371-375.
25. Gutteridge, J. (1980): FEBS Lett., 112:269-272.
26. Gutteridge, J. (1982a): Biochem. Soc. Trans., 10:72-73.
27. Gutteridge, J., Kerry, P., and Armstrong, D. (1982b): Biochim. Biophys. Acta, 711:460-465.
28. Halliwell, B. (1981): In: Age Pigments, edited by R. Sohal, pp. 1-62, Elsevier/North-Holland Biomedical Press, Amsterdam.
29. Harman, D. (1956): J. Gerontol., 11:298-299.
30. Hayes, K. (1974): Invest. Ophthalmol., 13:499-510.
31. Hegedus, Z., Altschule, M., and Nayak, U. (1982): Arch. Int. Physiol. Biochim., 90:55-60.
32. Hochstein, P., Kumar, K., and Forman, S. (1980): Ann. N.Y. Acad. Sci., 355:240-248.
33. Hochstein, P. and Jain, S. (1981): Fed. Proc., 40:183-188.
34. Horn, R., Friesen, E., Stephens, R., Hedrick, W., and Zimbrick, J. (1979): Cancer, 43:2392-2398.
35. Hothersall, J., El-Hassan, A., McLean, P., and Greenbaum, A. (1981): Enzyme, 26:271-276.
36. Hrycay, E. and O'Brien, P. (1971): Arch. Biochem. Biophys., 147:14-27.
37. Kalyanaramen, B. and Sivarajeh, K. (1984): In: Free Radicals in Biology, Vol. 6, edited by W. Pryor, pp. 149-198, Academic Press, New York.
38. Katz, M., Stone, W., and Dratz, E. (1978): Invest. Ophthalmol. Vis. Sci., 17:1049-1058.
39. Klinghart, L.G. (1974): Acta Neuropath., (Berl) 28:117-141.
40. Knowles, P., Marsh, D., and Rattle, H. (1976): Magnetic Resonance of Biomolecules, pp. 1-343, John Wiley & Sons, London.
41. Kochi, J. and Mains, H. (1965): J. Org. Chem., 30:1862-1872.
42. Kong, S. and Davison, A. (1980): Arch. Biochem. Biophys., 204:18-29.
43. Kosacic, P. and Kurz, M. (1966): J. Amer. Chem. Soc., 88:2068-2069.
44. Leung, H., Vang, M. and Mavis, R. (1981): Biochim. Biophys. Acta, 664:266-272.

45. Lloveras, J., Vincensini, P., Ribbes, G., Record, M., Ferre, G., Douste-Blazy, L., and Pescia, J. (1980): Radiat. Res., 82:45-54.
46. Manocha, S. and Sharma, S. (1978): Experientia, 34:377-378.
47. Mead, J. (1976): In: Free Radicals in Biology, edited by W. Pryor, pp. 51-68, Academic Press, New York.
48. Miyoshi, K. (1982): Rinsho Shinkeigaku, 22:1109-1111.
49. Nandy, K. (1978): Mech. Ageing Dev., 8:131-138.
50. Nandy, K. and Schneider, H. (1976): In: Neurobiology of Aging, edited by R. Terry and S. Gershin, pp. 245-264, Raven Press, New York.
51. Noda, Y., McGeer, P. and McGeer, E. (1982): Neurobiol. Aging, 3:173-178.
52. Nohl, H. and Wegner, D. (1978): Eur. J. Biochem., 82:563-567.
53. Nohl, H., Hegner, D., and Heinz-Summer, K. (1979): Mech. Ageing Dev., 11:145-151.
54. Nohl, H. and Jordan, W. (1980): Eur. J. Biochem., 111:203-210.
55. Nohl, H., Jordan, W., and Hegner, D. (1981): FEBS Lett., 123:241-244.
56. Nohl, H., Jordan, W., and Hegner, D. (1981): Hoppe-Seyler's Z. Physiol. Chem., 363:599-607.
57. Oldfors, A. and Sourander, P. (1981): Acta Neuropathol., 54:287-292.
58. Oliver, C. (1981): In: Age Pigments, edited by R. Sohal, pp. 335-353, Elsevier/North-Holland Biomedical Press, Amsterdam.
59. Osipov, A., Moravskii, A., Shuvalov, V., Azizova, O., and Vladimirov, Yu. (1980): Biophysics, 25:239-244.
60. Packer, J., Slater, T., and Willson, R. (1979): Nature, 278:737-738.
61. Pfeffer, S. and Swislocki, N. (1982): Mech. Ageing Dev., 18:355-367.
62. Player, T. and Horton, A. (1981): J. Neurochem., 37:422-426.
63. Pryor, W. (1978): Photochem. Photobiol., 28:787-801.
64. Robison, W., Kuwabara, T., and Bieri, J. (1982): Retina, 2:263-281.
65. Roubal, W. (1970): Lipids, 6:62-64.
66. Sagai, M. and Ichinose, T. (1980): Life Sci., 27:731-738.
67. Schaich, K. and Karel, M. (1976): Lipids, 11:392-400.
68. Sealy, R., Felix, C., Hyde, J., and Swartz, H. (1980): In: Free Radicals in Biology, Vol. 4, edited by W. Pryor, pp. 209-259, Academic Press, New York.
69. Sharma, S. and Manocha, S. (1977): Mech. Ageing Dev., 6:1-14.
70. Siakotos, A., Watanabe, J., Pennington, K., and Whitfield, M. (1973): Biochem. Med., 7:25-38.
71. Siakotos, A. and Munkrees, K. (1981): In: Age Pigments, edited by R. Sohal, pp. 181-202, Elsevier/North-Holland Biomedical Press, Amsterdam.
72. Siakotos, A. and Munkrees, K. (1982): In: Ceroid-Lipofuscinosis (Batten's disease), edited by D. Armstrong, N. Koppang and J. Rider, pp. 167-183, Elsevier Biomedical Press, Amsterdam.
73. Slater, T. (1978): Biochemical Mechanisms of Liver Injury, pp. 65-73, Academic Press, London.
74. Sohal, R. (1981): Age Pigments, pp. 1-394, Elsevier/North-Holland Biomedical Press, Amsterdam.

75. Spoerri, P. and Glees, P. (1974): Mech. Ageing Dev., 3:131-155.
76. Spoerri, P., Kelley, K., Armstrong, D., and Ellis, A. (1984): Ophthalmic Res., (in press).
77. Svingen, B., O'Neal, F., and Aust, S. (1978): Photochem. Photobiol., 28:803-809.
78. Takeuchi, N., Iritani, N., Fukuda, E., and Tanaka, F. (1978): In: Tocopherol, Oxygen and Biomembranes, edited by C. de Duve and O. Hayaishi, pp. 257-272, Elsevier/North-Holland Biomedical Press, Amsterdam.
79. Tappel, A. (1962): Vit. Hormones, 20:493-510.
80. Tappel, A., Fletcher, B., and Deames, D. (1973): J. Gerontol., 28:415-424.
81. Thomas, M., Mehl, K., and Pryor, W. (1978): Biochem. Biophys. Res. Comm., 83:927-932.
82. Tien, M., Morehouse, L., Bucher, J., and Aust, S. (1982): Arch. Biochem. Biophys., 218:450-458.
83. Totaro, E. (1981): Acta Neurologica, (Naples), 36:1-8.
84. Van Woert, M., Prasad, K. and Borg, D. (1967): J. Neurochem., 14:707-716.
85. Vistnes, A., Henriksen, T., Nicolaissen, B. Jr., and Armstrong, D. (1983): Mech. Ageing Dev., 22:335-345.
86. Wills, E. (1965): Biochim. Biophys. Acta, 98:238-251.
87. Willson, R. (1979): In: Oxygen Free Radicals and Tissue Damage, Ciba Foundation Symposium 65, pp. 19-42, Exerpta Medica, Amsterdam.
88. Winterbourne, C. (1981): Biochem. J., 198:125-131.
89. Zuckerman, B. and Geist, M. (1981): In: Age Pigments, edited by R. Sohal, pp. 283-302, Elsevier/North-Holland Biomedical Press, Amsterdam.

Possible Involvement of Iron and Oxygen Free Radicals in Aspects of Aging in Brain

Robert A. Floyd, Malgorzata M. Zaleska, and *H. James Harmon

*Oklahoma Medical Research Foundation, Oklahoma City, Oklahoma 73104; *Department of Zoology, Oklahoma State University, Stillwater, Oklahoma 74078*

The possible involvement of free radicals as etiological agents in aging has been debated for about 30 years now. Arguments implicating the involvement of free radicals in aging were recently summarized by Harman (9). There is ample evidence indicating that free radical events occur in normal as well as under pathological conditions in biological systems; and therefore in the context that all possible biological reactions contribute to aging in an organism, there is little doubt that free radicals are involved. However, from a critical point of view, it is clear that it has not yet been proven that a particular free radical event affects a rate limiting step in overall organism aging even though a few studies are suggestive. We present in this report an approach to critically testing some concepts in the relationship between free radicals and aging.

RATIONALE OF APPROACH

We have concentrated on testing the concept that aging in brain involves oxidative mechanisms with the participation of iron ions and oxygen free radicals. The reasons for our approach are outlined briefly below.

Lipofuscin Accumulation

Lipofuscin accumulates in cells, especially postmitotic cells such as brain neurons, with age and in general shows a linear increase with lifetime. This heterogenous polymer, often referred to as aging pigment, is apparently the result of aldehydes, such as malondialdehyde which is produced during lipid peroxidation, conjugating with primary amine groups of other lipids, nucleic acids, and proteins to form Schiff's base type compounds. The accumulation of lipofuscin with age is due probably to either its increased rate of formation or its decreased rate of decomposition or possibly the combination of both with age. Thus since oxidative mechanisms are involved in the formation of lipofuscin, and since its accumulation with age is about the only one parameter that consistently increases with age over a wide range of organisms, it seems reasonable to examine oxidative events during aging in more detail.

Iron and Oxygen Free Radicals

The recognition that oxygen free radicals participate in a wide range of biologically important reactions both in normal as well as under pathological conditions has become ever increasingly obvious as evidenced by the continously growing literature in this area. Fridovich and colleagues in early work demonstrated that superoxide production in the presence of hydrogen peroxide yielded a powerful oxidant that damaged biological compounds. It is now considered that reaction (3) below, which is the sum of reactions (1) and (2), explains their observations. Thus, the hydroxyl free radical, a very powerful oxidant, is produced in an iron-catalyzed reaction between superoxide and hydrogen peroxide.

$$O_2^{\overset{\bullet}{-}} + Fe(III) \longrightarrow Fe(II) + O_2 \qquad (1)$$

$$H_2O_2 + Fe(II) \longrightarrow Fe(III) + \overset{\bullet}{O}H + OH^- \qquad (2)$$

$$O_2^{\overset{\bullet}{-}} + H_2O_2 \longrightarrow O_2 + OH^- + \overset{\bullet}{O}H \qquad (3)$$

We have examined reaction 2 in some detail from a biologically relevant viewpoint and have been able to draw some important conclusions (3,4,5). First, the catalysis of $\overset{\bullet}{O}H$ formation from hydrogen peroxide requires ferrous ion, ferric being ineffective. We also demonstrated that the addition of ferrous ion to several buffer systems resulted in this ion rapidly becoming incapable of catalyzing $\overset{\bullet}{O}H$ formation from H_2O_2 except when certain synthetic ligands were present; DETAPAC (diethylenetriamine pentaacetic acid) was the most effective in this regard (3). This prompted us to examine many biologically relevant compounds to see if they were effective in ligating Fe(II) such that it would remain in a catalytically active state. Of the many compounds tested, only the di- and triphosphate ester nucleotides were found effective in this regard (5). The spin-trapping technique was utilized to assay for $\overset{\bullet}{O}H$ formation. When 2 mM ADP and 100 μM ferrous ion was added, the amount of $\overset{\bullet}{O}H$ spin-trapped was about 20% of the amount of ferrous ion added. Very little $\overset{\bullet}{O}H$ was spin-trapped when ADP was not present indicating that the nature of the ferrous ligand is very important. ATP was slightly more effective than ADP in allowing Fe(II) catalysis of H_2O_2 to form ȮH, but AMP was without effect. We have found that this was a consistent pattern with the guanosine, thymidine and cytosine nucleotides also, i.e. that the tri- and diphosphate ester nucleotides were quite effective, but the monophosphate nucleotides were no more effective than buffer itself (4). We have also studied the kinetics of Fe(II) effectiveness in catalyzing ȮH formation from H_2O_2 as a function of time after adding the metal ion (4). At zero time Fe(II) in the absence of ADP was more effective, but after 7.5 sec ferrous ion in buffer only had become about half as effective as at time zero and then after 15 sec had become completely ineffective. In contrast when ADP was present, ferrous ion effectiveness decayed much more slowly than in the absence of the nucleotide with a halftime of about 165 sec. We interpret this result as due to the rapid oxidation of Fe(II) to Fe(III) in the presence of buffer only and the much slower oxidation of this ion when ADP was present. Recent experiments demonstrated that this interpretation is correct (Zs.-Nagy, I. and Floyd, R.A., unpublished observations).

We have made several other observations regarding iron ligation by nucleotide that appears to be relevant to oxidative damage in biological systems. We have found that the ferric-nucleotide complex is reduced to the ferrous-nucleotide complex effectively by ascorbate (5). Also, we have found that Mg^{++}, the ion normally considered to be ligated to nucleotides intracellularly, will not prevent Fe(II) from binding to nucleotides even if magnesium is present at 100 times the level of ferrous ion (6). Thus from these results, it can be concluded that if iron is available for movement within the cell, then it may ligate with the di- and triphosphate nucleotides and complexed thus is easily reduced by ascorbate to the ferrous form which is very effective in catalyzing ȮH formation from H_2O_2. These results take on more significance when it is realized that ascorbate is about 2 mM in brain, and combined with the demonstration of Konopka et al, that di- and triphosphate nucleotides are very effective in moving iron from transferrin to ferritin (12) or mitochondria (13). Thus it is quite possible that freely mobile iron in the cell exists as a nucleotide complex. Also, it has been known since 1964 that iron-nucleotide complexes are very effective in accelerating lipid peroxidation (11).

Brain and Aging

There are many reasons for concentrating on brain as the primary tissue to study during aging. Due to its central role in the processing of information, any alteration in brain most likely will have consequences which are amplified in the whole organism. As was stated earlier, lipofuscin does accumulate in the brain of almost all organisms, and even if this compound is innocuous in so far as functional effects are concerned, its presence and accumulation does implicate that the tissue undergoes oxidative insults. It should be noted also that if oxidative damage is serious enough, death of neurons will occur; since neurons are postmitotic tissue, death of any neurons may have dire consequences. In addition, brain has a very high level of polyunsaturated fatty acids which are easily peroxidizable. The level of 22:6 and 20:4 fatty acids in membrane lipids of brain can account for more than 20% of the total fatty acids of the brain (2). Furthermore, susceptibility of brain to oxidative damage is enhanced by the fact that this tissue is generally low in oxidative protection when its potential for peroxidation is considered. The enzymes catalase, superoxide dismutase, and glutathione peroxidase are not elevated or in some cases lower than in other organs. For instance, Carmagnol et al. (1) report that human brain has about 3% of the total glutathione peroxidase of human liver and, interestingly, there was a complete absence of the non-selenium dependent enzyme. The total superoxide dismutase activity of brain is remarkably similar to that observed in heart and liver over a wide range of species differing greatly in length of maximum life span (17). Catalase levels of human brain are unusually low, less than 1% of that observed in liver or erthrocytes (15). Also, brain generally contains less vitamin E, a very good antioxidant, than most tissues. For instance Zaspel and Csallany (19) found a level of α-tocopherol of 27 µg/gm brain in rat which was about one-half of that obtained from liver. The glutathione level of brain tends to be about one-third the level of liver in rat (16). In contrast to vitamin E and glutathione, brain tends to be generally higher in ascorbate. The reason for this is not known, and the consequences of the higher levels of ascorbate upon oxidative damage are not known; at higher levels ascorbate is protective, yet at lower concentrations it enhances peroxidation of membranes. Another important factor to be considered regarding possible oxidative damage to brain is the fact that brain, especially certain areas of human brain, is high in total iron content. As will be discussed later, the basal ganglia of human brain is high in total iron. Iron in certain freely available forms accelerates oxidative damage. For all of the above reasons, we have concentrated upon examining oxidative damage in brain as a function of age.

METHODS

Animals

Female rats were utilized throughout this study. In the initial experiments, retired breeders of the Sprague-Dawley strain were utilized; but in most of the studies, Fisher 344 were used. These rats were purchased under contract from the National Institutes of Aging in groups of 15 at the ages of 3, 12 or 27 to

30 months. They were usually kept 2 weeks to 1 month in the animal facility before use.

Total Iron in Brain Areas

In experiments where total iron was determined, the animals were placed under light ether anesthesia; brains were perfused with cold saline, removed in about 15 to 30 sec, and then placed on an ice cold glass surface. The various areas of the brain were dissected according to Glowinski and Iversen (7) while the tissue was kept on an ice cold stage. The tissue was digested for 24 hrs at 50°C in Amersham tissue solubilizer (NCS). The total iron content of the clear digest was then determined on a Perkin Elmer atomic absorption unit with a programmable electric furnace. The internal standard method was utilized to obtain the amount of total iron present. Three separate determinations were conducted on each sample, and the average value used.

Isolation of Brain Mitochondria

Mitochondria from brain was prepared by a minor modification of the method of Lai and Clark (14). Briefly, the forebrain of 4 rats of the same age were homogenized in isolation media, and the homogenate, after centrifugation to remove cell debris, was then centrifuged on a discontinous 7.5%/13% ficoll gradient from which synaptosomes as well as non-synaptic mitochondria were obtained. The synaptosomes were lysed and synaptic mitochondria isolated from a discontinous 4.5%/6% ficoll gradient centrifugation. From 4 animals, about 1-3 mg of synaptic mitochondrial protein and about 10-12 mg of non-synaptic mitochondria protein was obtained.

Detection of Superoxide from Brain Mitochondria

The production of superoxide from intact synaptic and non-synaptic mitochondria was estimated using the oxygen electrode method. The rationale for the basic method is presented in Figure 1. Superoxide production is considered to originate from electron transfer from the semiquinone form of ubiquinone to oxygen. Superoxide dismutase (SOD) acts upon superoxide to form H_2O_2 which then is converted to oxygen and water by catalase. Superoxide is not detected by the oxygen electrode; in the presence of SOD and catalase, oxygen consumption will decrease if superoxide is produced and the two enzymes are present in sufficient quantity. The substrates utilized in the experiments presented were succinate and malate-glutamate in combination with the inhibitors rotenone, thenoyltrifluoroacetone (TTFA) and antimycin A in the specific orders of addition as indicated in the top portion of Figure 1. The point of action of each inhibitor in the electron transport chain is shown in the middle portion of Figure 1. Utilizing the stoichiometry presented in the bottom of Figure 1, it is possible to calculate the amount of superoxide produced by intact brain mitochondria using the electrode method. It should be noted that there is a certain amount of endogenous SOD and also catalase present; in fact, to the

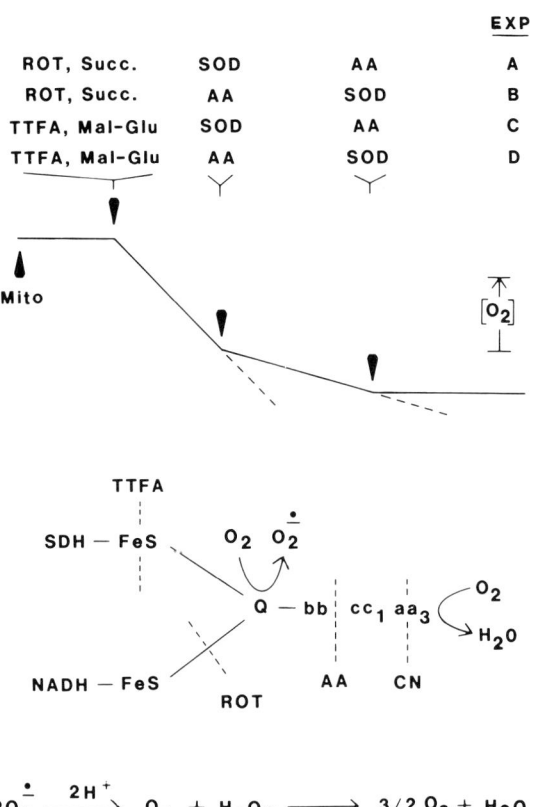

FIG. 1. Scheme to illustrate the rationale of the oxygen electrode method of superoxide production by brain mitochondria. The top portion illustrates a hypothetical change in oxygen consumption after the addition of various substrates and inhibitors at the indicated times. Four separate sets of experimental conditions, i.e. various substrates and inhibitors added in the order indicated, were performed on each set of synaptic and non-synaptic brain mitochondria at each age tested. The middle portion demonstrates the points of action of the various inhibitors in the mitochondria electron transport chain. The bottom portion presents the stoichiometry of the oxygen consumption during the production of superoxide and the action of added superoxide dismutase and catalase to yield oxygen.

extent that addition of external catalase produced no effect on the oxygen electrode values. It should also be noted that superoxide production estimated by the cytochrome $\underline{c}$ reduction method yielded similar values as was obtained by the oxygen electrode method. The cytochrome content of the mitochondria was determined by the method of Harmon (10).

Lipid Peroxidation of Brain Areas

Specific brain areas were isolated as described before and homogenized in 10 volumes of 50 mM pH 7.4 potassium phosphate buffer containing 15 mM Na^+ and 145 mM K^+ at ice temperature. Peroxidation was initiated by placing the homogenate in a shaking water bath at 37°C. Homogenate kept at ice temperature did not peroxidize. In certain experiments, the brain homogenates were exposed to high pressure oxygen. In these cases the high pressure oxygen chamber was placed in a 37°C shaking water bath. The amount of peroxidation was determined using the thiobarbituric acid method with 2-deoxyribose oxidized with periodic acid to yield malondialdehyde (18) as a standard.

RESULTS

Figure 2 demonstrates the amount of peroxidation obtained in specific brain areas of female Sprague-Dawley retired breeder rats as influenced by oxygen tension. The amount of peroxidation varied by a factor of two between the various brain areas and was increased significantly by increased oxygen pressure. Peroxidation was highest in cortex and cerebellum followed by hippocampus and then lowest in striatum, hypothalamus, and stem. In all brain areas, an increase in oxygen pressure caused a dramatic increase in peroxidation values with the percentage increase being larger from air to one atmosphere oxygen than from one atmosphere to five atmospheres of oxygen.

The differences in peroxidation between brain areas appear to be governed in part by the total iron content of that specific region. Figure 3 presents the amount of peroxidation achieved in 30 min on a per mg protein basis versus total iron content of each brain region examined. The lines drawn are the least squares fit of the data obtained under air as well as with one and five atmospheres of oxygen. The correlation coefficient varied from 0.82 to 0.89 for the data if the values for striatum, the points of which are designated by a superscript star, are excluded. There is in general an increased amount of peroxidation with increasing total iron content of the specific brain region. In addition, the higher the oxygen pressure, there is more dependence of peroxidation upon the total iron content.

Figure 4 emphasizes more clearly the important role of iron in the peroxidation of brain homogenate. In this case peroxidation was induced in forebrain under 5 atmospheres oxygen at 37°C for

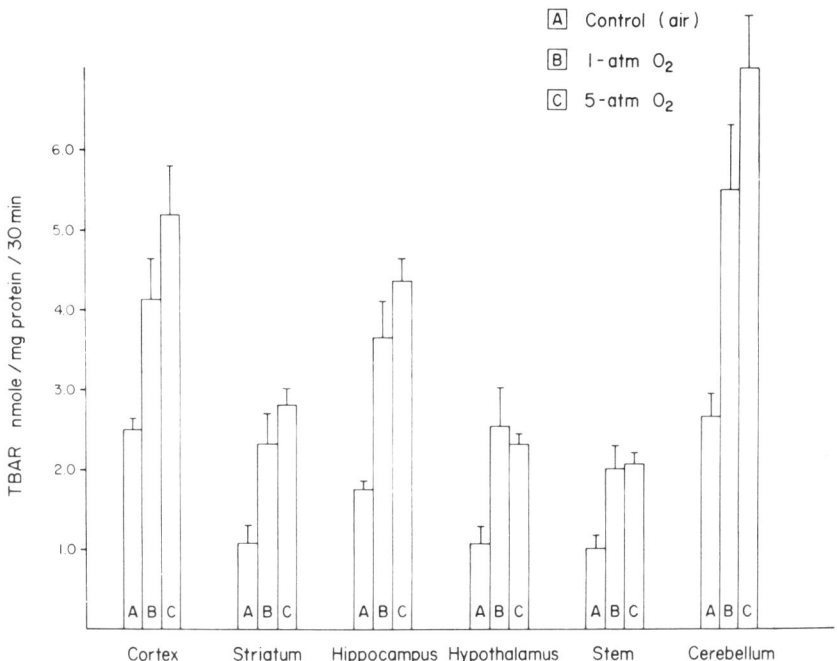

FIG. 2. The amount of peroxidation observed in specific brain areas of female Sprague-Dawley retired breeder rats as influenced by oxygen tension. The brain homogenate from each designated area was exposed to air, 100% oxygen or 5 atmospheres oxygen for 30 min at 37°C. Peroxidation was measured as outlined in the methods section.

60 min. The effect of the addition of various agents upon peroxidation was determined. Several agents inhibited peroxidation significantly, but the most effective were the iron chelators dipyrydyl, o-phenanthroline, and desferrioxamine. In addition, apo-transferrin as well as catechol were very effective in preventing peroxidation. It is interesting that thiourea, benzoate, and mannitol, all very effective hydroxyl free radical scavengers, were ineffective. Histidine and dimethylfuran, known singlet oxygen scavengers, inhibited slightly; but this may possibly be due to their iron ligating ability.

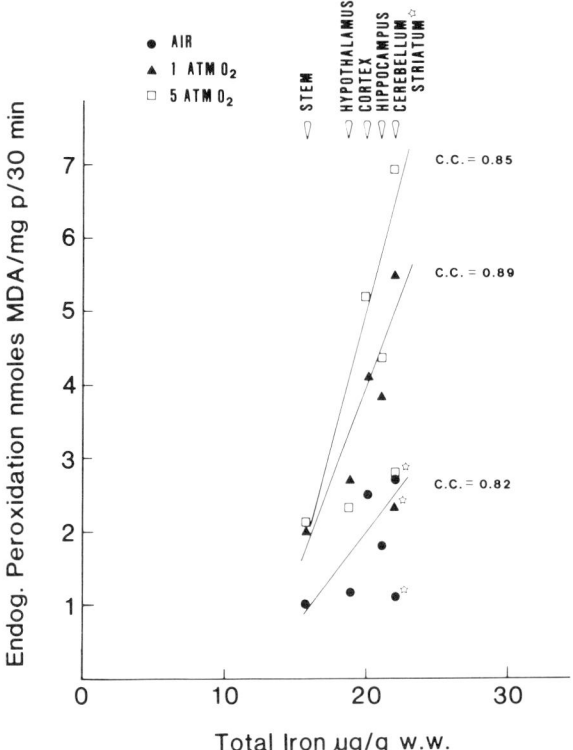

FIG. 3. The amount of peroxidation in specific brain areas as noted in Figure 2 is plotted versus total iron content on a wet weight basis. The values obtained in air, 100% oxygen and 5 atmospheres oxygen are designated by circles, triangles and squares, respectively. The total iron content of the brain area it corresponds to is designated in the top of the figure. The striatum and cerebellum has the same amount of total iron in these animals and the striatum values of peroxidation are designated by a superscript star.

The importance of iron in the processes involved in peroxidation of brain is further emphasized by the data presented in Figure 5. In this case iron and ascorbate, added to a final concentration of 10 µM and 250 µM, respectively, caused a drastic increase in peroxidation. The data in Figure 5 should be compared to the data in Figure 2 where the ordinate values are one-tenth the scale of the peroxidation observed in the presence of iron-ascorbate. The addition of iron-ascorbate caused a 10- to 20-fold increase in peroxidation of all brain areas with the largest increase observed in the striatum and hypothalamus.

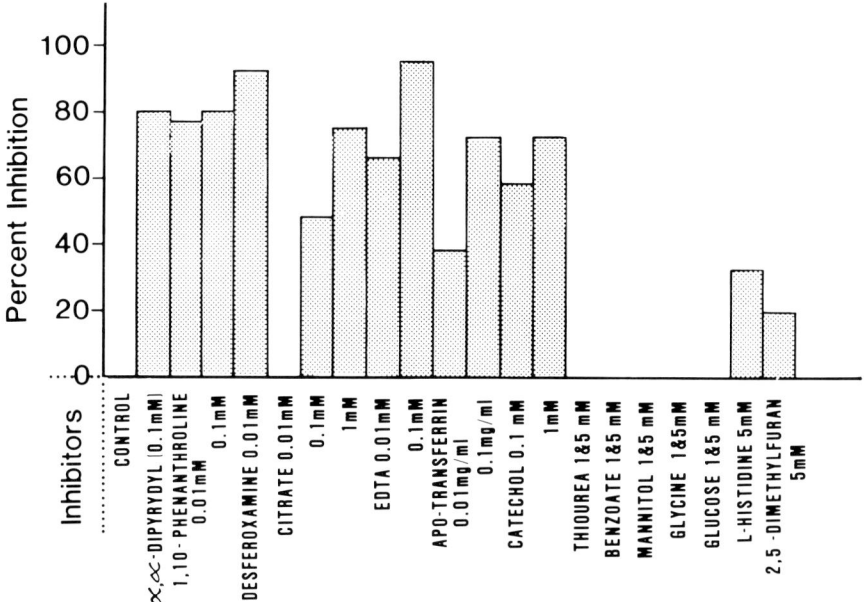

FIG. 4. The percent inhibition of peroxidation versus control of the forebrain homogenate of female retired breeder Sprague-Dawley rats as influenced by specific compounds added. The condition of peroxidation were 5 atmospheres oxygen at 37°C for one hour.

We have examined the effect of age on the total iron content of various brain areas in the female Fisher 344 rat. The results obtained are shown in Figure 6. It should be noted that in almost all areas of the brain there is an increase in total iron content on a wet weight basis as the age of the animal increases from 3 to 12 to 30 months old. The greatest and most consistent increase with age occurs in cerebellum, striatum, and hippocampus.

Table 1 presents the results obtained when the amount of lipid peroxidation in each brain area was determined at three age groups, 3, 12 and 30 months. The general overall pattern is a slight decrease in the amount of peroxidation in each brain area with increased age. This result when viewed with the previous results in mind, i.e. namely the demonstrated increase in total iron in brain areas with age and the close dependence of total iron with peroxidation, seems inconsistent, but this apparent contradiction only emphasizes the degree of complexity of the process and the fact that many factors including total iron control peroxidation in brain. For instance, we have noted that peroxidation of brain homogenate does not occur at ice temperature. This indicates that metabolism may play a role in the

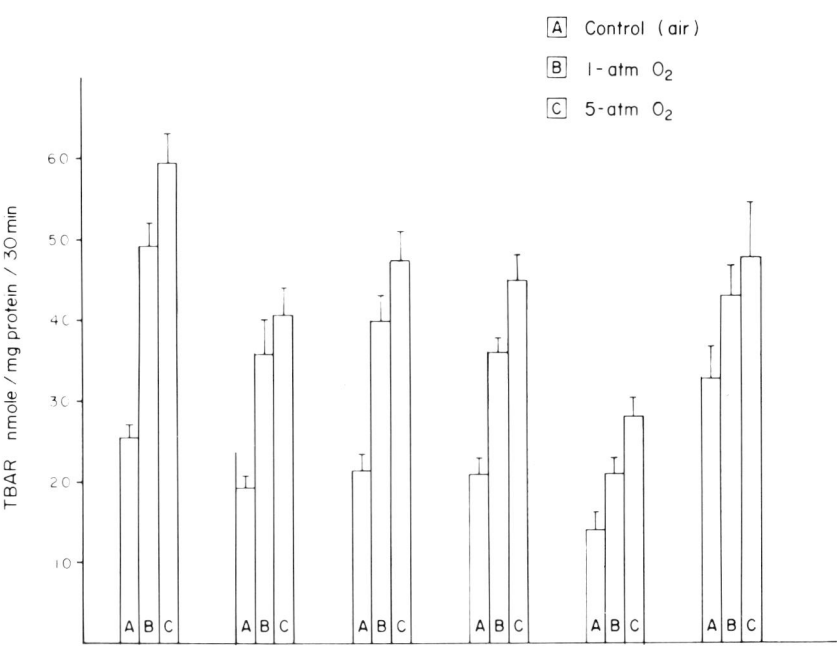

FIG. 5. The influence of added iron and ascorbate on the peroxidation as influenced by oxygen tension in various areas of brain of female Sprague-Dawley breeders. The final concentration of added iron and ascorbate was 10 µM and 250 µM, respectively.

peroxidation of brain. Mitochondria are known to produce superoxide, and thus it was considered necessary to examine the effects of age on the production of oxygen free radicals by brain mitochondria.

Table 2 shows that intact brain mitochondria produce substantial quantities of superoxide and that the amount of $O_2^{\bullet}$ produced varies with the age of the animal. There was a substantial decrease in the amount of superoxide produced with increasing animal age in synaptic (S) mitochondria when succinate was used as the substrate. There was also a decrease in $O_2^{\bullet}$ produced by S mitochondria when malate-glutamate was used as substrate yet the decrease was not as consistent as with succinate. In the non-

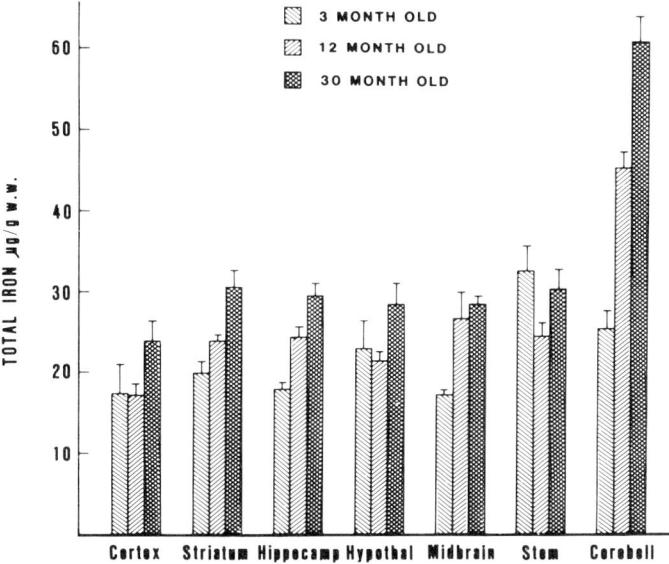

FIG. 6. The total iron content on a wet weight basis of various areas of female Fisher 344 rats 3, 12 and 30 months of age.

TABLE 1. Peroxidation of homogenate of various brain areas as influenced by age

Brain area	Age	Lipid peroxidation nmoles MDA/mg protein/60 min		
		3 mo	12 mo	30 mo
Cortex		1.91	1.57	1.72
Striatum		1.46	1.29	1.13
Hippocampus		2.03	1.53	1.31
Hypothalamus		1.67	1.32	1.44
Midbrain		1.58	1.33	1.30
Stem		1.04	0.70	0.77
Cerebellum		2.44	2.07	2.17

synaptic (NS) mitochondria there appeared to be some decrease of $O_2^{\pm}$ production with age yet the decrease is not as consistent and drastic as with the S mitochondria.

Table 3 shows the oxygen consumption on a heme a basis of S and NS mitochondria as influenced by age of the animal from which

TABLE 2. $O_2^{\bullet}$ generation in brain mitochondria as influenced by age

	Age	3 mo	12 mo	27 mo
I. SOD to complete respiration				
A. Non-synaptic				
Succinate		10.1[a]	0.07	0
Malate-glutamate		2.6	12.1	4.3
B. Synaptic				
Succinate		160	39	11
Malate-glutamate		33.5	9.9	14.7
II. SOD to antimycin A inhibited respiration				
A. Non-synaptic				
Succinate		17.2	0.6	10
Malate-glutamate		12.4	5.8	7.1
B. Synaptic				
Succinate		40.1	4	0
Malate-glutamate		131	25.7	0

[a] nmole $O_2^{\bullet}$/min-nmole heme $\underline{a}$.

TABLE 3. Oxygen consumption by isolated brain mitochondria as influenced by age

	Age	3 mo	12 mo	27 mo
Non-synaptic				
Succinate		158[a]	116	155
Malate-glutamate		86	74	60
Synaptic				
Succinate		305	157	145
Malate-glutamate		86	51	24

[a] nmole O_2/min-nmole heme $\underline{a}$.

they were isolated. In S mitochondria there was a large (50% or greater) decrease in oxygen consumption with age when either succinate or malate-glutamate was used as a substrate. There appeared to be a slight but a less consistent decrease in oxygen consumption in NS mitochondria with age when either succinate or malate-glutamate was used as a substrate. Using the data in Tables 2 and 3, it is possible to calculate the percentage of electron flow which is going to externally produced $O_2^{\pm}$ as a function of age and this is shown in Table 4. The percentage of electron flux going to $O_2^{\pm}$ in S mitochondria decreases with age when succinate is the substrate but appears not to change much or in fact perhaps increases when malate-glutamate was used as a substrate. With NS mitochondria, again the percentage of electron flux when succinate is used as the substrate decreases with age, but there appears to be an increase with age when malate-glutamate is used as substrate. The data in Table 4 also show that there was a striking difference in the percent of electron flux going to $O_2^{\pm}$ in S versus NS mitochondria at any age examined. In many cases 50% of the electron flux goes to $O_2^{\pm}$ in S mitochondria whereas in NS mitochondria this is usually less than 10%.

Utilizing the data in Tables 2, 3 and 4, it is possible to obtain values of oxygen consumption in brain mitochondria after excluding that consumed due to $O_2^{\pm}$. The data obtained is shown in Table 5. In general, oxygen consumption minus that due to $O_2^{\pm}$ in NS mitochondria was about the same with age, but in S mitochondria there was a difference depending upon the substrate; that supported by succinate stayed the same with age but with malate-glutamate there was a drastic decrease.

The cytochrome content of brain mitochondria is shown in Table 6. The values obtained are considerably higher (50-75%) than the mitochondria isolated by Lai and Clark (14) using essentially identical procedures. As shown in Table 6, the cytochrome content of S and NS mitochondria are fairly comparable in young

TABLE 4. Percent total electron flux used in $O_2^{\pm}$ generation

	Age	3 mo	12 mo	27 mo
Non-synaptic				
Succinate		6.4	0.06	--
Malate-glutamate		3.0	16.4	7.2
Synaptic				
Succinate		52.4	24.8	7.6
Malate-glutamate		39	19.4	61.3

TABLE 5. Respiratory oxygen consumption in brain mitochondria

Age	3 mo	12 mo	27 mo
Non-synaptic			
Succinate	148[a]	116	155
Malate-glutamate	83	62	56
Synaptic			
Succinate	145	118	134
Malate-glutamate	52	41	9

[a] nmole O_2/min-nmole heme a.

animals, but with increasing age S mitochondria change their content more than do NS mitochondria. The cytochrome b content of NS and S mitochondria does not change significantly with age in contrast to the content of c and a-type cytochromes. In NS mitochondria, cytochrome $c+c_1$ content does not decrease with age while in S mitochondria the cytochrome $c+c_1$ content decreased by 40% at 27 months of age. The content of cytochrome oxidase (aa_3) in NS mitochondria decreased less than 14% by 27 months while cytochrome oxidase levels in S mitochondria had decreased 50% at 27 months.

DISCUSSION

Factors Contributing to Brain Peroxidation

The results presented in this report demonstrate that there are several factors which contribute to peroxidation of rat brain. We presented clear evidence that the total iron content correlates with brain peroxidation. We interpret this finding as indicating that the total iron content is only one, but an important contributing factor, to brain peroxidation. It is highly likely that most of the iron of the tissue is in the form of ferritin which in this state really is not available for initiating peroxidation (M.M. Zaleska and R.A. Floyd, unpublished observations). Therefore, iron must be moved from ferritin into a mobile form such as perhaps a nucleotide complex which then is available to initiate and propagate membrane lipid peroxidation. Thus inherent in this argument is that operationally total iron and mobile iron are in an equilibrium and within the range of brain areas examined, excluding striatum, the higher the total iron the higher the amount of mobile or available iron for catalyzing peroxidation.

TABLE 6. Cytochrome content of brain mitochondria as influenced by age

Age	3 month		12 month		28 month	
Cytochrome	Non-Sy	Sy	Non-Sy	Sy	Non-Sy	Sy
$c+c_1$	0.56[a] ± 0.16	0.67 ± 0.22	0.63 ± 0.04	0.60 ± 0.06	0.58 ± 0.12	0.38 ± 0.12
b	0.23 ± 0.03	0.24 ± 0.07	0.20 ± 0.03	0.28 ± 0.08	0.20 ± 0.03	0.20 ± 0.11
aa_3	0.37 ± 0.07	0.46 ± 0.23	0.41 ± 0.04	0.38 ± 0.02	0.32 ± 0.14	0.22 ± 0.05

[a] nmole cytochrome/mg protein

Striatum is a special case, and perhaps the decreased level of peroxidation in this brain area is partially explainable by our previous observation that dopamine, the neurotransmitter which is high in striatum, ligates iron much more effectively than ADP and prevents this metal from catalyzing $\overset{\bullet}{O}H$ formation from H_2O_2 (6). Thus in striatum the available iron, which in fact may be very low in actual amounts, is most likely ligated with dopamine and as such this prevents its participation in events important to membrane peroxidation.

Our observations showing that hydroxyl free radical scavengers did not inhibit brain peroxidation under the conditions of five atmospheres oxygen for 60 minutes is interesting. We consider that this does not rule out the possibility that $\overset{\bullet}{O}H$ is an initiating species in peroxidation because the high oxygen pressure for such a long total time would perhaps allow enough of these radicals to be formed and perhaps bypass the scavengers to initiate peroxidation. In fact, substantial evidence points to the fact that oxygen free radicals are involved in peroxidation. Thus, one interpretation of our observation that brain homogenate does not peroxidize at ice temperatures but does very effectively at 37°C is most likely due to the superoxide production by brain mitochondria. Thus, if brain mitochondria are a potent source of oxygen free radicals, as our data indicates, then the amount of superoxide produced by these organelles may have important consequences in regard to brain oxidative damage. Our data shows that in general $O_2^{\pm}$ generation from brain mitochondria decreases with age of the animal (see table 2). Thus, we consider the decrease in $O_2^{\pm}$ production by brain mitochondria with age as a major contributory factor as to why there is a slight decrease in brain peroxidation with age even though there is an increase in total iron content (see Figure 6).

Brain Mitochondria Changes with Age

Modification of brain mitochondria with age is an important area that needs much more research. Our data clearly demonstrate that in female Fisher 344 rats, brain mitochondria change substantially with age. Brain mitochondria, especially the synaptic fraction, produces substantial amounts of $O_2^{\pm}$ in young rats, but the amount produced decreases drastically with age. Total oxygen consumption by synaptic mitochondria decreases with age, but this was not true of non-synaptic mitochondria. Interestingly, we noted a substantial decrease in the <u>a</u> and <u>c</u> cytochromes of synaptic mitochondria with age but no such change in the <u>b</u> cytochromes. In contrast, there appeared to be very little change in the cytochrome content of non-synaptic mitochondria with age. A decrease in cytochromes does indicate a possible loss in respiratory activity; yet on the basis of our present data, we cannot conclude that the actual decrease in cytochrome content was solely responsible for the loss in total oxygen consumption by mitochondria with age. It is interesting to note that $O_2^{\pm}$ is apparently formed from the reduction of oxygen by the semiquinone state of ubiquinone and that cytochromes c and a occur in the respiratory chain after ubiquinone; therefore, it seems unlikely that the decreased amount of these two cytochromes is exclusively responsible for the change in $O_2^{\pm}$ production with age especially since cytochrome b, the electron acceptor of ubiquinone, does not change with age. Alterations in dehydrogenases or ubiquinone content may occur and would contribute to the decreases in respiration and $O_2^{\pm}$ generation.

Overall Relevance to Aging in Brain

We consider the study of oxidative damage in brain of particular importance to aging for the following reasons: (A) Brain contains a high amount of unsaturated lipids and (B) is a tissue that utilizes about one-fifth of the total oxygen demand of the body. In addition, (C) it is not particularly enriched in any of the antioxidant enzymes (superoxide dismutase, catalase, and glutathione peroxidase) nor vitamin E; yet it does contain large amounts of ascorbate. All of the above facts would tend to indicate that brain should be a tissue that is highly susceptible to oxidative damage. Therefore, if oxidative damage in brain should occur, and if it is of a magnitude such that death or impairment of neurons occurs, then this would be expected to have important consequences because (D) brain is a post mitotic tissue.

Another little known fact that is very relevent is that human brain, especially certain areas such as the basal ganglia, contain large amounts of total iron. In fact, it has been noted that several areas of adult human brain have greater than 200 μg iron per gm wet weight (8). This is nearly 10 times the level observed in rat and several other laboratory animals. If total iron levels in human brain tend to have an influence on potential peroxidative damage as we have noted in rat brain homogenate, then in addition to all of the reasons listed above, the high level of iron should amplify the probability that oxidative damage may be important and highly relevant to aging in human brain.

If one speculates using the above facts, then it becomes important to ask are the areas most susceptible to oxidative damage the ones that contain the highest amount of iron such as basal ganglia in humans? This area exerts major controlling influence over movement; and thus it is possible that movement changes which are known to occur consistently with age could be based to a large extent upon oxidative damage. Parkinsonism, which occurs with increasing frequency in advanced age, may be particularly worth considering in the context of our studies. But, it should be noted that any brain change which occurs with age may have as a basis oxidative damage and/or death of neurons.

These conclusions remain speculative at the present time yet can be deduced based on the present facts. It is, however, perfectly clear that this area needs much more research.

ACKNOWLEDGMENTS

This work was supported in part by NIH grant No. R01 AG02599 and by a contract from the Department of the Air Force. We would like to thank Ms. Anita Hill and Sandra Nank for help regarding the preparation of the manuscript and mitochondria.

REFERENCES

1. Carmagnol, F., Sinet, P.M., and Jerome, H. (1983): Biochim. Biophys. Acta, 759:49-57.
2. Crawford, C.G., and Wells, M.A. (1979): Lipids, 14:757-762.
3. Floyd, R.A. (1982): Can. J. Chem., 60:1577-1586.
4. Floyd, R.A. (1983): Arch. Biochem. Biophys., 225:263-270.
5. Floyd, R.A., and Lewis, C.A. (1983): Biochemistry, 22:2645-2649.
6. Floyd, R.A., and Zaleska, M.M. (1983): In: Proceedings of The Third International Conference on Oxygen Radicals in Chemistry and Biology, edited by W. Bors, M. Saran, and D. Tait (in press). Walter De Gruyter and Co., Berlin.
7. Glowinski, J., and Iversen, L.L. (1966): J. Neurochem., 13:655-669.
8. Hallgren, B., and Sourander, P. (1958): J. Neurochem., 3:41-51.
9. Harman, D. (1981): Proc. Natl. Acad. Sci. USA, 78:7124-7128.
10. Harmon, H.J. (1982): J. Bioenerg. Biomembr., 14:377-383.
11. Hochstein, P., Nordenbrand, K., and Ernster, L. (1964): Biochem. Biophys. Res. Commun., 14:323-328.
12. Konopka, K., Mareschal, J-C., and Crichton, R.A. (1981): Biochim. Biophys. Acta, 677:417-423.
13. Konopka, K., and Romslo, I. (1980): Eur. J. Biochem., 107:433-439.
14. Lai, J.C.K., and Clark, J.B. (1976): Biochem. J., 154:423-432.
15. Marklund, S.L., Westman, N.G., Lundgren, E., and Roos, G. (1982): Cancer Res., 42:1955-1961.
16. Mikasa, H., Ageta, T., Mizoguchi, N., and Kodama, H. (1982): Anal. Biochem., 126:52-57.
17. Tolmasoff, J.M., Ono, T., and Cutler, R.G. (1980): Proc. Natl. Acad. Sci. USA, 77:2777-2781.
18. Waravdekar, V.A., and Saslaw, L.D. (1959): J. Biol. Chem., 234:1945-1950.
19. Zaspel, B.J., and Csallany, A.S. (1983): Anal. Biochem., 130:146-150.

Free Radicals in Molecular Biology, Aging, and Disease, edited by D. Armstrong et al. Raven Press, New York © 1984.

Potential Role of Autoxidation in Age Changes of the Retina and Retinal Pigment Epithelium of the Eye

Martin L. Katz, W. Gerald Robison, Jr., and *Edward A. Dratz

*National Eye Institute, Laboratory of Vision Research, National Institutes of Health, Bethesda, Maryland 20205; *Division of Natural Sciences, University of California, Santa Cruz, California 95064*

Lipofuscin Accumulation and Senescence

In seeking to determine the molecular and cellular processes that underlie mammalian senescence, one attempts to define phenomena which occur universally during aging in all members of a species. In addition, if a phenomenon is to play a role in senescence, it must contribute to the age-related decrease in the probability of survival; i.e. it must be deleterious. Primary events in the aging process must also account for the progressive nature of senescence; their effects must accumulate over a long period of time. Finally, if a cellular or molecular event is to play a primary role in the process of senescence, the occurrence of such an event must not be dependent on modifiable environmental factors. These four criteria for defining primary aging events (universality, deleteriousness, progressiveness, and intrinsicality), first suggested by Strehler (45), provide a framework with which to evaluate the potential role of various molecular and cellular phenomena in senescence.

The longest known manifestation of mammalian aging at the cellular level is the progressive accumulation of an autofluorescent pigment (lipofuscin) within the cytoplasm of a wide variety of post-mitotic cell types (5). Intracellular lipofuscin accumulation appears to occur universally in animals during senescence; pigment accumulation has been found to accompany aging in every animal species examined to date. The progressive nature of lipofuscin accumulation during aging has been demonstrated in a number of tissues, including heart (5, 46) and brain (26, 37). While the rate of lipofuscin accumulation can be modified by a number of factors (36, 43), no experimental manipulations have been successful in preventing the accumulation of lipofuscin during senescence. Thus, it seems likely that age-related lipofuscin accumulation is an intrinsic characteristic of mammalian species, and not simply an expression of pathology. On the basis of universality, progressiveness and intrinsicality, then, the events leading to lipofuscin accumulation are likely candidates for primary events in senescence. The question of whether lipofuscin accumulation is deleterious can be divided into two parts: (1) are the events leading to lipofuscin formation harmful? and (2) does the mere presence of lipofuscin impair cell function? While a number of experiments pertaining to these two questions have been performed, definitive answers to them are not yet available. As discussed below, however, data currently available suggest that at least the molecular events involved in lipofuscin formation may be deleterious.

Lipofuscin Accumulation and Aging of the Eye

The mammalian retina consists of a thin layer of neural tissue lining the posterior two-thirds of the eye. It is made up of a number of distinct cell layers (FIG. 1), the outermost of which is the photoreceptor cell layer. Photoreceptor cells mediate the transduction of light into nervous impulses. These impulses are transmitted to neurons in the inner nuclear layer of the retina, and then to the ganglion cells, the innermost neurons of the retina. The axons of the ganglion cells form the optic nerve which passes out the back of the eye and ends at the lateral geniculate nucleus in the brain. Some mammals, including man, have two distinct blood supplies to the retina: one connected to a capillary bed in the neural retina, and one serving a large bed of capillaries posterior to the retina (the choriocapillaris). In other mammals the latter blood supply alone supports the retina. In either case, the photoreceptor cells appear to be dependent mainly on the choriocapillaris for nutrition and waste removal (6). Between the photoreceptors and the choriocapillaris is a single layer of epithelial cells, the retinal pigment epithelium (RPE) (FIG. 1), which forms a blood-retinal barrier. Most exchange between the photoreceptors and the choriocapillis is mediated by the RPE. In addition, the RPE plays a crucial role in the turnover of the photosensitive membranes of the photoreceptor cells. The pigments responsible for light absorption and transduction are membrane proteins with a covalently bound vitamin A chromophoric group. These pigments are located in specialized regions of the photoreceptor cells called outer segments (FIG. 2), which are long and cylindrical in rod photoreceptors, the predominant photoreceptor type in rats. In rods, the outer segment consists of a stack of

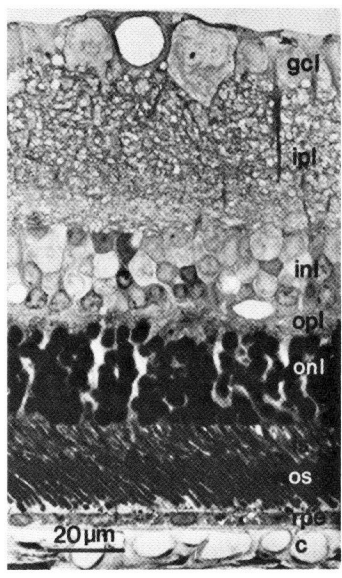

FIG 1. Overall morphology of the rat retina shown in a light micrograph. The retina is organized into a number of distinct layers which include the ganglion cell layer (gcl), inner plexiform layer (ipl), inner nuclear layer (inl), outer plexiform layer (opl), outer nuclear layer (onl), and photoreceptor inner segments (is) and outer segments (os). Also shown are the retinal pigment epithelium (rpe) and choriocapillaris (c).

flattened membranous sacs or disks surrounded by the plasma membrane. New disks are synthesized continuously, and added to the bases of the outer segments, while packets of old disks are shed periodically from the outer segments tips. The shed outer segment material is phagocytized and digested by the RPE. Impairment of phagocytosis by the RPE has been found to lead to photoreceptor cell degeneration (4, 19, 32), a fact which underlines the importance of this RPE function to photoreceptor cell maintenance. The RPE plays a number of other roles, including vitamin A transport and storage, which are crucial in maintaining photoreceptor cell integrity and function.

Age-related retinal degeneration is a major cause of serious visual impairment in man (16, 29, 30). In light of the importance of the RPE in maintaining retinal integrity, it is possible that retinal degeneration is secondary to primary age changes in the RPE. Like many other post-mitotic tissues, the RPE has been found to accumulate lipofuscin during senescence (12, 13, 25, 33, 48) (FIGS. 3 and 4). Thus it is possible that lipofuscin accumulation in the RPE plays a role in age-related retinal degeneration. Likewise, lipofuscin accumulation in other tissues could lead to an impairment of tissue function, which ultimately contributes to senescence as seen at the level of the whole organism.

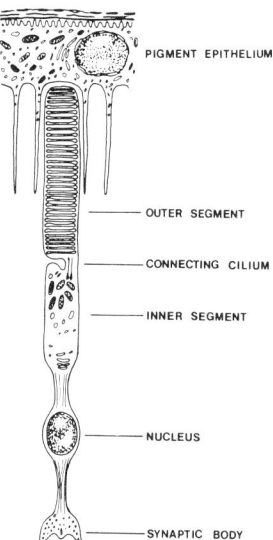

FIG. 2. Structure and organization of a rod photoreceptor cell. Light transduction takes place in the disk membranes of the outer segment. The outer segment is connected to the mitochondrion-rich inner segment by a connecting cilium. Interior to the inner segment is the rod cell nucleus. Finally, the rod cell terminates with a synaptic body which lies in the outer plexiform layer.

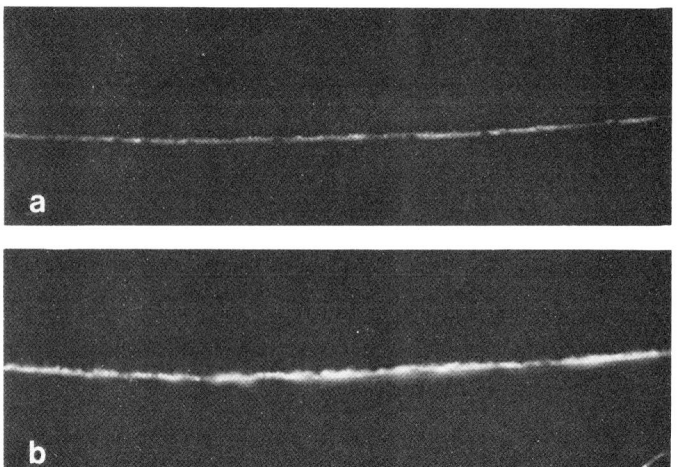

FIG. 3. Age-related increase in RPE lipofuscin content shown in fluorescence micrographs of frozen sections of the central retinas from 26-week-old (a) and 117-week-old (b) Sprague-Dawley rats.

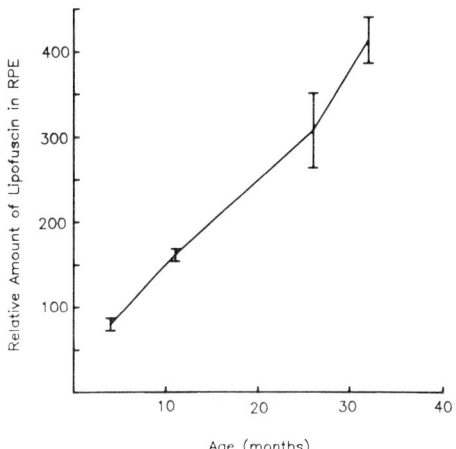

FIG. 4. Effect of age on the lipofuscin content of the central RPE. Lipofuscin in the RPE was quantitated from electron micrographs. Data from Katz and Robison (25).

Using the RPE as a model, we have attempted to determine whether lipofuscin accumulation is deleterious, and thus whether it may be a primary event underlying senescence. In order to determine whether lipofuscin accumulation is deleterious, one must first uncouple lipofuscin accumulation from senescence. We have used antioxidant nutrient deficiency as the primary means of doing this.

AUTOXIDATION AND LIPOFUSCIN ACCUMULATION

Lipofuscin Accumulation is Accelerated by Antioxidant Nutrient Deficiency

There is a growing body of evidence linking lipofuscin accumulation to autoxidative tissue damage, and thus suggesting that lipofuscin is at least partially a product of autoxidation of subcellular components. Vitamin E deficiency has long been known to accelerate lipofuscin accumulation in a variety of tissues (36, 43). It has been found that in animals on a diet high in polyunsaturated fatty acids, the RPE is particularly sensitive to vitamin E deficiency (FIG. 5); after 35 weeks of vitamin E deficiency there is about a 5-fold increase in the number of lipofuscin granules in the RPE relative to controls (40). While it has long been proposed that vitamin E acts as an antioxidant in vivo, the most compelling evidence for this hypothesis is the relatively recent finding that deficiency in this vitamin leads to an increase in the amount of ethane and pentane exhaled in the breath of experimental animals (10). These gases are autoxidation products of ω-3 and ω-6 fatty acids respectively. Evidence that lipofuscin is composed of products

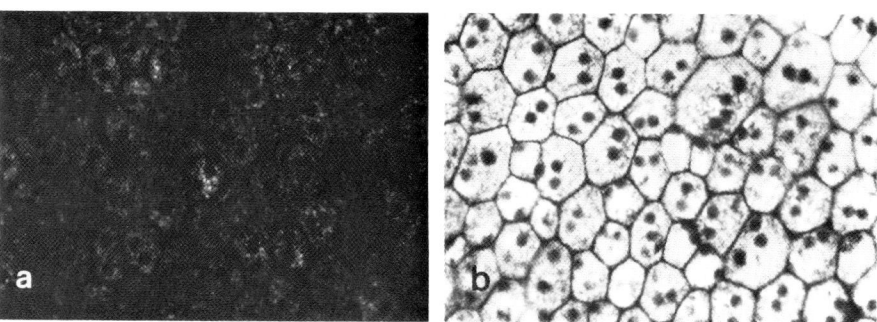

FIG. 5. Increase in RPE autofluorescence resulting from vitamin E deficiency. Fluorescence micrographs of RPE flat preparations from Sprague-Dawley rats which had been maintained for 44 weeks from weaning on diets supplemented with (a) or lacking (b) in vitamin E.

of tissue autoxidation comes from the finding by Chio and co-workers (8) that the products of in vitro autoxidation of microsomes have similar properties to lipofuscin formed in vivo during normal aging. Finally, we have recently shown that the pigment which accumulates in the RPE as a result of vitamin E deficiency has a fluorescence emission spectrum essentially identical to that of the pigment which accumulates during senescence (27) (FIG. 6). Thus "age-pigment" probably forms by the same mechanism as the pigment formed as a consequence of vitamin E deficiency.

Antioxidant Nutrient Deficiency Uncouples Lipofuscin Accumulation from Aging

Manipulation of dietary levels of vitamin E and other antioxidant nutrients (22, 28) provide us with the means of uncoupling lipofuscin accumulation from chronological aging and therefore of asking whether lipofuscin accumulation in the RPE is deleterious. It should be noted that not all tissues which accumulate lipofuscin during senescence are sensitive to vitamin E deficiency, and likewise not all tissues which accumulate lipofuscin as a consequence of vitamin E deficiency show lipofuscin accumulation during aging (27). Therefore antioxidant nutrient deficiency is not a useful tool for evaluating the role of lipofuscin accumulation in senescence at the level of the whole organism, but only in certain tissues, such as the RPE.

Autoxidation of subcellular components resulting from antioxidant nutrient deficiency has the potential for directly causing irreversible cell damage (17, 49). Therefore, if antioxidant nutrient deficiency has deleterious effects on the RPE, these effects could either be the direct result of autoxidative damage, or an indirect result of such damage acting through the build-up of lipofuscin. As discussed later, we have

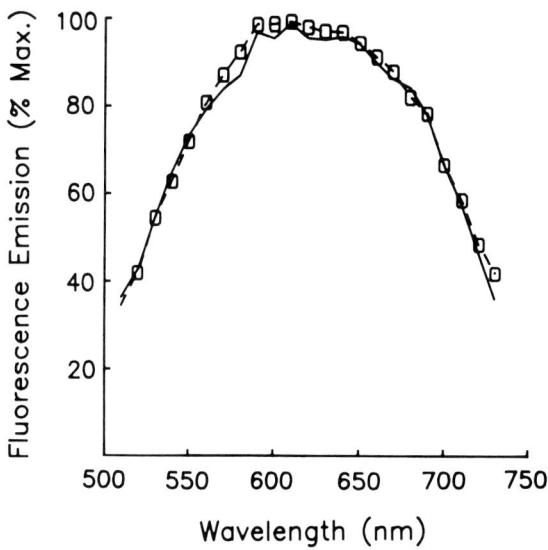

FIG. 6. Fluorescence spectra of RPE lipofuscin from young vitamin E deficient rats, _____, and from senescent rats which had been fed a diet containing adequate vitamin E, 0-0-0-0. Data from Katz, et al. (27).

evaluated a potential means of distinguishing between these possibilities. Even if the mere presence of lipofuscin within cells is not deleterious, the amount of lipofuscin in a cell appears to reflect the amount of direct autoxidative damage that has occurred within that cell.

Factors Potentiating Autoxidative Damage to the Retina and RPE

A number of factors suggest that photoreceptor cells and the RPE may be particularly susceptible to autoxidative damage. In mammals, 35 to over 50% of the fatty acids of rod outer segment membrane phospholipids are docosahexaenoic acid (15), which contains six double bonds. The more double bonds a fatty acid has, the more susceptible it is to autoxidation (21). On the basis of their lipid composition, therefore, rod outer segment membranes would be expected to be easily oxidized. The photoreceptor outer segments, as well as the RPE, contain large amounts of vitamin A in various forms. This vitamin is highly susceptible to autoxidation, particularly in the presence of light. Thus its presence may potentiate autoxidative damage to the retina and RPE. In addition, the retina has a very high rate of oxygen consumption (1) indicating that there must be a high flux of oxygen across the photoreceptors and RPE from the choriocapillaris to the neural retina.

DELETERIOUS EFFECTS OF ANTIOXIDANT NUTRIENT DEFICIENCY

Antioxidant Nutrient Deficiency Effects on Retina and RPE

In rats fed diets deficient in antioxidant nutrients, the disk membranes at the apical ends of the rod outer segments become highly disrupted (3, 22, 40) (FIG. 7). This suggests that

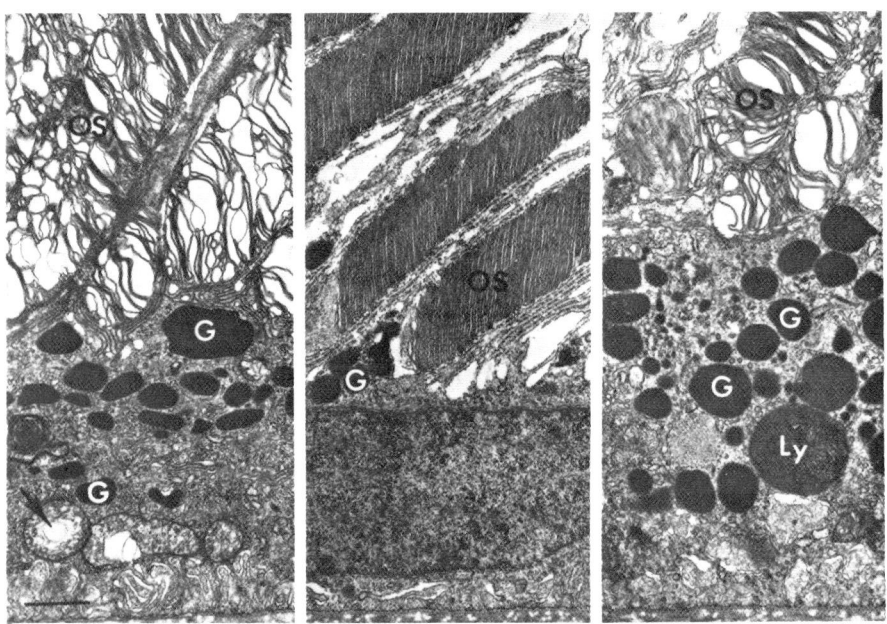

FIG. 7. Effect of vitamin E deficiency on rat retinal pigment epithelium and rod outer segment (OS) tips. Animals were fed a normal diet for 34 weeks (middle) or a vitamin E-free diet for 20 weeks (left) or 34 weeks (right). Retinas from vitamin E-deprived rats showed disruption of outer segment membranes (OS), increased RPE cell height, altered mitochondria (arrow), and increased numbers of lysosomes (Ly) and lipofuscin granules (G). From Robison, et al. (40).

there may be direct autoxidative damage to the photoreceptor disk membranes, and that autoxidized components of these membranes may become incorporated into RPE lipofuscin granules after being phagocytized. Recent experiments indicate, however, that antioxidant deficient animals accumulate large amounts of lipofuscin in the RPE even in the absence of photoreceptor cells (42) (FIG. 8). Amemiya (3) has suggested that the disruption of the photoreceptor outer segments seen in vitamin E deficient animals may result at least to some degree from the release of lysosomal enzymes from the RPE. Thus the RPE may be the primary

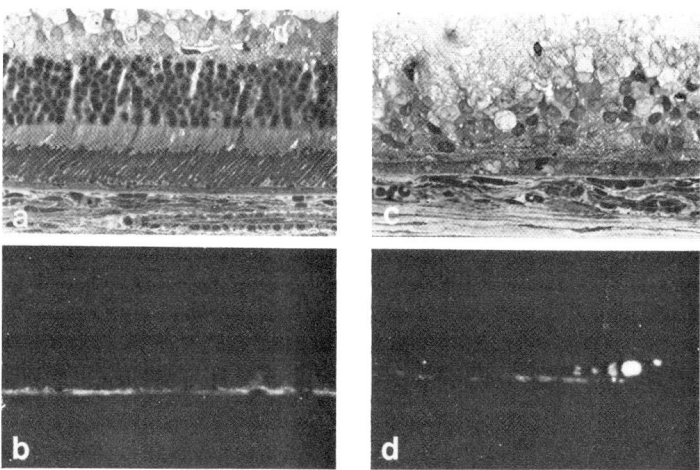

Fig. 8. Effect of absence of photoreceptors on age-related lipofuscin accumulation. In normal rats [retinal structure shown in (a)], there is a significant accumulation of autofluorescent pigment in the RPE at 40 weeks of age (b). In RCS rats, whose photoreceptors degenerate very early in life [retinal structure shown in (c)], the amount of lipofuscin in the RPE at 40 weeks of age is slightly less than in normal rats and its distribution is more irregular (d).

site of autoxidative damage in vitamin E deficient rats, and damage to the photoreceptors could be secondary. Photoreceptor cells actually die in response to antioxidant nutrient deficiency (22, 41); the number of photoreceptor cells per unit retinal area is significantly reduced in animals on diets deficient in antioxidant nutrients (FIG. 9). Again, it is not known whether photoreceptor cell death results from direct autoxidative damage to these cells, or is secondary to damage of the RPE cells which play important roles in maintaining the photoreceptors.

Since lipofuscin accumulates in the RPE of antioxidant deficient animals even if these animals lack photoreceptor cells, autoxidized disk membrane components cannot be the sole source of RPE lipofuscin. Another potential substrate for lipofuscin formation in the RPE is vitamin A, which is highly susceptible to oxidation. It has been found that the rate of lipofuscin accumulation in the RPE is highly dependent on dietary vitamin A levels (39, 41, 42). There appears to be a linear relationship between the logarithm of dietary vitamin A intake and the amount of lipofuscin in the RPE of both vitamin E deficient and supplemented animals (FIG. 10). The dependence of lipofuscin accumulation on vitamin A intake is greater in vitamin E deficient than in vitamin E supplemented animals. This suggests that vitamin E may protect vitamin A from autoxidation and thus

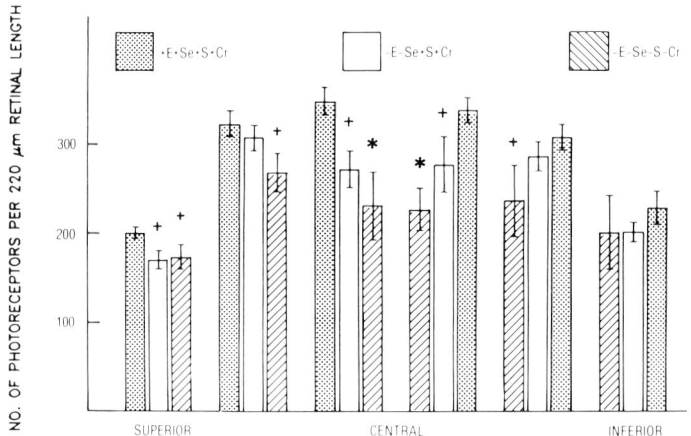

FIG. 9. Loss of photoreceptors due to dietary deficiency in antioxidant nutrients. Dietary levels of vitamin E (E), selenium (Se), sulfur amino acids (S), and chromium (Cr) were varied, and different regions of the retina were examined for photoreceptor loss. Symbols above the bars indicate statistical significance from control values: +, $p<0.05$; *, $p>0.01$. Adapted from Katz, et al. (22).

reduce the rate of incorporation of vitamin A oxidation products into lipofuscin granules. The protective effect of vitamin E on vitamin A _in vivo_ was first noted by Davies and Moore (9), who found that liver levels of vitamin A were reduced in vitamin E deficient rats. Robison and co-workers (40) reported that plasma vitamin A levels were reduced in vitamin E deficient rats as well. Recently we have found that both vitamin E and selenium deficiencies lower the vitamin A levels in the RPE of dark-adapted rats (23). Since vitamin A deficiency alone leads to photoreceptor cell degeneration (7, 11), it is possible that the loss of photoreceptor cells from the retinas of animals deficient in antioxidant nutrients is at least partially the result of an induced local deficiency of vitamin A in the eye.

As noted previously, the RPE plays a crucial role in the turnover of photoreceptor outer segment membranes. In the RCS rat strain, the photoreceptor cells undergo degeneration shortly after they develop (4, 19). The genetic defect in these animals apparently leads to an inability of the RPE to phagocytize photoreceptor outer segment membranes (4, 19, 32). Thus phagocytosis by the RPE appears to be crucial to the survival of the photoreceptors. We have found that the number of phagosomes in the RPE is greatly reduced in animals deficient in antioxidant nutrients (22) (FIG. 11). The reduction in phagosome number is greater than can be accounted for by photoreceptor cell loss alone. Therefore it is possible that the loss of

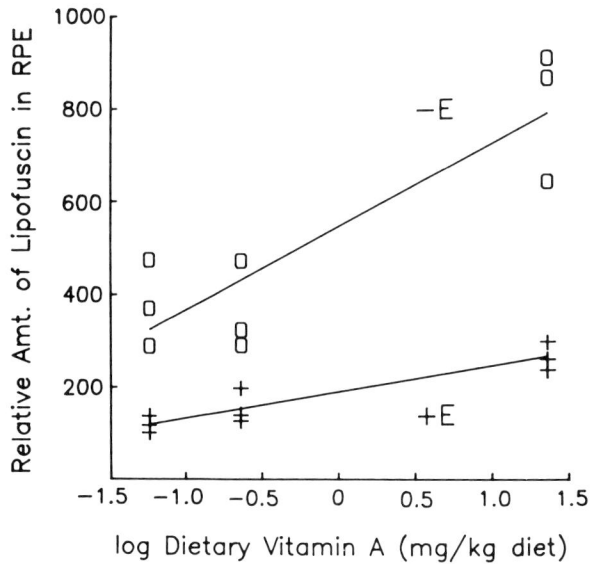

FIG. 10. Effect of dietary levels of vitamin A on RPE lipofuscin content of vitamin E supplemented (+E) and deficient (-E) rats. Each data point represents a determination from an individual animal.

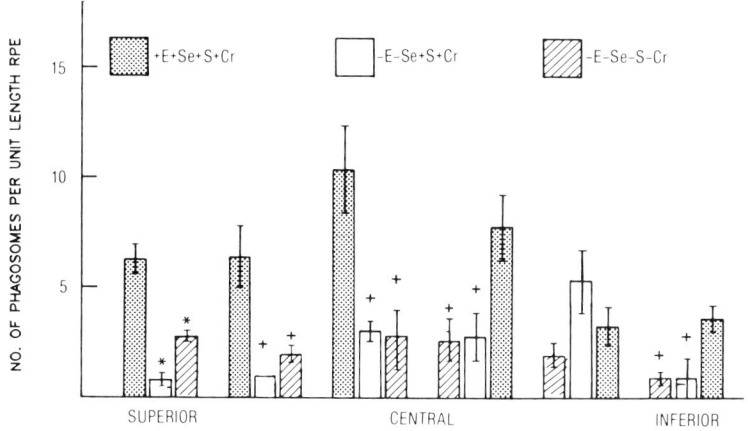

FIG. 11. Reduction in number of RPE phagosomes due to dietary deficiency in antioxidant nutrients. For designation of diets and symbols, see FIG. 9. Adapted from Katz, et al. (22).

photoreceptors associated with antioxidant nutrient deficiency may result at least to some degree from an impairment of phagocytosis by the RPE. This hypothesis is supported by the finding that outer segment debris accumulates at the apical surface of the RPE in rats deficient in antioxidant nutrients (22), suggesting that there is an imbalance between the rate of disk membrane production by the photoreceptors and phagocytosis by the RPE.

Additional evidence that antioxidant nutrient deficiency may have deleterious effects on the RPE comes from the finding that cells, apparently of RPE origin, are found in the space between the photoreceptor outer segments and the RPE of antioxidant deficient animals (22, 42) (FIG. 12). These cells

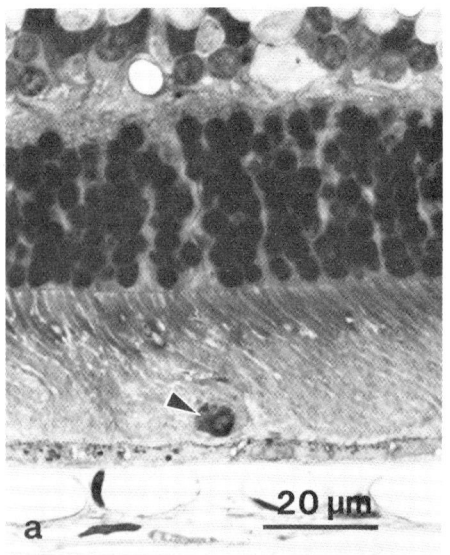
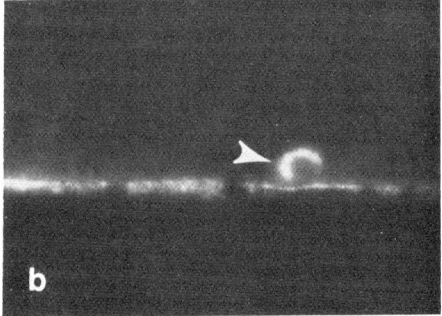

FIG. 12. Appearance of cells in the outer segment region of the rat retina in response to antioxidant nutrient deficiency [arrow in (a)]. These cells appear to be of RPE origin, since they contain large amounts of autofluorescent pigment similar to that present in the underlying RPE [arrow in (b)]. Adapted from Katz, et al. (22).

are filled with an autofluorescent pigment apparently identical to that in the underlying RPE. In addition, RPE cells can occasionally be seen in the process of detaching from their basal lamina. The presence of these detached cells in the subretinal space is likely to interfere with retina-RPE interactions. Since RPE cells have not been found to undergo mitosis in mature animals, and the RPE appears to form a complete layer between the retina and choroid, the remaining RPE cells apparently expand to fill the spaces left by those which detach from the basal lamina. This would lead to an increase in the metabolic load to which each RPE cell is exposed.

In response to antioxidant nutrient deficiency, the RPE undergoes a number of other changes which are not obviously deleterious. Average RPE cell height increases significantly as a result of deficiencies in a variety of antioxidant nutrients (22, 40) (Fig. 7). Whereas RPE cell height was quite regular in any particular region of the retina in rats receiving adequate levels of antioxidant nutrients, cell height became quite irregular in the peripheral retinas of rats deficient in antioxidant nutrients (22). These deficiencies also resulted in the appearance of a large number of lipid droplets in the RPE, particularly in the peripheral regions (22). In addition, lysosomal enzyme activity is elevated in the RPE of antioxidant deficient animals; Amemiya (3) found increased levels of acid phosphatase in the RPE of vitamin E deficient rats, while Katz and associates (23) found that acid lipase activities were elevated by vitamin E deficiency, or by a combined deficiency in vitamin E and selenium. The significance of the changes in RPE cell height, increases in RPE lipid content, and increases in RPE lysosomal enzyme activity in terms of RPE cell function is unknown.

Does the Presence of Lipofuscin Impair Cell Function?

While antioxidant nutrient deficiency has a number of deleterious effects on RPE cells, it is not known whether these effects are a direct result of autoxidative damage, or are due to the presence of increased amounts of lipofuscin in the RPE. In order to address this question, one must be able to manipulate the amount of lipofuscin in RPE cells in a manner that will not affect the amount of autoxidative stress to which these cells are subjected, or conversely, one must be able to manipulate the amount of autoxidation occurring in cells without affecting their lipofuscin content. A potential tool for evaluating the direct effect of lipofuscin accumulation on RPE cell function is the drug centrophenoxine (dimethylaminoethyl-p-chlorophenoxyacetate). Nandy and Bourne (35) reported that intramuscular or intraperitoneal injection of senile guinea pigs with this drug for 4 to 8 weeks resulted in an apparent reduction in the lipofuscin content of various neurons in the central nervous system. The effectiveness of centrophenoxine in reducing the lipofuscin content of certain cells of the central nervous system has been confirmed subsequently by a number of researchers (18, 34, 38). Centrophenoxine treatment has also been reported to alter the morphology of remaining lipofuscin granules in cells affected by this drug (44). In light of these findings, centrophenoxine seemed like a promising tool for determining whether the mere presence of lipofuscin within the RPE was deleterious to these cells.

Robison and co-workers (42) reported that centrophenoxine failed to reduce the lipofuscin content of the RPE in rats which had accumulated large amounts of this pigment due to vitamin E deficiency. In addition, no effect of centrophenoxine on RPE lipofuscin morphology was seen. Recently Katz and Robison (26) have reported that centrophenoxine also has no affect on the lipofuscin which has accumulated in the RPE during senescence (FIG. 13). Thus it appears that centrophenoxine will not be a useful tool for evaluating the effect of large amounts of lipofuscin on RPE cell function.

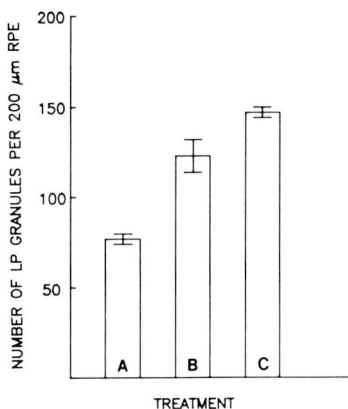

FIG. 13. Lack of an effect of centrophenoxine on RPE lipofuscin content in senescent rats. Data shown is from 28-week-old untreated rats (A), saline-injected 117-week-old rats (B), and 117-week-old rats which had been injected with centrophenoxine for 11 weeks. Centrophenoxine treatment had no significant effect. From Katz and Robison (26).

An alternative approach to separating at least some of the effects of autoxidation from the effects of lipofuscin accumulation on RPE cell function is to slow the rate of autoxidation in animals which have already accumulated large amounts of lipofuscin. This approach will of course only separate the acute effects of autoxidation from irreversible effects combined with indirect effects due to lipofuscin accumulation. Robison and co-workers (42) found that if rats were supplemented with vitamin E for 14 weeks after having been fed a vitamin E-free diet for 28 weeks, the lipofuscin content of the RPE was about equal to that which one would see in rats which had simply been on a vitamin E-free diet for 28 weeks. In other words, supplementation with vitamin E greatly slowed but did not reverse lipofuscin accumulation in the RPE. Vitamin E supplementation did, however, reverse the vesiculation and disruption of the photoreceptor outer segment disk membranes (2). The latter therefore appears not to be the result of the presence of large amounts of lipofuscin in the RPE. The ability of antioxidant nutrient supplementation to reverse other effects of antioxidant deficiency on the retina and RPE have yet to be examined.

EFFECTS OF AUTOXIDATION RESEMBLE EFFECTS OF AGING ON RETINA AND RPE

If autoxidation plays a role in senescence at the cellular level, one would expect antioxidant nutrient deficiency to accelerate at least some of the normal age changes that occur in the retina and RPE. As mentioned previously, lipofuscin, which appears to be an end-product of tissue autoxidation, accumulates in both the human and rat RPE during aging (12, 13, 25, 33, 48), and the rate of its accumulation is vastly accelerated as a consequence of dietarily induced antioxidant deficiency. Another age-related change reported to occur in the human RPE is the development of an irregularity in cell size and shape (33), similar to that which developed in the RPE of rats raised on diets which produced physiological antioxidant deficiencies (22). It has also been reported that lipid droplets or "vacuoles" accumulate in the human, monkey, and rat RPE during aging (14, 20, 25, 33). The similar build-up of lipid which occurs in the RPE as a consequence of antioxidant deficiency in the rat (22) suggests another parallel with the aging process. Phagocytosis of photoreceptor disk membranes by the RPE appears to be retarded both as a consequence of senescence (25), and as a result of antioxidant nutrient deficiency (22). Finally, a pronounced loss of photoreceptor cells, similar to that which occurs as a consequence of antioxidant deficiency, has been reported to occur in the rat retina during aging (24, 25, 31, 47). It therefore appears that antioxidant nutrient deficiency accelerates many of the changes which occur in the retina and RPE during normal aging. This finding suggests that senescence of these tissues may be at least partially a consequence of imperfect antioxidant protection, and thus of the accumulation of the effects of autoxidative damage.

Not all of the effects of antioxidant nutrient deficiency appear to occur during senescence. The vesiculation and disorganization of the apical ends of the photoreceptor outer segments, which is so striking in antioxidant deficient rats, does not appear to occur during normal aging (25). Likewise, the elevation of lysosomal enzyme activity reported to occur as a result of antioxidant deficiency, does not seem to occur during aging (25). Whatever the mechanism of damage to the photoreceptor disk membranes in antioxidant deficient animals, it is possible that the same damage occurs during senescence, but at such a slow rate as not be grossly apparent. Likewise the factors leading to elevated RPE lysosomal enzyme activity in antioxidant deficient rats may also be present in aging animals, but since their effects are spread over a much longer period, they may not be readily evident. Overall, the effects of antioxidant nutrient deficiency and of aging on the retina and RPE are similar enough to suggest that autoxidation probably plays a significant role in senescence of these tissues.

SUMMARY AND CONCLUSIONS

Strehler (45) has suggested four criteria which must be fulfilled by any event for it to play a primary role in senescence. These are: (a) universality, (b) intrinsicality, (c) progressiveness, and (d) deleteriousness. A variety of evidence suggests that intracellular lipofuscin accumulation probably meets at least the first three of these criteria. It also appears that autoxidation, particularly of polyunsaturated fatty acids and vitamin A, is involved in lipofuscin formation, and that autoxidation in vivo has deleterious effects. Many of the effects of antioxidant nutrient deficiency on the retina and retinal pigment epithelium resemble changes which occur normally in these tissues during aging. Therefore the autoxidative reactions leading to lipofuscin formation probably play a primary role in the development of at least some senescent changes in animals.

ACKNOWLEDGMENTS

We are glad to acknowledge the following collaborators who have participated in our research in this area: J.G. Bieri, C.C. Farnsworth, A.B. Groome, G.J. Handelman, K.R. Parker, D. Sisk, W.L. Stone, and M.E. Truckenmiller. Our thanks also to P. Boyer for typing the manuscript.

REFERENCES

1. Adler, F.M. (1965): Physiology of the Eye, C.V. Mosby Co., St. Louis.
2. Amemiya, T. (1981): Int. J. Vitam. Nutr. Res., 51:114-118.
3. Amemiya, T. (1981): Albert v. Graefes Arch. Klin. Ophthalmol., 216:103-109.
4. Bok, D. and Hall, M.O. (1971): J. Cell Biol., 49:664-682.
5. Brizzee, K.R. and Ordy, J.M. (1981): In: Age Pigments, edited by R.S. Sohal, pp. 102-156. Elsevier/North-Holland, Amsterdam.
6. Brown, K.T. (1969): In: The Retina: Morphology, Function, and Clinical Characteristics, edited by B.R. Straatsma, M.O. Hall, R.A. Allen, and F. Crescitelli, pp. 319-379.
7. Carter-Dawson, L., Kuwabara, T., O'Brien, P.J. and Bieri, J.G. (1979): Invest. Ophthalmol. Vis. Sci., 18:437-446.
8. Chio, K.S., Reiss, U., Fletcher, B. and Tappel, A.L. (1969): Science, 166:1535-1536.
9. Davies, A.W. and Moore, T. (1941): Nature, 147:794-796.
10. Dillard, C.J., Dumelin, E.E. and Tappel, A.L. (1977): Lipids, 12:109-114.

11. Dowling, J.E. and Wald, G. (1958): Proc. Nat. Acad. Sci.USA, 44:648-661.
12. Feeney, L., Grieshaber, J.A., and Hogan, M.J. (1963): In: The Structure of the Eye II Symposium, edited by F.K. Schattauer Verlag, Stuttgart.
13. Feeney, L. (1978): Invest. Ophthalmol. Vis.Sci., 17:583-600.
14. Fine, B.S. and Kwapien, R.P. (1978): Invest. Ophthalmol. Visual Sci., 17:1059-1069.
15. Fliesler, S.J. and Anderson, R.E. (1983): Prog. Lipid Res., 22:79-131.
16. Ganley, J.P. and Roberts, J. (1983): DHHS Publication, No. (PHS)83-1678.
17. Halliwell, B. (1981): In: Age Pigments, edited by R.S. Sohal, pp. 1-62.
18. Hasan, M., Glees, P. and Spoerri, P.E. (1974): Cell Tiss. Res., 150:369-375.
19. Herron, W.L., Riegel, B.W., Myers, O.E., and Rubin, M.L. (1969): Invest. Ophthalmol., 8:595-604.
20. Hogan, M.J. (1972): American Acad. Ophthalmol. Otolarynol. Trans., 76:64-80.
21. Holman, R.T. and Elmer, O.C. (1947): J. Am. Oil Chem. Soc., 24:127-129.
22. Katz, M.L., Parker, K.R., Handelman, G.J., Bramel, T.L. and Dratz, E.A. (1982): Exp. Eye Res., 34:339-369.
23. Katz, M.L., Parker, K.R., Handelman, G.J., Farnsworth, C.C., and Dratz, E.A. (1982): Ann. N.Y. Acad. Sci., 393:196-197.
24. Katz, M.L. and Robison, W.G. (1982): Invest. Ophthalmol. Vis. Sci., (ARVO suppl.) 24:181.
25. Katz, M.L. and Robison, W.G. (1983): Exp. Eye Res., (in press).
26. Katz, M.L. and Robison, W.G. (1983): J. Gerontol., 38:525-531.
27. Katz, M.L., Robison, W.G., Herrmann, R.K., Groome, A.B. and Bieri, J.B. (1983): Mech. Age. Dev., (in press).
28. Katz, M.L., Stone, W.L. and Dratz, E.A. (1978): Invest. Ophthalmol. Vis. Sci., 17:1049-1058.
29. Kini, M.M., Liebowitz, H.M., Colton, T., Nickerson, R.J., Ganley, J., and Dawber, T.A. (1978): Am. J. Ophthalmol., 85:28-34.
30. Kornzweig, A.L., Feldstein, M. and Schneider, J. (1957): Am. J. Ophthalmol., 44:29-37.
31. Lai, Y., Jacoby, R.O. and Jonas, A.M. (1978): Invest. Ophthalmol. Vis. Sci., 17:634-678.
32. Mullen, R.J. and LaVail, M.M. (1976): Science, 192:799-801.
33. Mishima, H., Hasebe, H. and Kondo, K. (1978): Jpn. J. Ophthalmol., 22:476-485.
34. Nandy, K (1968): J. Geront., 23:82-92.
35. Nandy, K. and Bourne, G.H. (1966): Nature, 210:313-314.
36. Porta, E.A. and Hartroft, W.S. (1969): In: Pigments in Pathology, edited by H. Wolman, pp. 192-236, Academic Press, N.Y.
37. Reichel, W., Hollander, J., Clark, J.M. and Strehler, S.L. (1968): J. Geront., 23:71-81.

38. Riga, S. and Riga, D. (1974): Brain Res., 265-275.
39. Robison, W.G., Katz, M.L. and Bieri, J.G. (1983): Invest. Opthalmol. Vis. Sci. (ARVO suppl.), 24:181.
40. Robison, W.G., Kuwabara, T. and Bieri, J.G. (1979): Invest. Ophthalmol. Vis. Sci., 18:683-690.
41. Robison, W.G., Kuwabara, T. and Bieri, J.G. (1980): Invest. Ophthalmol. Vis. Sci., 19:1030-1037.
42. Robison, W.G., Kuwabara, T. and Bieri, J.G. (1982): Retina, 2:263-281.
43. Sohal, R.S., editor (1981): Age Pigments, Elsevier/North Holland, Amsterdam.
44. Spoerri, P.E. and Glees, P. (1974): Exp. Geront., 10:225-228.
45. Strehler, B.L. (1977): Time, Cells and Aging, 2nd edition, Academic Press, N.Y.
46. Strehler, B.L., Mark, D.D., Mildvan, A.S. and Gee, M.V. (1959): J. Gerontol., 14:430-439.
47. Weisse, I., Stotzer, H. and Seitz, R. (1974): Virchows Arch. Path. Anat. and Histol., 362:145-156.
48. Wing, G.L., Blanchard, G.C. and Weiter, J.J. (1978): Invest. Ophthalmol. Vis. Sci., 17:601-607.
49. Yau, T.M. (1979): Mech. Age Dev., 11:137-144.

Free Radicals in Molecular Biology, Aging, and Disease, edited by D. Armstrong et al.
Raven Press, New York © 1984.

Dietary Restriction and the Aging Process

Richard Weindruch

Department of Pathology, University of California at Los Angeles School of Medicine, Los Angeles, California 90024

Dietary restriction (abbr. DR) increases both mean and maximum survival times in several animal species. These include the protozoan Tokophyra (74), rotifers (23), Daphnia (42), fish (16), rats (3, 51, 61, 68, 115) and mice (11, 89, 103). DR is the only strategy yet tested in homeotherms which convincingly reduces mortality rates (77) and increases maximum survival times (17, 70). DR appears to be the best available method to interfere with aging in mammals. A growing number of laboratories are using DR to study the biology of decelerated aging (for reviews see 1, 4, 48, 102). Yet, the precise way DR inhibits aging remains unknown. The DR model can be used to test theories of aging, allowing biologic phenomena influenced by DR to be separated from uninfluenced ones. Thus, the proposed role of free radicals and mitochondria in aging (37, 38) can be examined in animals on DR. The goals of this review are: 1) to summarize results from recent biogerontologic studies done on rodents subjected to DR and, 2) to consider findings from DR studies which may tell us more about the role of free radicals in aging.

Weaning-initiated DR (abbr. WDR) of rodents amounting to a 20-60% decrease in food intake starting at ~4 weeks of age is associated with 10-300% increases in mean and maximum survival times (3, 10, 11, 51, 61, 68, 89, 115). The largest relative increases were reported in rats by Ross (68) and appear due, in part, to negative effects of certain diets fed ad lib. Most evidence indicates that the feeding strategies which prolong life do so via caloric restriction combined with adequate intakes of other diet essentials (undernutrition without malnutrition). Spontaneous late life diseases occur at later ages or in lower incidences in rodents subjected to WDR (2, 24, 71-73, 86, 115).

The vast majority of DR studies have tested WDR and not adult-initiated DR (abbr. ADR). The impact of gradually imposed ADR on longevity, cancer incidence, and age-sensitive immunologic functions in mice is a major interest of our laboratory.

The diet strategies used in DR studies differ from laboratory to laboratory. We have fed control and mice on DR the semipurified diets shown in Table 1. Note that mice on DR are fed on an intermittent basis a diet enriched in protein (casein), salts and vitamins such that per week intakes of these nutrients are quite similar for all mice. Mice on DR eat less carbohydrate (cornstarch and sucrose), fat and fiber per week than do controls. The restricted diet fed to rats by Masoro and colleagues is also a semipurified, vitamin-enriched one and is fed at 60% of the ad lib intake level (5). DR is carried out at the Gerontology Research Center by giving rats free access to a laboratory chow diet but only every other day (27). Merry and Holehan (52) feed their rats on DR a pelleted breeder diet at 50% of the ad lib level.

TABLE I. Composition of Diets (g/kg diet)

Constituents	Diet 1[a]	Diet 2[b]
Casein, vitamin-free	200.0	350.0
Cornstarch	260.8	157.6
Sucrose	260.8	157.6
Corn oil	135.0	135.0
Non-nutritive fiber	56.4	40.0
Salt mixture, USP XIV	60.0	110.0
Vitamin mixture	23.0	42.0
Brewers' yeast	4.0	7.4
Zinc oxide	0.05	0.1

[a] Diet fed to control mice. Fed as seven ~3.0 g feedings (one daily feeding on Monday through Thursday, three feedings on Friday) per week, providing ~85 kcal/week.

[b] Diet fed to restricted mice. 35% casein diet, enriched also in salt and vitamin mixtures, Brewers' yeast, and zinc oxide. Fed as four ~3.0 g feedings (one daily feeding on Monday and Wednesday, two feedings on Friday) per week, providing ~50 kcal/week.

RECENT FINDINGS FROM OTHER LABORATORIES

University of Texas at San Antonio

Experiments on diverse physiologic aspects of aging in male Fischer 344 rats were reported by Masoro and coworkers. Rats put on DR at 6 weeks of age showed a mean lifespan of 32 months (vs. 23 months for ad lib fed controls) and a maximum lifespan of 47 months (vs. 32 months for controls) (115). Renal and other lesions occurred earlier in life in controls and progressed more rapidly. Body mass, lean body mass and adipose mass all fell in late life and each of these declines was greatest in controls (5, 115). The decline in mass of the gastrocnemius muscle began at ~18 months for controls versus ~24 months for DR rats (115). DR was reported (47) to delay: 1) age related increases in serum concentrations of cholesterol and phospholipids and, 2) age-related decreases in postabsorptive serum free fatty acid concentration. DR started at 6 months of age was found to be as effective as DR started at 6 weeks of age in modulating age changes in serum lipids (50). Age-related losses of tension development in vascular smooth muscle were retarded by DR (41). DR delayed and reduced age-related losses in adipocyte responsiveness to the lipolytic hormones glucagon and epinepherine (6, 114). Later studies indicated that DR prevents the loss in glucagon responsiveness through its effects on receptor-plasma membrane events (94). DR was recently reported (43) to prevent senile bone loss and suppress age-related increases in serum levels of parathyroid hormone.

Masoro's group reported an observation quite relevant to free radical views on aging. They found that rats on DR consumed a *greater* number of calories per gram of body weight during their lifetimes (134 kcal/g lifetime) than did control rats (92 kcal/g lifetime) (49). This increase in kcal/g lifetime for DR rats led the authors to conclude that DR does not slow the rate of aging by slowing metabolic rate. These findings differ from those of Sacher (77) who elegantly discussed relations between longevity and metabolism and then calculated kcal/g lifetime using data from two of Ross' prior studies (67, 69). Sacher found that kcal/g lifetime was almost the same (~100) for each of five diet groups which varied hugely in daily caloric intake from 18-75 kcal/day. He concluded that this near constancy of kcal/g lifetime indicated that DR prolongs life by reducing the metabolic rate per gram of rat and allowing a longer time to reach 100 kcal/g lifetime (and death). This view fits with the conclusion advanced in 1928 by Pearl (62) from studies on _Drosophila_: "In general the duration of life varies inversely as the rate of energy expenditure during life."

In my view, Sacher's calculations are confusing since he combined data from two of Ross' studies (67, 69) and differences existed between those populations. A basic problem seems that Ross' data were not appropriate for this type of calculation, a view also expressed by Masoro, Yu and Bertrand (49).

Other concerns cloud the meaning of the kcal/g lifetime value as an index of metabolic potential in WDR vs. control rodents. Perhaps rodents on WDR differ from controls in the efficiency of caloric absorption. WDR could also influence the percent of ingested calories used for either fuel or maintenance. One might also ask if dividing energy consumption by body weight is misleading in comparisons of control and DR animals. Do rodents on DR consume more calories per animal over the lifetime? Using the data of Masoro et al (49), I find that the average rat on DR ate ~38,000 kcal/lifetime vs. ~42,000 for ad lib controls. Sacher (77) found that Ross' rats on WDR ate ~20,000-30,000 kcal/lifetime vs. ~60,000 kcal/lifetime for ad lib controls.

Gerontology Research Center, National Institute on Aging (Baltimore)

Several DR studies have been carried out on rats and mice at the Gerontology Research Center. Three longevity studies initiated by the late Charles L. Goodrick and completed by Donald K. Ingram and colleagues were reported (27, 29, 30). In one (27), male Wistar rats were fed ad lib or subjected to WDR (given free access to food but only every other day). Mean lifespan was higher for WDR rats (32 vs 17 months) as was maximum lifespan (41 vs 23 months). In a second study (29), Wistar rats were fed ad lib until 10.5 or 18 months of age and then were housed in either standard laboratory cages or in activity wheel cages. Part of each group was then subjected to ADR (every other day feeding). The other rats were not restricted. ADR increased survival times in both age groups but exercise failed to yield a clear-cut effect on longevity. Maximum lifespans did not exceed 37 months. Analysis of mortality rates suggested that the rate of aging was slower for the ADR groups. In a third study (30), male Wistar rats were subjected to WDR (every other day feeding). These rats were compared to ad lib controls for spontaneous activity and longevity. Restricted rats showed less activity early in life and higher activity later in life. Overall, little support was found for the view that DR increases longevity via increasing activity.

Other work from the Gerontology Research Center shows that Wistar rats on WDR (again via every other day feeding) show slower losses with age of brain striatal dopamine receptors (46). The concentration of dopamine receptors in 24 month old rats on WDR was ~50% higher than in age-matched controls and was comparable to levels in 3-6 month old controls.

Three other DR studies from the Gerontology Research Center have recently been reported (as meeting abstracts), and indicate that: 1) genetic background can modulate the life-extending effects of WDR and ADR in mice (28, 31) and, 2) DR started at 6 months of age retards age-related losses in thermoregulation (85).

University of Hull (England)

Merry and Holehan (53) imposed WDR (50% of ad lib) on male Sprague-Dawley rats. Survivorship was very similar to that reported by the San Antonio group (115). Age changes in several hormones were studied. Puberty onset in males (as judged by a pubertal peak in serum testosterone levels) occurred at ~2 months of age in controls but was delayed 10-20 days by WDR. Peak levels of testosterone were ~3 fold lower for rats on WDR. Serum FSH levels were very low in WDR rats between 30-70 days but major effects of DR were not apparent later in life. Reproductive lifespan for male rats was not increased by WDR.

These investigators also studied influences of WDR on puberty onset and the duration of fertility in female Sprague-Dawley rats (52). Sexual maturation was delayed but not prevented by underfeeding. Vaginal opening and the first estrus occurred in 75% of control rats between 36-45 days of age. In contrast, only ~50% of the restricted rats at 143 days of age were sexually mature. By 227 days of age, all restricted rats were mature. Reproductive lifespan was clearly increased by WDR. About 80% of the rats on WDR could conceive and wean pups at 510 days, which is 100 days beyond when breeding ends for controls. Further, 25% of the rats on WDR could breed at 840-930 days.

University of Sydney

A.V. Everitt and colleagues have studied effects of hypophysectomy and DR on aging (longevity, diseases, collagen) in male Wistar rats. Since hypophysectomized rats eat only ~40% the amount of food as do intact rats, the DR group is needed to serve as pair-fed, intact controls. All rats ate a commercial chow diet. In one study (20), 70 day old rats subjected to either DR or hypophysectomy outlived ad lib controls (mean survival of 28-30 months for treated rats vs 26 months for controls; maximum survival of 42-44 months for treated vs 37 months for controls). Both treatments retarded the rate of aging of tail tendon collagen fibers and inhibited certain late-life diseases (tumors, renal, heart enlargement, hind limb paralysis). In a second study (21), rats were fed either 12.5, 25, 50, or 75 kcal/day. Mean and maximum survivorships were greatest for rats eating 25 or 50 kcal/day. No rat lived beyond ~41 months. DR led to a later onset of proteinuria and a lower incidence of glomerular lesions. Acute DR (25 kcal/day) imposed on 26 month old rats lowered proteinuria by 40% in one week with no further reduction in the next week. Hypophysectomy had a greater effect than DR in retarding collagen aging when food intakes were kept equal (22).

Other Investigators

A group from Bombay, India reports (13) that WDR (one-half of ad lib intake) imposed on Swiss albino female mice reduces: (1) in vitro lipid peroxidation (malondialdehyde formation) in liver homogenates, (2) lipofuscin accumulation in brain and heart and, (3) the percentage of lysosomal enzyme activity which is free (i.e. not inside the lysosome) in liver, brain and intestinal homogenates. Mice were studied up to 12 months of age. The in vitro rate of lipid peroxidation for 12 month old mice on WDR was ~70% that of age-matched controls. These workers also report (19) that low protein diets (6 or 12% protein) fed ad lib for up to 12 months produce similar effects. The low protein diets reduced growth rate but not final body weight. No survivorships were reported in either study. The authors view these results as supporting the free radical theory of aging since mice on the restricted diets (which presumably would live longer) showed less signs of free radical mediated damage.

These very interesting data beg a basic question: What are the best indexes of in vivo free radical mediated damage? Are measures such as malondialdehyde formation in vitro or lipofuscin accumulation providing a good gauge?

Two recent reports (75, 76) describe sharply higher activities of hepatic microsomal drug metabolizing and NADPH-generating enzymes in male Sprague-Dawley rats on WDR (45-50% of ad lib). Feeding did not go beyond 7 weeks. At this time, the activities of three NADPH-generating enzymes (malic enzyme, glucose-6-phosphate dehydrogenase, 6-phosphogluconate dehydrogenase) were ~2x higher in liver and ~3-5x higher in adipose tissue of rats on WDR as compared to controls (76). Activities of hepatic drug metabolizing enzymes (e.g. aniline hydroxylase, p-nitrobenzoate reductase) were ~20-50% higher in restricted rats. It seems important to next determine what role increases in drug-metabolizing and NADPH-generating capacities may have in determining health and longevity.

Interestingly, dehydroepiandrosterone (DHEA, a steroid hormone reported to have anti-cancer and anti-obesity effects when fed to rodents [80, 81, 113]) has also been found to raise the activity of liver malic enzyme in rats (14, 15). We recently observed that mice fed a diet containing DHEA ate ~30% less food than did mice on a control diet (110). Perhaps a decreased food intake may contribute to certain biologic effects attributed to feeding DHEA.

Eve and Gerald Reaven and colleagues have studied pancreatic structure and function, insulin action, and serum triglyceride levels in male Sprague-Dawley rats on WDR. Restriction was carried out by combining standard laboratory chow and cellulose in a 1:2 ratio and feeding this diet ad lib. Controls were fed the standard chow ad lib. Rats were not studied beyond 12 months of age. At this age, rats on WDR and another group of rats which were fed the control diet ad lib but allowed to exercise on a running wheel showed a 3-fold reduction in serum levels of triglycerides and insulin compared to sedentary rats on the control diet (65). Body weights for both treated groups at 10-12 months averaged ~500g versus ~750g for controls.

Calorie intakes were ~125 kcal/d for the exercised rats, ~110 kcal/d for the sedentary controls, and ~80 kcal/d for rats on the calorie-diluted diet. Pancreatic islets from 12 month old sedentary controls appeared enlarged, multi-lobulated and fibrotic; in contrast, restricted or exercised rats of this age did not show pancreas pathology (64). Compared to 12 month old rats fed ad lib, WDR rats showed: (1) lower plasma insulin levels after an oral glucose load and, (2) less insulin resistance (66). Glucose-stimulated insulin secretion per volume of islet fell with age but was not influenced by WDR.

RECENT FINDINGS FROM OUR LABORATORY

Since 1976, I have been investigating aging in mice from long-lived strains subjected to DR. This work has been carried out with Roy Walford in his laboratory at UCLA. We have mainly studied influences of WDR and ADR on longevity, cancer incidence and age-sensitive immunologic functions. Also, we have examined other effects of WDR (e.g. liver mitochondrial respiration, eye lens aging, body temperature). We feel it is best to use long-lived rodent strains in aging studies. The mouse strains we study show a mean lifespan of ~32-38 months when fed semipurified control diets; in contrast, in the oft-cited paper of Stuchlikova et al (84), the mean survival for control and WDR mice (strain not given) averaged only ~21-24 months. Maximum survival did not exceed 36 months. In this review, findings on WDR mice are first discussed followed by results from ADR studies.

Weaning-Initiated Dietary Restriction

Lifespan extension. Two longevity studies involving mice on WDR were reported by Kay Cheney and others in our laboratory. In the first study (10), C57BL/6J mice (females were mostly studied) on WDR showed a greater maximum lifespan (but not mean lifespan) and lower incidence of lymphoma than did controls. Mice on WDR tended to show higher mortality in early life and lower mortality later. A semipurified diet was used with controls eating ~105 kcal/wk and mice on WDR eating ~60 kcal/wk. Four cohorts (each comprised of control and WDR mice) were studied. In only one cohort did mice on WDR eat a salt- and choline-enriched diet and these mice lived the longest of all tested. In the second study (11), (C57BL/10Sn x C3H/HeDiSn)F_1 female mice (abbr. B10C3F_1) were fed semi-purified diets similar to those of the preceding study except that all mice on DR ate a diet doubled in salt and vitamin content. Pre-weaning DR was tested by suckling 9 mouse pups per mother (versus 5 per mother in the control group) and separating the mothers from the litter every other day starting at one week of age. Mice subjected to life-long DR of this sort showed a median lifespan of ~45 months (vs. ~38 months for controls) and a maximum 10% survival time of ~50 months (vs. ~43 months for controls). Mice on DR showed lower late-life mortality rates than controls. Mean lifespan of tumor-bearing mice tended to be greater in DR than in non-DR groups.

We recently completed a study of longevity and cancer incidence using female (C3H.SW/Sn x C57BL10.RIII/Sn)F_1 mice (abbr. C3B1ORF$_1$). Preliminary results were reported (108) and a full report is being prepared. This parental combination was selected because of its potential for generating a very long-lived hybrid which develops a variety of cancers (82). Six diet groups were set up at weaning: Group 1 - fed Purina Lab Chow ad lib; Group 2 - fed Diet 1 as described in Table 1 (~85 kcal/wk); Group 3 - fed Diet 2 as described in Table 1 (~50 kcal/wk); Group 4 - As per Group 3, but also restricted pre-weaning by separating pups from mother every other day one week after birth, Group 5 - fed ~50 kcal/wk but gradually restricted in protein intake (35% casein diet from weaning → 4 months; 25% casein from 4 months → 12 months; 20% from 12 months → 24 months; 15% from 24 months → death); Group 6 - fed ~40 kcal/wk of Diet 2. Each group consisted of 49-71 individually housed mice. The amount of food given Group 2 mice was less than the ad lib amount as reflected by their adult body weights of ~35-40g versus ~45-50g for Group 1 (and ~20-25g for restricted mice). Mean lifespan was shortest in Group 1 (27.4 $\pm$ 0.9 [SE] months), longer in Group 2 (32.7 $\pm$ 0.7 months), even longer in Group 5 (39.7 $\pm$ 0.9 months), and longest in the other three restricted groups (42.3 - 42.9 $\pm$ 0.9 months for Groups 3 and 4, 45.1 $\pm$ 0.9 months for Group 6). The maximum lifespans (calculated as mean lifespan of the longest lived 10% for each group) were: 35.1 $\pm$ 0.3 months for Group 1, 39.7 $\pm$ 0.6 months for Group 2, 48.5 $\pm$ 0.5 months for Group 5, 50.5 - 51.1 $\pm$ 0.2 months for Groups 3 and 4, and 53.0 $\pm$ 0.3 months for Group 6. To my knowledge, no prior report describes mice living as long as did the mice in Group 6. These results indicate: (1) As the severity of WDR increased, so did longevity. The 40 kcal intake is about the lowest amount of food we can feed to C3B1ORF$_1$ female mice and still get healthy looking mice (without excess early life mortality). (2) Limiting preweaning food intake failed to further enhance longevity for mice put on WDR. (3) Survival times for mice fed Purina Lab Chow ad lib were ~15% less than those for mice fed a semi-purified diet in slightly less than ad lib amounts. (4) Mice on both protein restriction and WDR lived somewhat shorter lifespans (~6%) than did mice on WDR eating a high protein diet.

Cancer incidence was assessed via gross autopsy and histopathologic study. Overall incidence was lowest for Group 6 (38% of mice were tumor-bearing) and highest for Group 2 (79%). Lymphoma was the most common tumor and was seen most frequently in Group 1 (29% incidence) and Group 2 (44%). In contrast, mice on the high-protein DR showed less lymphoma (13-23% incidence for Groups 3, 4 and 6). Average survival for lymphoma-bearing mice was greatest for those on DR (~40 months vs. ~30 months for Groups 1 and 2). Hepatoma was the next most common tumor. The incidence of hepatoma was not affected by WDR; however, longevity for mice with hepatoma was greater in all four DR groups (41-45 months) than in the two control groups (29-34 months).

It seems instructive to see how the results from this longevity series fit with a contrarian view of DR expressed by Cherkin (12) and Cutler (18). They suggest that a major problem with DR studies is the use of ad lib fed controls. Cherkin (12) writes: "I believe it is timely to reexamine this general acceptance of a causal relationship between dietary restriction and increased life span. The reason is that the experimental evidence equally supports an opposite interpretation, namely, that dietary excesses causes reduced life span." He identifies "the unstated assumption that an ad libitum diet is 'normal' or 'optimal' for longevity" as the source of this problem. I agree that it is probably incorrect to view ad lib as "normal" (despite the fact that most humans in developed nations eat ad lib). For this reason, we did not allow mice in Group 2 to eat ad lib. The lifespan of these mice exceeded that of mice fed Purina Lab Chow ad lib but was much less than that of mice on WDR. Cutler (18) also argues that overeating accelerates aging. He believes that DR "acts to bring the animal back to the aging rate it would normally have in its natural ecological niche." He also writes: "Life span extension is most frequently thought to be a process that prolongs life beyond what is the normal genetic potential for the animal. Calorie restriction and/or intermittent fasting does not appear to be such a process." In order to accept this view, one must know what the "normal genetic potential" is for longevity in a particular animal model. To my knowledge, such values are not known with any certainty. Does the 53 month average lifespan for the six longest lived mice in Group 6 exceed the "maximum genetic potential" for longevity in this F_1 hybrid? We also do not know how long mice can live in the wild since so few seem to have the good luck to reach the advanced ages attained by other mice involved in longevity research. Quite possibly, times of low food intake along with ample exercise could provide wild mice with a long life.

Lifetime Energy Dissipation. As discussed earlier, Masoro's group reported (49) that rats on DR averaged 134 kcal/g lifetime. This value exceeded the average value (92 kcal/g lifetime) for rats fed ad lib. Using the general method followed by Masoro et al (49), I find that this value for $C3B10RF_1$ mice in Groups 2, 3 and 6 is about 330, 410 and 400 kcal/g lifetime. Thus, mice on WDR showed about a 20-30% greater lifetime energy dissipation per gram of animal than did control mice (fed less than ad lib). This compares with a 45%

increase in this value for rats on DR (compared to ad lib rats) in the report of Masoro and colleagues (49). The large difference between rats (low lifetime energy dissipation values) and the C3B10RF$_1$ mice (high values) is due to daily adult calorie intakes per gram of rat being ~0.15 kcal/g/d versus ~0.30 kcal/g/d for the mice. Calculation of calories consumed per animal presents a different picture with mice from Groups 3 and 6 consuming ~8000-9000 kcal/lifespan vs. ~12,000 kcal/lifespan for Group 2 controls.

Mitochondria. Very little is known about effects of long-term DR on mitochondrial aging. A growing literature suggests that mitochondria may play a major role in the aging process. Certain of these findings are reviewed below followed by a discussion of the limited information on effects of DR on mitochondria.

Mitochondrial dysfunction with age may be due to deficits or defects in the organelle. As discussed in this Symposium, mitochondria produce free radicals as a consequence of oxidative metabolism. These free radicals can inflict damage to cells and extracellular molecules. Table 2 summarizes evidence for an age-dependent loss of mitochondria in mammals. Also, mitochondria fractionated from old rat heart (55) or liver (112) are more prone to appear degenerate or to become lost during centrifugation.

Age-dependent alterations in the in vitro respiratory activity of mitochondria have been observed in some, but not all, investigations. This is usually studied by measuring mitochondrial O_2 consumption. Oxidation and phosphorylation supported by β-hydroxybutyrate was reduced 40% in mitochondria from 24 month old rat liver compared to preparations from 3-4 month old rats (98). Beyond 20 months of age, rat heart mitochondria showed reduced ADP-stimulated (state 3) respiration supported by β-hydroxybutyrate, palmitylcarnitine, glutamate + malate, and glutamate + pyruvate (9). Rat heart mitochondria also exhibited age-related falls in state 3 rates of O_2 uptake due to fatty acid oxidation (33). Other reports do not describe lower respiratory rates for mitochondria from old rodents (26, 101). Hansford (34) carefully reviewed this area.

Nohl's group examined heart mitochondria from young (3 month) and old (23-24 month) rats for free radical production, activities of enzymes, and membrane composition. The formation rate of O_2 radicals by old mitochondria was ~25% greater than in young preparations (58). A constant fraction (~20%) of these radicals escaped quenching by mitochondrial superoxide dismutase. Peroxidative degradation of mitochondrial membrane lipids paralleled increases with age in free radicals. This occurred despite activities of mitochondrial catalase and glutathione peroxidase increasing with age (59). Several inner membrane enzymes from old rats showed decreased specific activities. These decreases disappeared after solubilization

TABLE 2. Evidence For A Loss Of Mitochondria In Old Mammals[a]

	SPECIES	TISSUE	METHOD	OBSERVATION	REFERENCE
1.	Human	Liver	EM[a]	↓ no. of mito. & ↑ size of mito.	87
2.	Monkey	Brain	EM	↓ no. of mito. in capillary endothelium	8
3.	Rat	Brain	EM	↓ no. of mito. in Purkinje cells	63
		Liver	EM	↓ no. of mito.. & ↑ size of mito.	88
		Liver	Biochem.	↓ amount of mito. DNA & protein	83
		Liver	Biochem.	↓ amount of mito. protein	91
		Liver	Biochem.	↓ sp. act. of cyt. c. ox.	92
		Heart	Biochem.	↓ sp. act. of cyt. c. ox.	79
		Heart, soleus diaphragm	Biochem.	↓ sp. act. of cyt. c. ox.	35
4.	Mouse	Heart, Liver	EM	↓ no. of mito.	39
		Liver	Biochem.	↓ sp. act. of cyt. c. ox.	101
		Brain	EM	↓ no. of mito.	78

[a] Abbreviations used: mito. = mitochondria; EM = electron microscopy; cyt. c. ox. = cytochrome c oxidase; sp. act. = specific activity; no. = number; ↑ = increase; ↓ = decrease.

of the surrounding phospholipids by detergent (56). Inner membranes from old mitochondria were less fluid and contained lower amounts of polyunsaturated fatty acids. Nohl suggests (56) that aging does not alter intrinsic properties of membrane-bound enzymes but influences the association of lipids and proteins via radical-induced peroxidation of membrane lipids. Later work (57, 60) showed that mitochondria from old rat hearts were ~40% less active than young mitochondria in translocating adenine nucleotides across the inner membrane. Endogenous levels of adenine nucleotides fell with age by ~25% because of falls in ATP. Also, mitochondrial membranes from old rats contained less negatively charged phospholipids.

Lipids from rat liver mitochondria were studied as a function of age by Vorbeck and coworkers (93). Total phospholipid content fell and cholesterol increased with age. Evidence was presented for lower lipid fluidity in old mitochondria.

Hansford and Castro (35) reported that muscle homogenates from 24 month old rats showed overt falls (20-50% lower than for 6 month old rats) in mitochondrial enzyme activities for both the tricarboxylate cycle and for lipid oxidation. Soleus, diaphragm and heart muscles were studied. Hansford and Castro (36) also reported that rates of uptake and release of Ca^{2+} in and out of heart mitochondria were 25-35% lower in 24 month old as compared with 6 month old rats. Mitochondrial Ca^{2+} content was not affected by age.

The literature on metabolic rate for rodents on DR does not yield a consensus view. Heat production in 850 day old rats on WDR in McCay's colony was higher than for controls when expressed per unit body weight but slightly lower per body surface area (111). Young C57BL/6J mice on a 50% DR for 6 weeks produced less heat than controls when expressed per animal (kcal/hr/animal) but about the same amount per unit of metabolic body weight (kcal/hr/kg$^{0.75}$) (90). Seven week old Sprague-Dawley rats fed ~60% of ad lib for a month showed lower heat production than controls on either a per rat or per metabolic body weight basis (25). A 30% drop in basal metabolic rate was found for these rats on DR. Other workers measured O_2 consumption to evaluate metabolic rate. A 12 day restriction at 50% of ad lib intake imposed on adult Sprague-Dawley rats reduced O_2 consumption per animal or per metabolic body weight by 10-20% (7). A 50% restriction imposed on very young (~40 g) Wistar rats for 8 weeks did not influence O_2 consumption per metabolic body weight (54). Thus, from these data, it is difficult to find strong support for the notion that DR increases lifespan via a metabolic slowdown.

We studied effects of aging and WDR on mitochondrial recovery and respiratory capacities in mice (101). Male C57Bl/6J mice were used for the aging study (old = 23-26 months, adult = 9-12 months) and female C3B1ORF$_1$ mice for the WDR study (3-7 months old). Old mice did not differ from adults in amounts of protein recovered from mitochondrial fractions of liver, brain and spleen, but did show a decline in specific activity of cytochrome c oxidase in liver and spleen. Age effects on in vitro respiration by mitochondria occurred in liver and spleen. In liver, only one substrate (ß-hydroxy-butyrate) of four tested was respired at a slower rate by old than by young mitochondria. Effects of WDR were studied in liver and brain. WDR reduced recovery of liver mitochondrial protein (Table 3) and liver cyt. c ox. specific activity (Table 4). Liver mitochondria from mice on WDR generally showed increased state 3 rates with no differences from controls in state 4 rates for respiration supported by glutamate or pyruvate + malate (Table 5), resulting in an increased RCI in the WDR group. Rates of 2,4-dinitrophenol-uncoupled respiration were also raised by WDR. Electron microscopy of liver mitochondrial preparations revealed more

TABLE 3. Effects of WDR on Mitochondrial Recovery from Liver and Brain.[a]

	Control	WDR	P
Liver (n = 13)			
Wet wt. (g)	1.22 ± .06	1.10 ± .05	<.05
Total mito. protein (mg)	33.4 ± 1.5	26.6 ± 1.3	<.01
Mito. protein/wet wt. (mg/g)	27.3 ± 1.1	23.4 ± 1.3	<.01
Brain (n = 5)			
Wet wt.	0.45 ± .01	0.39 ± .01	<.05
Total mito. protein	11.0 ± 0.4	10.4 ± 0.5	NS
Mito. protein/wet wt.	24.7 ± 0.7	27.0 ± 0.9	NS

[a] Values are $\bar{X}$ ± SE for 3-7 month old mice. Adapted from (101).

TABLE 4. Effects of WDR on Cytochrome c Oxidase Activity of Liver Mitochondria.[a]

	Control	WDR	P
Activity/mg mito. protein	2.1 ± 0.1	1.8 ± 0.1	NS(<0.1)
Activity/g liver	51.2 ± 2.6	37.8 ± 4.0	<.05

[a] Values are $\bar{X}$ ± SE for n = 7. Activity equals µmoles of cytochrome c oxidized per minute. Adapted from (101).

TABLE 5. Effects of WDR on Liver Mitochondrial Respiration Supported by Malate Plus Pyruvate.[a]

	Control	WDR	P
State 4	2.0 ± 0.1	1.9 ± 0.1	NS
State 3	4.7 ± 0.3	6.2 ± 0.5	<.01
2,4 DNP	5.3 ± 0.4	6.5 ± 0.6	<.01
RCI	2.5 ± 0.1	3.3 ± 0.1	<.01

[a] Values are $\bar{X}$ ± SE for n = 14 and represent µg atom O/min./100 mg protein. RCI (respiratory control index) is defined as the ratio of respiration in the presence of ADP to that after its use (State 3/4). Adapted from (101).

non-mitochondrial contaminants for old mice and larger mitochondria for mice on WDR. These findings are compatible with the notion of age-dependent losses of liver mitochondria which can be influenced by DR.

A higher RCI suggests better coupling of oxidative phosphorylation to electron transport. Perhaps the better coupling shown by liver mitochondria from mice on WDR results in reduced free radical generation and less mitochondrial damage. This could postpone age-related losses of mitochondria and explain higher kcal/g lifetime values observed for rats and mice on WDR. It is also possible that WDR lowers the number of mitochondria an organism needs and this might increase a restricted animal's capacity to synthesize and assemble new mitochondria in late life.

Immune system.
The immune systems of mammals change overtly upon maturation and senescence. Decreases in immune response capacities to exogenous stimuli (e.g., T cell mitogens, viruses) occur with aging along with increases in autoimmunity (32, 107). Walford proposed that immune system aging may contribute to the pathogenesis of aging (95-97). Studies in mice indicate that age changes in the immune system are influenced by WDR (99). We put B10C3F$_1$ females on WDR and reported that young (6 weeks to 8 months old) restricted mice showed lower spleen weights, a dampening of thymus growth, higher proliferative responses by splenic T lymphocytes to mitogen or alloantigen stimulation, and a lowering of body temperature (99). Similar results were obtained in this same F$_1$ hybrid by Cheney et al (11) using a different diet and feeding protocol. We further investigated thymus structure in B10C3F$_1$ mice on WDR via quantitative histologic methods and observed a "younger" appearance of thymuses from 6 month old WDR mice than for age-matched controls (100). In other experiments, C3B10RF$_1$ mice were put on WDR and studied at between 3-15 months of age for T lymphocyte proliferation induced by mitogens (105). The increased splenic T cell response to phytohemagglutinin for mice on WDR appeared due (at least in part) to an increased proportion of responsive T cells. We also studied natural killer (abbr. NK) cell activity in mice from this strain on WDR. It has been suggested (40) that NK cells defend against cancer. Lower basal NK responses were seen for restricted mice (2-33 months old) than for age-matched controls; however, after injection with the interferon inducer polyinosinic:polycytidylic acid (which boosts NK activity), old mice on WDR showed higher responses than old, injected controls (109). In addition, mice on WDR showed higher in vitro generation of cytotoxic T lymphocytes to allogeneic tumors than did age-matched controls. The suggestion was made that mice on WDR may better resist cancer via an NK system very responsive to induction signals coupled with higher levels of T cell cytolysis.

Lens proteins.
With colleagues at UCLA's Jules Stein Eye Institute, we have studied aging of the eye lens in mice on WDR. It is well known that with aging and cataract development in rodents and humans a loss occurs in the amount of soluble protein in the gamma crystallin fraction. Using high performance liquid chromatography and other technics, we found that WDR retards the loss

with age in gamma crystallins in C3B10RF$_1$ mice (44,45). To our knowledge, no other report describes a deceleration of this loss.

Adult-Initiated Dietary Restriction

Gerontologists have not studied ADR nearly as intensively as WDR. We became interested in the ADR model because the available findings did not answer the question of whether or not ADR is an effective anti-aging strategy. The longevity data were not at all conclusive (see 103 for discussion) and data on other age-sensitive phenomena were very limited. Another motivation to study ADR is that only it seems potentially applicable to humans. For these reasons, we carried out the studies described below.

Lifespan extension.

Male mice from two strains (B10C3F$_1$ and C57BL/6J [abbr. B6]) eating commercially available chow ad lib from weaning until 12-13 months of age were studied. At this age, food intake was gradually restricted via nutrient enriched, semi-purified diets. Controls were fed a non-enriched companion diet in amounts adequate to maintain body weights. ADR mice consumed 60-70% of the calories eaten by controls. Many mice were studied (B10C3F$_1$: n=67 ADR, n=68 control; B6: n=29 ADR, n=24 control). Body weights and survivorship are shown in Figure 1. The mean lifespan for B10C3F$_1$ controls was 33.0 $\pm$ 0.7 (SE) months vs. 36.9 $\pm$ 0.7 months for mice on ADR (12%↑). Mean survival for the longest lived 10% of each B10C3F$_1$ group (n=7) was 40.6 $\pm$ 0.5 months for controls vs. 45.1 $\pm$ 0.6 months for ADR mice (11%↑). Mean survival for B6 controls was 24.9 $\pm$ 0.9 months vs. 29.9 $\pm$ 1.4 months for ADR mice (20%↑). Mean survival for the longest-lived 10% of each group (n=3) was 31.5 $\pm$ 0.5 months for controls vs. 38.2 + 1.4 months for ADR mice (21%↑). The incidence of spontaneous lymphoma was lower for mice in both strains on ADR. Also, for mice bearing a lymphoma, those on ADR tended to live longer than controls.

Cheney and coworkers (11) found that female B10C3F$_1$ mice put on ADR at 14 months of age lived 5-10% longer (average and maximum lifespans) than did normally fed controls. Mice on ADR did not live as long as did mice on WDR. Tumor incidence was reduced by ADR.

The results indicate that appropriate ADR can inhibit cancer and extend the average and maximum life span.

Lifetime energy dissipation.

Approximate values were calculated for male B10C3F$_1$ from our ADR longevity study (103). Mice on ADR consumed ~575 kcal/g over their lifespan. Control mice consumed a slightly lower amount (~530 kcal/g). B10C3F$_1$ mice ate ~0.50

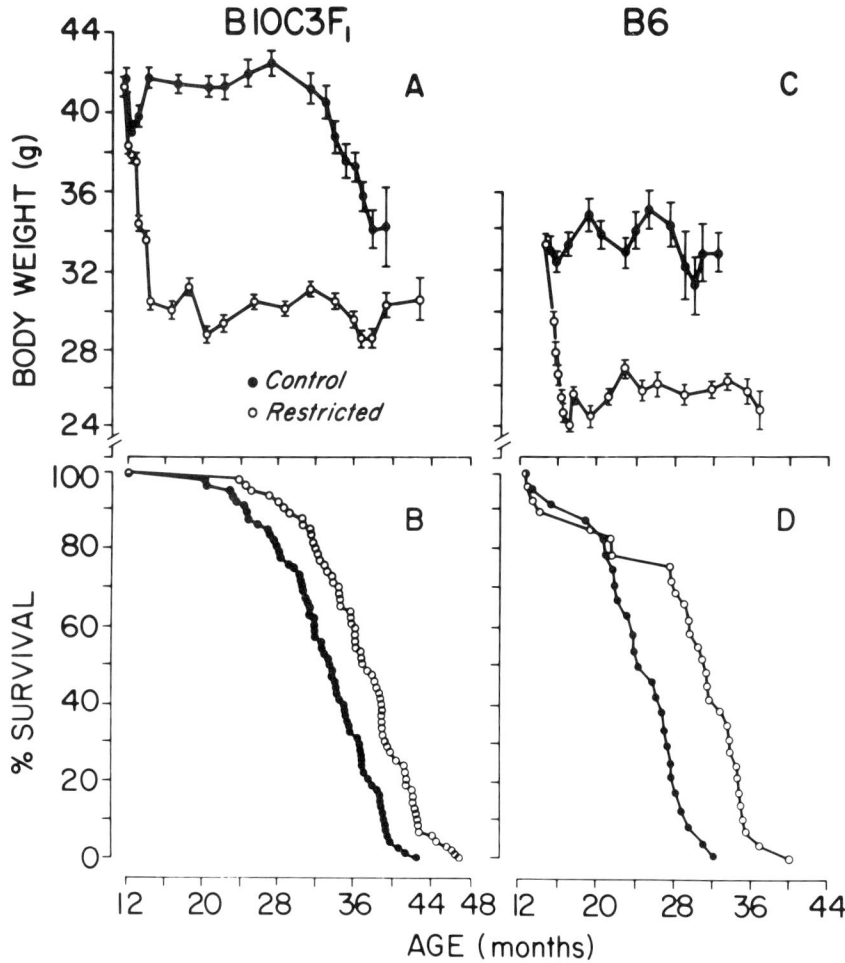

FIG. 1. Body weights and survival of B10C3F$_1$ mice (A and B) and B6 mice (C and D) fed control or restricted diets. Each point in the survival curves represents one mouse. (From Weindruch and Walford, *Science* 215: 1415-1418. Copyright 1982 by the AAAS.)

kcal/g/day irrespective of diet. This value exceeds that for female C3B10RF$_1$ mice (~0.30) discussed previously but appears similar to that for the control B10C3F$_1$ females studied by Cheney and coworkers (11). Apparently, lifetime energy dissipation values may show large intraspecies variations. The average mouse on ADR ate ~18,000 kcal/lifetime, which was less than the ~21,000 kcal/lifetime eaten by controls.

Immune system.
Our first experiment on ADR involved male B10C3F$_1$ mice restricted from 12 months until 16.5 months of age. After this feeding period, mice on ADR showed more robust splenic lymphocyte proliferative responses to T cell mitogens (99). We next studied C3B10RF$_1$ males put on gradual ADR starting at 12, 17 or 22 months of age (104). Underfeeding lowered the numbers of nucleated cells per spleen but increased the percentages of T cells. For mice restricted at 12, 17, or 22 months and tested at various ages thereafter, the proliferative response of spleen cells after phytohemagglutinin stimulation exceeded that for age-matched controls. ADR did not, however, alter splenocyte responses to concanavalin A or to B cell mitogens. In the splenic plaque-forming cell response to injected sheep erythrocytes, ADR and control mice differed more clearly in response kinetics than in peak levels. The splenic cell mediated lymphocytotoxic response to alloantigens was comparable in old mice (27-29 months) on ADR since 12 months of age to that of young (5-6 months) controls, and was greater than that of age-matched controls. In a third study (106), B10C3F$_1$ mice were put on ADR when 12 months old. A small blood sample was taken at 13 and 23 months of age and analyzed for IgG and immune complex levels. Both measures fell after long term ADR. These results indicate that ADR influences immunosenescence in a fashion not unlike WDR.

CONCLUSIONS

Prior to the last ten years or so, gerontologic studies on rodents subjected to DR were largely limited to examining lifespan extension and disease inhibition. Over the preceding decade, the DR studies have turned increasingly mechanistic and have provided endocrinologic, immunologic, physiologic and biochemical insights. Yet, the precise way that DR influences survivorship and disease remains unknown. This most likely is due both to the diversity of changes brought on by DR and by the mysterious nature of biologic aging. Several findings were discussed which implicate free radical generation and mitochondrial dysfunction (and losses) in the aging process. Only a cluster of findings are available regarding influences of DR on free radicals/mitochondria; however, this limited information seems compatible with the possibility that DR could act by reducing free radical generation and mitochondrial losses. Studies aimed at testing this notion may well clarify both the role that free radicals play in aging and how DR is able to retard the aging process.

ACKNOWLEDGEMENTS

I thank Drs. Roy Walford and Sheldon Ball for their comments and the Word Processing Center personnel for their assistance.

My research is supported by U.S. Public Health Service Research Grants AG-00424 and CA-26164.

REFERENCES

1. Barrows, C.H. Jr., and Kokkonen, G.C. (1982): In: Nutritional Approaches to Aging Research, edited by G.B. Moment, pp. 219-243. CRC Press, Boca Raton, Florida.
2. Berg,B.N., and Simms, H.S. (1960): J. Nutr., 71:255-263.
3. Berg,B.N., and Simms, H.S. (1961): J. Nutr., 74:23-32.
4. Bertrand, H.A. (1983): In: Review of Biological Research in Aging (Volume 1), edited by M. Rothstein, pp. 359-378. Liss, New York.
5. Bertrand, H.A., Lynd, F.T., Masoro, E.J., and Yu, B.P. (1980): J. Gerontol., 35:827-835.
6. Bertrand, H.A., Masoro, E.J., and Yu, B.P. (1980): Endocrinology, 107:591-595.
7. Boyle, P.C., Storlien, L.H., Harper, A.E., and Keesey, R.E. (1981): Am. J. Physiol., 241:R392-397.
8. Burns, E.M., Kruckerberg, T.W., Comerford, L.E., and Buschmann, M.T. (1979): J. Gerontol., 34:642-650.
9. Chen, J.C., Warshaw, J.B., and Sanadi, D.R. (1972): J. Cell. Physiol., 80:141-148.
10. Cheney, K.E., Liu, R.K., Smith, G.S., Leung, R.E., Mickey, M.R., and Walford, R.L. (1980): Exp. Gerontol., 15:237-258.
11. Cheney, K.E., Liu, R.K., Smith, G.S., Meredith, P.J., Mickey, M.R., and Walford, R.L. (1983): J. Gerontol., 38:420-430.
12. Cherkin, A. (1979): Age, 2:51.
13. Chipalkatti, S., De, A.K., and Aiyar, A.S. (1983): J. Nutr., 113:944-950.
14. Cleary, M.P., Shepherd, A., Zisk, J., and Schwartz, A. (1983): Nutr. Behav., 1:127-136.
15. Cleary, M.P., Billheimer, J., Finan, A., Sartin, J.L., and Schwartz, A.G. (1984): Horm. Metab. Res., (in press).
16. Comfort, A. (1963): Gerontologia, 8:150-155.
17. Cutler, R.G. (1981): In: Aging: Biology and Behavior, edited by J.L. McGaugh and S.B. Kiesler, pp. 31-76. Academic Press, New York.
18. Cutler, R.G. (1982): In: Testing the Theories of Aging, edited by R.C. Adelman and G.S. Roth, pp. 25-114. CRC Press, Boca Raton, Florida.
19. De, A.K., Chipalkatti, S., and Aiyar, A.J. (1983): Mech. Ageing Dev., 21:37-48.
20. Everitt, A.V., Seedsman, N.J., and Jones, F. (1980): Mech. Ageing Dev., 12:161-172.
21. Everitt, A.V., Porter, B.D., and Wyndham, J.F. (1982): Gerontology, 28:168-175.
22. Everitt, A.V., Wyndham, J.F., and Barnard, D.L. (1983): Mech. Ageing Dev., 22:233-251.
23. Fanestil, D.D., and Barrows, C.H. Jr. (1965): J. Gerontol., 20:462-469.

24. Fernandes, G., Yunis, E.J., and Good, R.A. (1976): Nature, 263:504-507.
25. Forsum, E., Hillman, P.E., and Nesheim, M.C. (1981): J. Nutr., 111:1691-1697.
26. Gold, P.H., Gee, M.V., and Strehler, B.L. (1968): J. Gerontol., 23:509-512.
27. Goodrick, C.L., Ingram, D.K., Reynolds, M.A., Freeman, J.R., and Cider, N.L. (1982): Gerontology, 28:233-241.
28. Goodrick, C.L., Ingram, D.K., Reynolds, M.A., Freeman, J.R., and Cider, N.L. (1982): Gerontologist, 22:95 (abstract).
29. Goodrick, C.L., Ingram, D.K., Reynolds, M.A., Freeman, J.R., and Cider, N.L. (1983): J. Gerontol., 38:36-45.
30. Goodrick, C.L., Ingram, D.K., Reynolds, M.A., Freeman, J.R., and Cider, N.L. (1983): Exp. Aging Res., 9:203-209.
31. Goodrick, C.L., Ingram, D.K., Reynolds, M.A., Freeman, J.R., and Cider, N.L. (1983): Age, 6:145 (abstract).
32. Gottesman, S.R.S., and Walford, R.L. (1982): In: Testing the Theories of Aging, edited by R.C Adelman and G.S. Roth, pp. 233-279. CRC Press, Boca Raton, Florida.
33. Hansford, R.G. (1978): Biochem. J., 170:285-295.
34. Hansford, R.G. (1981): In: CRC Handbook of Biochemistry in Aging, edited by J.R. Florini, pp. 137-162. CRC Press, Boca Raton, Florida.
35. Hansford, R.G., and Castro, F. (1982): Mech. Ageing Dev., 19:191-201.
36. Hansford, R.G., and Castro, F. (1982): Mech. Ageing Dev., 19:5-13.
37. Harman, D. (1981): Proc. Natl. Acad. Sci. USA, 78:7124-7128.
38. Harman, D. (1983): Age, 6:86-94.
39. Herbener, G.H. (1976): J. Gerontol., 31:8-12.
40. Herberman, R.B. and Ortaldo, J.R. (1981): Science, 214:24-30.
41. Herlihy, J.T., and Yu, B.P. (1980): Am. J. Physiol., 238:H652-655.
42. Ingle, L., Wood, T.R., and Banta, A.M. (1937): J. Exp. Zool., 76:325-352.
43. Kalu, D.N., Yu, B.P., and Norling, B.K. (1983): Age, 6:141-142 (abstract).
44. Leveille, P., Weindruch, R., Bok, D., Walford, R., and Horwitz, J. (1982): Age, 5:132 (abstract).
45. Leveille, P., Weindruch, R., Bok, D., Walford, R., and Horwitz, J. (submitted).
46. Levin, P., Janda, J.K., Joseph, J.A., Ingram, D.K., and Roth, G.S. (1981): Science, 214:561-562.
47. Liepa, G.U., Masoro, E.J., Bertrand, H.A., and Yu, B.P. (1980): Am. J. Physiol., 238:E253-257.
48. Masoro, E.J., Yu, B.P., Bertrand, H.A., and Lynd, F.T. (1980): Fed. Proc., 39:3178-3182.

49. Masoro, E.J., Yu, B.P., and Bertrand, H.A. (1982): Proc. Natl. Acad. Sci. USA, 79:4239-4241.
50. Masoro, E.J., Compton, C., Yu, B.P., and Bertrand, H.A. (1983): J. Nutr., 113:880-892.
51. McCay, C.M., Crowell, M.F., and Maynard, L.A. (1935): J. Nutr., 10:63-79.
52. Merry, B.J., and Holehan, A.M. (1979): J. Reprod. Fertil., 57:253-259.
53. Merry, B.J., and Holehan, A.M. (1981): Exp. Gerontol., 16:431-444.
54. Mohan, P.F., and Narasinga Rao, B.S. (1983): J. Nutr., 113:79-85.
55. Murfitt, R.R., and Sanadi, D.R. (1978): Mech. Ageing Dev., 8:197-201.
56. Nohl, H. (1979): Z. Gerontol., 12:9-18.
57. Nohl, H. (1982): Gerontology, 28:354-359.
58. Nohl, H., and Hegner, D. (1978): Eur. J. Biochem., 82:563-567.
59. Nohl, H., Hegner, D., and Summer, K.-H. (1979): Mech. Ageing Dev., 11:145-151.
60. Nohl, H., and Kramer, R. (1980): Mech. Ageing Dev., 14:137-144.
61. Nolen, G.A. (1972): J. Nutr., 102:1477-1494.
62. Pearl, R. (1928): The Rate of Living. Knopf, New York.
63. Porta, E.A., Nitta, R.T., Kia, L., Joun, N.S., and Nguyen, L. (1980): Mech. Ageing Dev., 13:319-355.
64. Reaven, E.P., and Reaven, G.M. (1981): J. Clin. Invest., 68:75-84.
65. Reaven, G.M., and Reaven, E.P. (1981): Metabolism, 30:982-986.
66. Reaven, E.., Wright, D., Mondon, C.E., Solomon, R., Ho., H. and Reaven, G.M. (1983): Diabetes, 32:175-180.
67. Ross, M.H. (1959): Fed. Proc., 18:1190-1207.
68. Ross, M.H. (1961): J. Nutr., 75:197-210.
69. Ross, M.H. (1969): J. Nutr., 97(Suppl. 1):563-602.
70. Ross, M.H. (1978): In: The Biology of Aging, edited by J.A. Behnke, C.E. Finch and G.B. Moment, pp. 173-189. Plenum, New York.
71. Ross, M.H., and Bras, G. (1965): J. Nutr., 87:245-260.
72. Ross, M.H., and Bras, G. (1971): J. Natl. Cancer Inst., 47:1095-1113.
73. Ross, M.H., and Bras, G. (1973): J. Nutr., 103:944-963.
74. Rudzinska, M.A. (1952): J. Gerontol., 7:544-548.
75. Sachan, D.S. (1982): Biochem. Biophys. Res. Commun., 104:984-989.
76. Sachan, D.S., and Das, S.K. (1982): J. Nutr., 112:2301-2306.
77. Sacher, G.A. (1977): In: Handbook of the Biology of Aging, edited by C.E. Finch and L. Hayflick, pp. 582-638. Van Nostrand Reinhold, New York.

78. Samorajski, T., Friede, R.L., and Ordy, J.M. (1971): J. Gerontol., 26:542-551.
79. Sanadi, D.R. (1977): In: Handbook of the Biology of Aging, edited by C.E. Finch and L. Hayflick, pp. 73-98. Van Nostrand Reinhold, New York.
80. Schwartz, A.G. (1979): Cancer Res., 39:1129-1132.
81. Schwartz, A.G., and Tannen, R.H. (1981): Cancinogenesis, 2:1335-1337.
82. Smith, G.S., and Walford, R.L. (1978): In: Genetic Effects on Aging, edited by D. Bergsma and D.E. Harrison, pp. 281-312. Liss, New York.
83. Stocco, D.M. and Hutson, J.C. (1978): J. Gerontol., 33:802-809.
84. Stuchlikova, E., Juricova-Horakova, M., and Deyl, Z. (1975): Exp. Gerontol., 10:141-144.
85. Talan, M., and Ingram, D.K. (1983): Gerontologist, 23:79 (abstract).
86. Tannenbaum, A. (1942): Cancer Res., 2:460-467.
87. Tauchi, H., and Sato, T. (1968): J. Gerontol., 23:454-461.
88. Tauchi, H., and Sato, T. (1980): Mech. Ageing Dev., 12:7-14.
89. Tucker, M.J. (1979): Int. J. Cancer, 23:803-807.
90. Vander Tuig, J.G., Trostler, N., Romsos, D.R., and Leveille, G.A. (1979): Proc. Soc. Exp. Biol. Med., 160:266-271.
91. Vorbeck, M.L., and Martin, A.P. (1976): J. Cell. Biol., 70:36a (abstract).
92. Vorbeck, M.L., Martin, A.P., Park, J.K.J., and Townsend, J.F. (1982): Arch. Biochem. Biophys., 214:67-79.
93. Vorbeck, M.L., Martin, A.P., Long, J.W. Jr., Smith, J.M., and Orr, R.R. Jr. (1982): Arch. Biochem. Biophys., 217:351-361.
94. Voss, K.H., Masoro, E.J., and Anderson, W. (1982): Mech. Ageing Dev., 18:135-149.
95. Walford, R.L. (1969): The Immunologic Theory of Aging. Munksgaard, Copenhagen.
96. Walford, R.L. (1974): Fed. Proc., 33:2020-2027.
97. Walford, R.L. (1982): J. Am. Geriatr. Soc., 30:617-625.
98. Weinbach, E.C., and Garbus, J. (1959): J. Biol. Chem., 234:412-417.
99. Weindruch, R.H., Kristie, J.A., Cheney, K.E., and Walford, R.L. (1979): Fed. Proc., 38:2007-2016.
100. Weindruch, R.H., and Suffin, S.C. (1980): J. Gerontol., 34:525-531.
101. Weindruch, R.H., Cheung, M.K., Verity, M.A., and Walford, R.L. (1980): Mech. Ageing Dev., 12:375-392.
102. Weindruch, R.H., and Makinodan, T. (1981): In: Nutrition in the 1980s. Constraints on Our Knowledge, edited by N. Selvey and P.L. White, pp. 319-325. Liss, New York.
103. Weindruch, R., and Walford, R.L. (1982): Science, 215:1415-1418.

104. Weindruch, R., Gottesman, S.R.S., and Walford, R.L. (1982): Proc. Natl. Acad. Sci. USA, 79:898-902.
105. Weindruch, R., Kristie, J.A., Naeim, F., Mullen, B., and Walford, R.L. (1982): Exp. Gerontol., 17:49-64.
106. Weindruch, R., Chia, D., Barnett, E.V., and Walford, R.L. (1982): Age, 5:111-112.
107. Weindruch, R., and Walford, R.L. (1982): In: The Reticuloendothelial System: A Comprehensive Treatise. Volume 3: Phylogeny and Ontogeny, edited by N. Cohen and M.M. Sigel, pp. 713-748. Plenum, New York.
108. Weindruch, R.H., Fligiel, S., Mullen, B., and Walford, R.L. (1982): Gerontologist, 22:167 (abstract).
109. Weindruch, R.H., Devens, B.H., Raff, H.V., and Walford, R.L. (1983): J. Immunol., 130:993-996.
110. Weindruch, R.H., McFeeters, G., and Walford, R.L. (1984): Exp. Gerontol., (in press).
111. Will, L.C., and McCay, C.M. (1943): Arch. Biochem., 2:481-485.
112. Wilson, P.D., and Franks, L.M. (1975): Gerontologia, 21:81-94.
113. Yen, T.T., Allan, J.A., Pearson, D.V., Acton, J.M., and Greenberg, M.M. (1977): Lipids, 12:409-413.
114. Yu, B.P., Bertrand, H.A., and E.J. Masoro. (1980): Metabolism, 29:438-444.
115. Yu, B.P., Masoro, E.J., Murata, I., Bertrand, H.A., and Lynd, F.T. (1982): J. Gerontol., 37:130-141.

Adrenocortical Cultures as Model Systems for Investigating Cellular Aging

Peter J. Hornsby, Kathy A. Aldern, and Sandra E. Harris

Department of Medicine, University of California at San Diego, La Jolla, California 92093

Because of the continued interest in the role of oxygen-centered radicals in the aging process (19), it is important to establish whether there is a role for such free radicals in the phenomena of cell culture senescence. In view of the lack of consensus on the relevance of in vitro aging to in vivo aging (6, 21, 45, 55), it is essential to acquire as much information as possible on the mechanisms involved in the cell culture aging process, in order to determine its relevance to in vivo aging. If there are features common to both the cell culture and the in vivo aging processes, cell culture aging systems are validated as models of in vivo aging, and those common elements should be studied.

Several investigators have investigated whether lowering the cellular level of oxygen free radicals in cultured cells by supplementation with vitamin E or lowering the oxygen tension is accompanied by an increase in the total number of population doublings achieved ("culture life span") (5, 21, 40, 41, 44, 52). There has been disagreement as to whether these alterations of the cell culture environment do have any effect on culture life span. Vitamin E supplementation was found to be effective in extending the replicative potential of human fibroblasts only with one serum lot (40). Whereas the increase of life span in this one experiment was unequivocal, the reason for the inability of vitamin E to extend life span with other serum lots remained undetermined.

Our interest in this question was initiated by the finding that adrenocortical cells in long-term culture, in the presence of 10% fetal bovine serum, were essentially completely deficient in both vitamin E and selenium (23). In previous experiments using vitamin E supplementation in culture, the selenium status of the cells does not appear to have been taken into account. It is well established that in vivo it is difficult to observe effects of vitamin E in the presence of an adequate supply of selenium, and vice versa (11, 34, 46). The protective effects of vitamin E and selenium against oxidative damage in cells are very similar. Vitamin E is a potent chain-breaking phenolic antioxidant, and so acts directly to prevent peroxidative damage to membrane lipids (1, 2). Selenium acts indirectly as a biological antioxidant, as an essential component of the enzyme glutathione peroxidase. Glutathione peroxidase acts to protect against lipid peroxidation by removing peroxides, which can provide a major source of reinitiation of peroxidation (13). Any possible extension of culture life span by vitamin E can be adequately investigated only under conditions of established deficiency of selenium and low glutathione peroxidase activity; and possible effects of supplementation with selenium on replicative potential, which do not appear to have been investigated previously, may only be observable under conditions of established deficiency of vitamin E.

We have used the adrenocortical cell culture system in previous experiments on cellular aging and experiments on the possible effects of oxidative damage on differentiated functions in this cell type (7, 8, 15, 22, 23, 25, 27, 29, 30, 48). The system will be introduced here with a description of previous work, and then our recent experiments on the effects of vitamin E and selenium on cell culture life span will be presented. We feel that the use of differentiated cell types in cellular aging research offers some advantages over the use of the standard fibroblast systems, in that the ability to monitor changes in differentiated functions as well as replicative potential may provide valuable insights into the mechanisms of cellular senescence. An additional advantage is the ability to obtain large number of cells for preparation of primary cultures; thus, the earliest passages of the culture life span, during which several changes in the functions of adrenocortical cells occur, may readily be studied.

Aging of adrenocortical cells in culture: Introduction

The adrenocortical cell culture system has been used in several investigations of cellular aging. Earlier work has been reviewed in a previous publication (30). Most of the work has been performed on bovine adrenocortical cells, with some preliminary work having been done on the aging of human adrenocortical cells in culture (31). As has been discussed elsewhere, bovine cells, being derived from a long-lived non-rodent species, show a finite life span phenomenon in culture, and very rarely transform spontaneously (36). They therefore generally provide a good model for human cells, with the advantages of more ready availability in the case of the differentiated cell types and in many cases considerably simpler growth conditions. An overview of the life history of bovine adrenocortical cells in culture is provided in Fig. 1.

Fig. 1 shows that several important changes in the function of adrenocortical cells occur over the very early portion of the

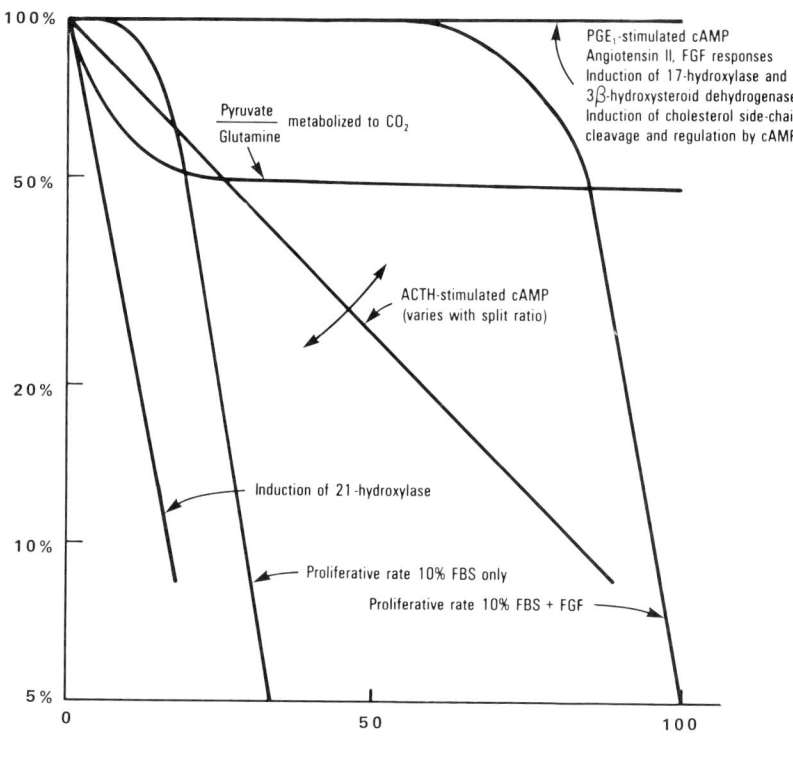

FIG. 1. General scheme of the life history of bovine adrenocortical cells in culture. The ordinate shows the measured level of the indicated functions after various numbers of divisions in culture, taking the level in the primary culture as 100%. In addition to those functions shown, 11β-hydroxylase induction has been shown to be maintained to mid-life span, when reinduction is carried out under conditions that prevent oxidative damage to the enzyme (see text).

culture life span. Adrenocortical cells offer some advantages over fibroblast systems in the investigation of such early changes in the function of cultured cells. Fibroblasts pass through a "Phase I" period during which the cells migrate out from the initiating explant tissue (21). Changes that may occur during this period are not readily observable. Furthermore, it is not possible to establish whether any cell selection may have taken place during this period, and thus whether the cells that form the mid-life span "Phase II" proliferating population are representative of the initial population in culture. Adrenocortical cells may be shown not to undergo such a selection process. Fig. 2 shows that all of the cells in the initial cell suspension from the adrenal cortex that attach to the culture dish will incorporate [^{3}H]thymidine into their nuclei, indicating that all the cells enter the proliferating population, without selection of a minority of cells that is capable of division in culture (26). This has enabled study of both short- and long-term alterations occurring in this culture system.

In mid- and late-life span, bovine adrenocortical cells show an increase in the concentration of ACTH required to stimulate steroidogenesis, and a change in the type of steroids secreted, with a loss of synthesis of the normal final product, cortisol (27, 30, 48). At the time of the earlier review of this system, two functional alterations had been observed to occur over the life span that appeared to account for the changes in differentiated function: a loss of ACTH-stimulated cyclic AMP production, without losses of cellular responsiveness to other agents acting at the cell surface (prostaglandin E_1, angiotensin II, and fibroblast growth factor (FGF)); and a loss of reinduction of one of the steroid hydroxylases, 11β-hydroxylase, by agents that raise intracellular cyclic AMP (30). The loss of responsiveness to ACTH appears to result from loss of ACTH receptors in a process that occurs at low cell density and is only partially reversible (29).

We have now found that the loss of inducibility of the 11β-hydroxylase enzyme is a function of the conditions used for its reinduction, and is not an absolute loss. On exposure to certain steroids that act as pseudosubstrates, 11β-hydroxylase is subject to a process of oxidative damage that causes loss of enzymatic activity (22). In a process that is dependent on the presence of oxygen, these steroid pseudosubstrates greatly shorten the half-life of the enzyme. When reinduction of the enzyme is attempted, by exposure of the cultures to agents that stimulate cyclic AMP, such as ACTH, prostaglandins, or cholera toxin, enzyme activity is not increased under "standard" culture conditions of 19% oxygen and 10% fetal bovine serum (22, 48). Reinduction may be accomplished, however, by the use of a lowered oxygen tension (22), or by reinduction under serum- and lipoprotein-free conditions, where the destructive pseudosubstrate steroids are not synthesized (Hornsby, Aldern, and Harris, unpublished observations). Steroidogenesis in bovine adrenocortical cells is absolutely dependent on an exogenous supply of cholesterol (33). When reinduction of 11β-hydroxylase was investigated in serum-free medium at 5% oxygen, no loss of reinducibility was seen, at least until mid-life span (Hornsby, Aldern, and Harris, unpublished observations).

The mechanism of loss of 11β-hydroxylase on exposure to steroid pseudosubstrates has been discussed elsewhere (25). The loss appears to result from the release of superoxide from the enzyme when attempted metabolism of pseudosubstrates occurs and normal hydroxylation cannot take place. Superoxide may initiate lipid peroxidation which destroys the enzyme. Antioxidants, such as butylated hydroxyanisole, dimethyl sulfoxide, and ascorbic acid,

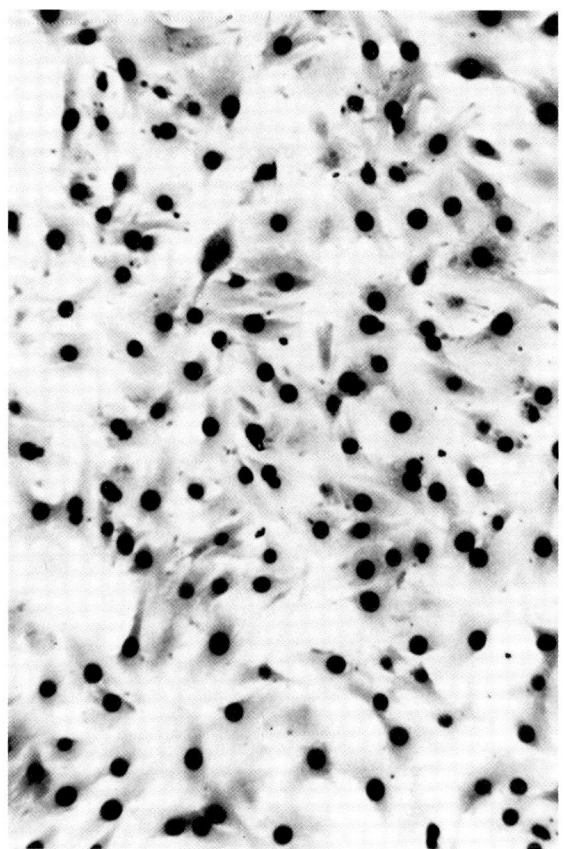

FIG. 2. Autoradiograph of bovine adrenocortical cells, plated from the primary cell suspension, after exposure to [^{3}H]thymidine. Bovine adrenocortical cells were enzymatically dissociated from the intact gland and placed in culture. After 48 hours, [^{3}H]thymidine was added for an additional 24 hours, after which the culture was processed for autoradiography. Almost all the cells that attach to the dish commence DNA synthesis, thus demonstrating an absence of initial cell selection in this system. (From ref. 26).

prevent the loss of enzyme activity (22). Thus such antioxidants stabilize the differentiated properties of adrenocortical cells, under these particular conditions (9).

The loss of the reinducibility of the 11β-hydroxylase under "standard" culture conditions only partly accounts for the observed lack of synthesis of the expected final steroid product, cortisol, in the later part of the life span. Another steroid hydroxylase, the 21-hydroxylase, is not reinducible under any conditions thus far investigated (Hornsby, Aldern, and Harris, unpublished observations). Other investigators have shown a lack of reinduction of 21-hydroxylase activity in primary cultures of bovine adrenocortical cells despite induction of mRNA coding for the enzyme after exposure of the cells to ACTH (14). Although 21-hydroxylase, like 11β-hydroxylase, is subject to loss of activity on incubation with steroid pseudosubstrates (24), there is much less evidence that this loss is due to oxidative damage, as appears to be the case for 11β-hydroxylase. It is to be noted that the 21-hydroxylase is a microsomal enzyme, in contrast to the mitochondrial 11β-hydroxylase (12). In both cases there are other cytochrome P-450 enzymes that are still induced normally in late passage cultures; the cholesterol side-chain cleavage enzyme in mitochondria, and 17-hydroxylase in the endoplasmic reticulum (30, 48).

In the changes in differentiated function that have been observed in this cell system, the alterations occur in the early phase of the culture life span; little or no change is observed at the end of the life span. This has been discussed previously in relation to the concept that the cessation of division at the end of the life span is the result of "further differentiation" (28). This hypothesis is unlikely in this cell type which begins its culture life span as fully differentiated, and in which there is no dichotomy between proliferation and expression of differentiated function (28).

Vitamin E, selenium, and mitochondrial function

The early alterations in differentiated functions occurring in bovine adrenocortical cells in culture encouraged us to examine "household" cellular functions that might change as a function of population doubling level and possibly as a function of oxidative damage. The finding that loss of the mitochondrial 11β-hydroxylase enzyme involved oxidative damage suggested that it would be worthwhile to examine other mitochondrial functions. Oxidation of tricarboxylic acid (TCA) cycle substrates was examined (29). Pyruvate was used because it is the major mitochondrial substrate in vivo (56) and glutamine was used because it is extensively oxidized by cells in culture, often in preference to other TCA cycle substrates (37, 32). The ratio of oxidation of the two substrates was examined as a function of the population doubling level in culture (Fig. 3). The ratio was shown to rise over the early portion of the life span reaching a new, steady value which was maintained to the end of the life span. The change in this ratio was accompanied by an increasing susceptibility to the toxic effects of an inhibitor of aminotransferases, aminooxyacetate (Fig. 3). In adrenocortical cells, as in some other cell types, glutamate from glutamine is oxidized in mitochondria by a pathway dependent on aminotransferase activity (50). Aminooxyacetate was shown to specifically inhibit oxidation of glutamine (29) and this

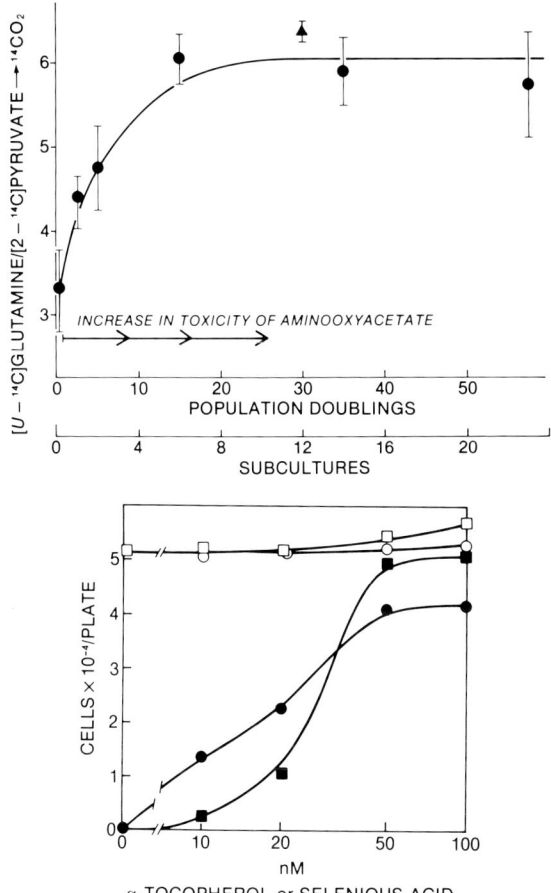

FIG. 3 (top). Change in the glutamine/pyruvate oxidation ratio as a function of population doublings in bovine adrenocortical cells. The measurement and significance of this ratio are described in detail in refs. 29 and 23. In the early portion of the life span, over the period indicated by the arrows on this figure, the toxic effect of aminooxyacetate increases with the rise in the ratio. (From ref. 29).

FIG. 4 (bottom). Prevention of toxicity of aminooxyacetate in mid-life span bovine adrenocortical cells with vitamin E or selenium. In the absence of added vitamin E or selenium, almost all cells were killed by 72 hours exposure to 2 mM aminooxyacetate. In the presence of the indicated concentrations of vitamin E or selenium, cell death was partially or almost completely prevented. Solid symbols, cells treated with aminooxyacetate; open symbols, control cells; squares, addition of vitamin E; circles, addition of selenium. (From ref. 23).

presumably accounts for its toxic effects. The involvement of oxidative damage in the change in the glutamine/pyruvate oxidation ratio and the increasing toxicity of aminooxyacetate was indicated by the increase in the ratio and in aminooxyacetate resulting from the 11β-hydroxylase/steroid pseudosubstrate interaction (29). Apparently, the oxidative damage resulting from this interaction spreads from its site of origin to affect other components of the inner mitochondrial membrane. We have hypothesized that the component affected is ubiquinone; this is discussed in detail elsewhere (23).

In a survey of antioxidant compounds that might prevent the toxic effects of aminooxyacetate, it was found that vitamin E and selenite were both effective at very low concentrations in completely preventing cell death caused by treatment with aminooxyacetate (23) (Fig. 4). A variety of synthetic antioxidants were also completely effective at preventing aminooxyacetate-induced cell death, but 2- to 1000-fold higher concentrations were required. Compounds were effective in the order of their potency as antioxidants, i.e. as reactants for peroxy radicals (23; see ref. 1). This is in contrast to other data on the prevention of peroxidative damage to 11β-hydroxylase with antioxidants. In this case, small phenolic compounds such as BHA (butylated hydroxyanisole) were much more effective (22). Subsequently this has been shown to be most likely the result of the fact that these phenols may bind at or close to the active site of the cytochrome P-450 enzyme involved, and so have privileged access to destructive radicals (Hornsby, Aldern, and Harris, unpublished observations). For overall antioxidant protection of the cell, it appears that the biological antioxidants, vitamin E, and selenium as glutathione peroxidase, are superior to any of the synthetic antioxidants (25). However, the fact that synthetic antioxidants could completely substitute for vitamin E in the prevention of aminooxyacetate toxicity indicates that these compounds are not restricted from access to the inner mitochondrial membrane, if the actual target for damage is ubiquinone, as we have hypothesized (23).

Deficiency of vitamin E and selenium in cultured adrenocortical cells: Effects on long-term growth rate

The experiments with aminooxyacetate suggested that adrenocortical cells were essentially completely deficient in vitamin E and selenium after long-term growth in culture. The rise in glutamine/pyruvate oxidation and increasing sensitivity to aminooxyacetate toxicity observed over the early phase of the life span could indicate development of deficiencies of vitamin E and selenium, assuming that cells are not deficient in these nutrients on isolation from the animal. Presumably, the medium used and the particular lot of fetal bovine serum used did not supply adequate bioavailable amounts of either nutrient. This was of interest in that although vitamin E has not normally been considered to be an essential growth factor for mammalian cells in culture (18), selenium has been shown to be an essential factor for growth and survival in serum-free media (38, 49). It was possible that either trace amounts of the nutrients were being supplied by the medium (but not sufficient for protection against aminooxyacetate toxicity) or else that these biological antioxidants were not essential in the presence of other components of serum. This was tested by examining growth of adrenocortical cells over 25 days in medium with 10% untreated fetal bovine serum or 10% fetal bovine serum that had been depleted of residual vitamin E and dialyzable selenium compounds by extraction with diisopropyl ether and butanol followed by extensive dialysis (3) (Fig. 5). The serum lot was that used in the experiments on aminooxyacetate described above. Fig. 5 shows that long-term growth was little affected either by the readdition of selenium and vitamin E to untreated serum or to extracted, dialyzed serum. The maximal increase in growth rate was about 10%. Some decrease (about 30%) in the growth rate was observed in cells in the extracted, dialyzed serum, but this was not corrected by the addition of either vitamin E or selenium. Although fetal bovine serum is usually reported to contain substantial amounts of selenium by elemental analysis (43, 49), certain lots of serum may be deficient in selenium, since selenium deficiency is common in cattle in certain geographical areas (53). Additionally, much of the selenium in serum detected by elemental analysis may be in a protein-bound form (46), of uncertain bioavailability to cells in culture. Selenium present as selenite, known to be bioavailable to cells in culture (38), is removed during the dialysis procedure. Since it has been demonstrated that selenite has a marked effect on adrenocortical cell growth under serum-free, defined culture conditions (49), it is likely that other factors in serum are able to compensate for the effects of deficiencies in selenium and vitamin E on growth rate.

Effect of added selenium on glutathione peroxidase activity

Although the dialysis procedure used in the previous experiment removes low molecular weight selenium compounds, it is still possible that selenium may be obtained from macromolecular sources in the serum. To assess the bioavailable selenium in medium (with or without serum) it is necessary to assay the activity of a selenium-dependent cell function. The only known selenium-dependent enzyme in animal cells is glutathione peroxidase (13, 46). Thus, selenium available to cultured cells may be assessed by their glutathione peroxidase activity. This method is used in whole animal studies as the most reliable estimate of selenium bioavailability (35). Using this method, we have shown that the lot of serum used in the previous experiment is essentially completely deficient in bioavailable selenium, even before dialysis, thus confirming that adrenocortical cells can grow for long periods under conditions of selenium deficiency. Bovine adrenocortical cells grown to the third passage in medium containing 10% of this fetal bovine serum lot had only 18% of the activity of fresh bovine adrenocortical tissue (Table 1).

TABLE 1. Glutathione peroxidase activity in adrenocortical tissue before and after growth in culture[a]

	Glutathione peroxidase activity (U/mg protein)
Fresh adrenocortical tissue	0.045±0.008
Third passage adrenocortical cells	0.008±0.001
After 72 h exposure to 20 nM Se	0.063±0.007
After 72 h exposure to serum-free medium	0.007±0.002
After 72 h exposure to serum-free medium with 20 nM Se	0.050±0.011

[a]Glutathione peroxidase activity was determined in homogenized fresh adrenocortical tissue or cultured adrenocortical cells that had been grown for three passages in culture in 10% fetal bovine serum, followed by the indicated treatments. (From Hornsby, Pearson, Aldern, Harris, and Autor, submitted for publication).

Readdition of 20 nM selenite increased glutathione peroxidase activity over a period of 48-72 hours to a level comparable to that in fresh adrenocortical tissue (Fig. 6). Similar results were obtained when cells were treated with selenium in the serum-free defined medium formulated for adrenocortical cells (49), rather than serum-containing medium (Table 1). The present experiments using cultured cells confirm experiments in vivo demonstrating increases in glutathione peroxidase activity on administering selenium (34, 42, 51, 53). Effects of selenium on glutathione peroxidase activity in cultured mammalian cells have not been extensively investigated. It has been reported that selenium supplementation of mouse mammary cells in culture increased glutathione peroxidase specific activity, but only at 5 μM, and not at 50 nM (39). The concentrations of selenium used here to increase glutathione peroxidase activity and to protect against the toxicity of cumene hydroperoxide (see below) are similar to those reported to be necessary for growth and survival under serum-free conditions (38). The time-course of increase in glutathione peroxidase implies synthesis of new glutathione peroxidase molecules, rather than addition of selenium to apoenzyme molecules. This is consistent with recent reports on the mechanism of synthesis of this enzyme (20, 54).

Effect of added selenium and vitamin E on toxicity of cumene hydroperoxide

The increase in glutathione peroxidase occurring on supplementation of adrenocortical cells with selenium was shown to have functional significance by demonstrating that the cells

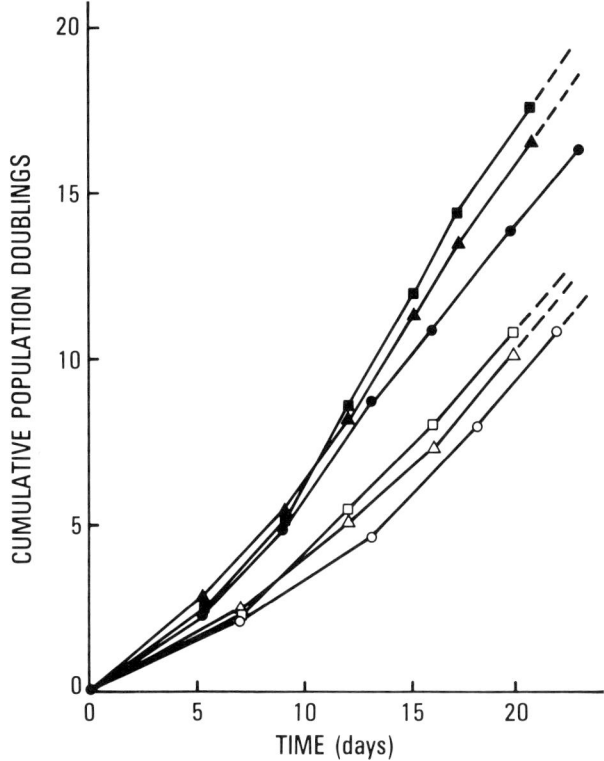

FIG. 5. Effect of selenium and vitamin E on bovine adrenocortical cell growth in culture. Primary cultures of bovine adrenocortical cells were grown in untreated (solid symbols) or ether-extracted, dialyzed (open symbols) 10% fetal bovine serum. Cells were subcultured using a 1:5 split ratio and cumulative population doublings were calculated as described in the text. Cultures were supplemented with 20 μM selenite (triangles), or 1 μM dl-α-tocopherol (squares), or received no supplementation (circles). (From Hornsby, Pearson, Aldern, Harris, and Autor, submitted for publication).

acquired resistance to the toxic effects of an added peroxide, cumene hydroperoxide, which is a substrate for the enzyme (47). Cell death was measured by the reduction in cloning efficiency after 24 h exposure to cumene hydroperoxide. Fig. 7 shows that on readdition of selenium, resistance of the cells to cell death from hydroperoxide treatment increased with approximately the same time-course as that for restoration of glutathione peroxidase activity shown in Fig. 6, thus providing circumstantial evidence that glutathione peroxidase was the protective factor involved.

The selenium-containing medium was removed at the time of addition of cumene hydroperoxide, and no added selenium was present during clonal growth. No protection was conferred by selenium added at the time of addition of the cumene hydroperoxide or added during clonal growth after peroxide treatment. Vitamin E supplementation as dl-α-tocopherol provided similar protection against cumene hydroperoxide. However, no time lag in its action was observed, in contrast to supplementation with selenium.

The general conclusions that may be drawn from this study are that neither vitamin E nor selenium is an essential nutrient for adrenocortical cell growth in culture, at least in the presence of other serum components. The deficiency becomes apparent, however, when the cells are challenged with a toxic peroxide, and there may be certain other culture conditions which are suboptimal for health of mammalian cells, such as serum-free medium or when glutamine oxidation is blocked with aminooxyacetate, where the deficiencies become critical for continued cell growth and survival.

Vitamin E, selenium and cell culture life span

Although deficiencies of vitamin E and selenium appeared compatible with normal cell growth and survival, effects of these deficiencies on replicative potential (life span) were still possible. Since the experiments using cumene hydroperoxide indicate a lowered level of protection against oxidative damage in deficient cells, an effect on life span might be observed if the finite life span phenomenon were associated with oxidative damage to the cell. We carried out an experiment to test this hypothesis. Bovine adrenocortical cells were passaged continuously, starting at the primary culture, in 10% fetal bovine serum of the same lot used in the experiments reported above, now established to be deficient in bioavailable selenium and vitamin E. This is the same lot used in our previous experiments on the replicative life span of bovine adrenocortical cells (27), experiments now known to have been performed under conditions of selenium and vitamin E deficiency. Some cultures were supplemented from the outset with 1 μM vitamin E (dl-α-tocopherol) and some with 20 nM selenite, concentrations found to provide optimal protection against peroxide toxicity. Although conditions for this experiment were substantially the same as in our previously published work, two differences were (a) the use of fibronectin coating of the dishes, which we have shown increases the growth rate of adrenocortical cells in culture (31, 49), and (b) the use of bacterial protease for subculturing rather than trypsin; this increases plating efficiency and reduces the lag before resumption of growth after replating (29).

The results of this experiment, as of the date of writing, are shown in Fig. 8. For the first 70 population doublings, little difference among the different lines was noted, with the exception that a slightly increased growth rate was seen in cultures treated with vitamin E or with selenium. The difference in growth rates is

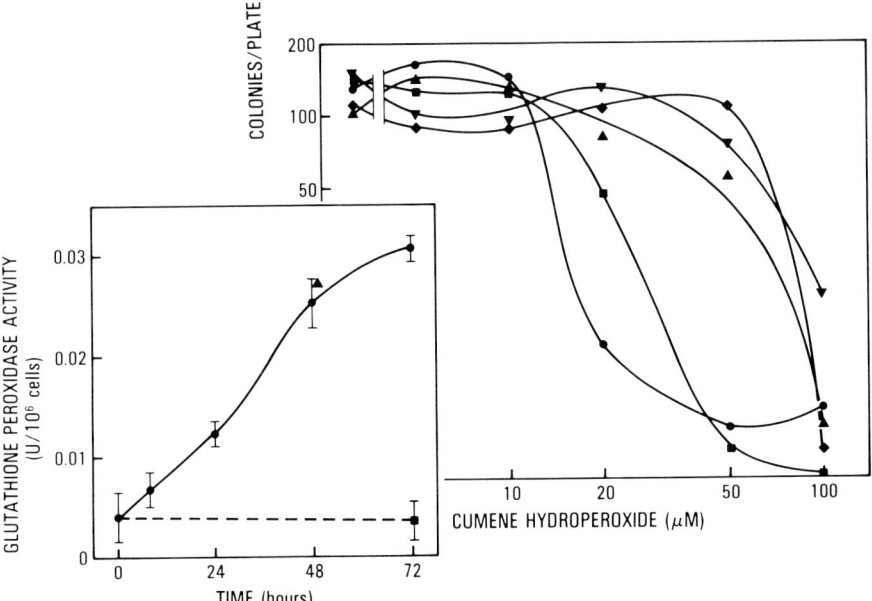

FIG. 6 (left). Increase in glutathione peroxidase activity on addition of selenium to cultured bovine adrenocortical cells. Third passage bovine adrenocortical cells were grown in 10% untreated fetal bovine serum with 20 nM selenite (circles), 200 nM selenite (triangle), or no added selenium (square). At the indicated times cells were harvested for assay of glutathione peroxidase activity. (From Hornsby, Pearson, Aldern, Harris, and Autor, submitted for publication).

FIG. 7 (right). Cell killing by cumene hydroperoxide before and after treatment with selenium. Third passage bovine adrenocortical cells were grown in 10% untreated fetal bovine serum. The medium was changed with the addition of the indicated concentration of cumene hydroperoxide. After 24 hours, the peroxide-containing medium was removed and cells were subcultured for estimation of cloning efficiency as a measure of cell killing. Cloning efficiency is shown as colonies formed per plate. Additionally, prior to addition of cumene hydroperoxide, cells were exposed to 20 nM selenite for 24 hours (squares), 48 hours (triangles), or 96 hours (inverted triangles); or were exposed to 1 µM dl-α-tocopherol for 1 hour (diamonds); or received no additional treatment (circles). Selenium and vitamin E were not added to the medium during cumene hydroperoxide treatment. (From Hornsby, Pearson, Aldern, Harris, and Autor, submitted for publication).

primarily noticeable at the beginning of the growth period, as more clearly shown in Fig. 5; at later time points the growth curves were approximately parallel. At about the 70th population doubling (28th passage with the 1:5 split ratio used) the curves began to diverge. At the time of writing, the control cells have undergone 105 doublings (41 passages); the vitamin E-treated cells have undergone 120 doublings (49 passages); and the selenium-treated cells have undergone 115 doublings (46 passages). All three lines have a time between subcultures of 7 days and do not reach as high a saturation density as earlier passages. Although this experiment is not currently complete, it does appear that there is a slight extension of culture life span with both vitamin E and selenium supplementation. However, since finite cell lines may be clonal at late passages (57), it is not clear that the differences in life span in this experiment are not due simply to clonal variation. The total life span in controls is somewhat higher than that previously reported (27). This may be the result of the use of fibronectin coating of the culture dishes and the use of bacterial protease for subculture instead of trypsin.

Effect of FGF on culture life span

The significance of any extension of life span by vitamin E and selenium must be viewed in the context of extension of life span by other factors, principally growth factors such as FGF. We have previously demonstrated that in the absence of FGF long-term growth of bovine and human adrenocortical cells is poor and clonal growth or growth from low densities does not occur (17); and that cells that have been grown to mid- or late-life span cease growth within a few doublings when FGF is withdrawn (48). Thus, apparently, there are major effects on life span of factors such as FGF, as has been reported for other cell types such as granulosa cells (16). An experiment to test this directly is shown in Fig. 9. Primary bovine adrenocortical cells were grown and sequentially subcultured in the presence or absence of FGF (in the absence of vitamin E and selenium). The rate of proliferation was much less in cells not grown in the presence of FGF, and by the 5th passage (12 population doublings) the time between subcultures had already lengthened to 7 days, the same as for cells at the 100th population doubling when grown with FGF. At this stage, the cells appear senescent and reach only low saturation densities. Whether such cells are truly senescent and can no longer respond to the addition of FGF with increased growth will be determined in future

---

FIG. 8 (opposite page). Effects of vitamin E and selenium on replicative life span in cultured bovine adrenocortical cells. Upper left, control cells; upper right, cells supplemented with 1 µM dl-α-tocopherol; lower left cells supplemented with 20 nM selenite. The lower right diagram shows the same data plotted together. Solid line, control; dashed line, vitamin E; dotted line, selenium.

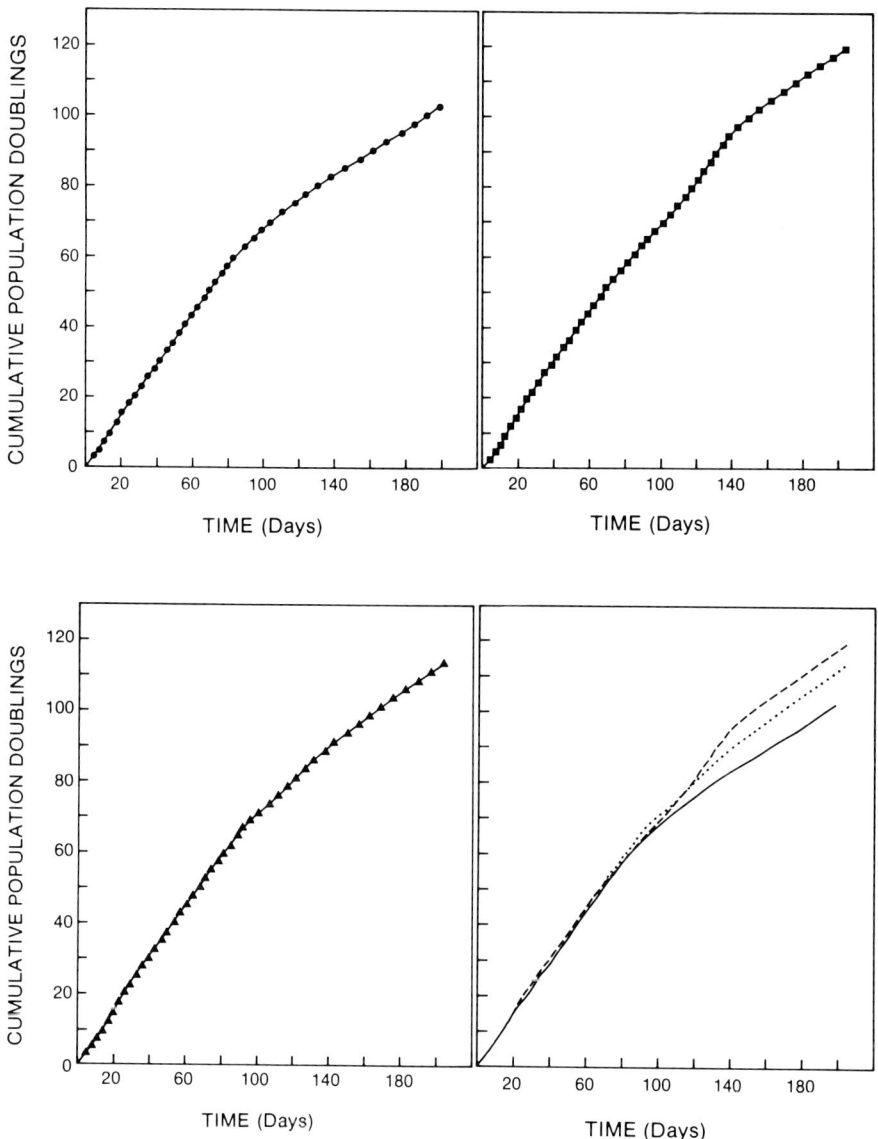

FIG. 8

experiments. It has been reported that growth factors had no effect on the life span of keratinocytes when cumulative population doublings were calculated by a method that measured the number of cells forming part of the proliferating population after each subculture (10). A similar calculation method was used in these experiments. The increase in cell number over one passage was calculated as

$$\frac{[\text{Cell number at confluence}]}{[\text{Cells distributed into plate}] - [\text{Cells unattached at medium change}]}$$

Thus, cells that may fail to attach to the dish and not form part of the growing population after subculture were taken into account. Nevertheless, a large effect on life span of FGF supplementation is apparent from Fig. 9.

Summary and Conclusions

These experiments do not support the concept that the finite culture life span of normal diploid adrenocortical cells in culture results from oxidative damage, or, at least, not from oxidative damage that can be prevented by the major biological antioxidant nutrients, vitamin E and selenium. This conclusion is in agreement with those from previously published studies (5, 21). However, differences from previous investigations are that (a) the selenium status of the cells was taken into account in these studies; and (b) experiments were performed to demonstrate that cells were deficient in both vitamin E and selenium, that these nutrients were supplied to the cells in a bioavailable form, and that protection against oxidative damage was increased by supplementation with these substances. Lack of bioavailable selenium was confirmed by demonstrating very low levels of glutathione peroxidase. Adrenocortical cells can survive and grow over their entire culture life span in a condition of deficiency of both vitamin E and selenium. It may be that either standard culture conditions with 19% oxygen in the gas phase do not produce oxidative stress in adrenocortical cells, or that the cells can survive and grow under conditions of chronic oxidative stress. The antioxidant deficiency becomes apparent when oxidative stress is increased with addition of a hydroperoxide or when glutamine oxidation is blocked with aminooxyacetate. Additionally, oxidative damage has effects on the differentiated functions of adrenocortical cells, in particular the mitochondrial 11β-hydroxylase enzyme. However, this oxidative damage differs in that it is observable in the primary culture as well as in later passages, and that it may be prevented by some synthetic antioxidants, but not by vitamin E or selenium. In the primary culture, the cells are probably not deficient in vitamin E and selenium. Oxidative damage to this enzyme may occur in an environment to which the major cellular biological antioxidants do not have access.

The major determinant of life span in adrenocortical cells is the supply of powerful growth factors such as FGF. The small

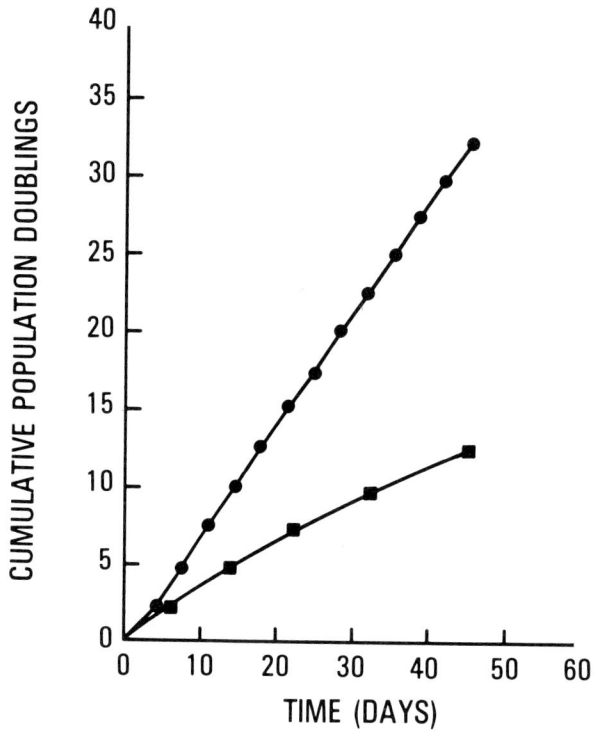

FIG. 9. Effect of FGF on long-term growth of bovine adrenocortical cells. Circles, cells grown with FGF; squares, no FGF.

effects on life span of supplementation with vitamin E or selenium may result from the slight increases in growth rate that are noted on supplementation with these nutrients. If oxidative damage is a feature of the aging process in vivo, and cell cultures are to be used as models of this feature of the aging process, it may be necessary to study processes in culture other than the finite replicative life span phenomenon. We have shown that, regardless of an effect of oxidative damage on life span, there are effects of oxidative damage in cultured adrenocortical cells on cytochrome P-450 hydroxylases, mitochondrial energy metabolism, and cell survival in the presence of hydroperoxides. The validity of culture systems as models for in vivo aging depends on the recognition of features of the in vivo aging process that may be studied in culture, and the development of suitable model systems for their investigation.

Acknowledgments

This work was supported by NIH research grants AG 00936 and CA 32468 to P.J.H.

References

1. Burton, G.W., and Ingold, K.U. (1981): J. Am. Chem. Soc. 103:6472-6477.
2. Burton, G.W., Le Page, Y., Gabe, E.J., and Ingold, K.U. (1980): J. Am. Chem. Soc. 102:7791-7792.
3. Cham, B.E., and Knowles, B.R. (1976): J. Lipid Res. 17:176-181.
4. Cornelis, R., and Versieck, J. (1980): In: Trace Element Analytical Chemistry in Medicine and Biology, edited by Bratter, P., and Schramel, P., pp. 587-591. Walter de Gruyter, Berlin.
5. Cristofalo, V.J., and Balin, A.K. (1981): In: Aging: A Challenge to Science and Society, Vol. 1, edited by Danon, D., Shock, N.W., and Marois, M., pp. 99-124. Oxford University Press, Oxford.
6. Cristofalo, V.J., and Stanulis-Praeger, B.M. (1982): Adv. Cell Culture 2:1-68.
7. Crivello, J.F., Hornsby, P.J., and Gill, G.N. (1982): Endocrinology 111:469-479.
8. Crivello, J.F., Hornsby, P.J., and Gill, G.N. (1983): Endocrinology 113:235-242.
9. Cutler, R.G. (1982): In: Testing the Theories of Aging, edited by Adelman, R.C., and Roth, G.S., pp. 25-114. CRC Press, Boca Raton.
10. Didinsky, J.B., and Rheinwald, J.G. (1981): J. Cell. Physiol. 109:171-179.
11. Draper, H.H. (1980): In: Vitamin E. A Comprehensive Treatise, edited by Machlin, L.J., pp. 272-300. Marcel Dekker, New York.
12. Finkelstein, M., and Shaefer, J.M. (1979): Physiol. Rev. 59:353-406.
13. Flohe, L. (1982): In: Lipid Peroxides in Biology and Medicine, edited by Yagi, K., pp. 149-159. Academic Press, New York.
14. Funkenstein, B., McCarthy, J.L., Dus, K.M., Simpson, E.R., and Waterman, M.R. (1983): J. Biol. Chem. 258:9398-9405.
15. Gill, G.N., Crivello, J.F., Hornsby, P.J., and Simonian, M.H. (1982): Cold Spring Harbor Conf. Cell Proliferation 9:461-482.
16. Gospodarowicz, D., and Bialecki, H. (1978): Endocrinology 103:854-865.
17. Gospodarowicz, D., Delgado, D., and Vlodavsky, I. (1980): Proc. Natl. Acad. Sci. U.S.A. 77:4094-4098.
18. Ham, R.G., and McKeehan, W.L. (1979): Meth. Enzymol. 58:44-93.
19. Harman, D. (1982): In: Free Radicals in Biology, Vol. 5, edited by Pryor, W.A., pp. 255-275. Academic Press, New York.
20. Hawkes, W.C., and Tappel, A.L. (1983): Biochim. Biophys. Acta 739:225-234.

21. Hayflick, L. (1980): In: <u>Annual Review of Gerontology and Geriatrics, Vol. 1</u>, edited by Eisdorfer, C., Besdine, R.W., Birren, J.E., Cristofalo, V., Lawton, M.P., Maddox, G.L., and Starr, B.D., pp. 26-67. Springer, New York.
22. Hornsby, P.J. (1980): <u>J. Biol. Chem.</u> 255:4020-4027.
23. Hornsby, P.J. (1982a): <u>J. Cell. Physiol.</u> 112:207-216.
24. Hornsby, P.J. (1982b): <u>Endocrinology</u> 111:1092-1101.
25. Hornsby, P.J., and Crivello, J.F. (1983): <u>Molec. Cell. Endocrinol.</u> 30:1-20 and 123-147.
26. Hornsby, P.J., and Gill, G.N. (1977): <u>J. Clin. Invest.</u> 60:342-352.
27. Hornsby, P.J., and Gill, G.N. (1978): <u>Endocrinology</u> 102:926-936.
28. Hornsby, P.J., and Gill, G.N. (1980): <u>Science</u> 208:1481-1483.
29. Hornsby, P.J., and Gill, G.N. (1981): <u>J. Cell. Physiol.</u> 109:111-120.
30. Hornsby, P.J., Simonian, M.H., and Gill, G.N. (1979): <u>Int. Rev. Cytol. Suppl. 10</u> :131-162.
31. Hornsby, P.J., Sturek, M., Harris, S.E., and Simonian, M.H. (1984): <u>In Vitro</u> 20: (in press).
32. Kovacevic, Z., and McGivan, J.D. (1983): <u>Physiol. Rev.</u> 63:547-605.
33. Kovanen, P.T., Faust, J.R., Brown, M.S. and Goldstein, J.L. (1979): <u>Endocrinology</u> 104:599-609.
34. Levander, O.A. (1982): <u>Annals N.Y. Acad. Sci.</u> 393:70-80.
35. Levander, O.A. (1983): <u>Federation Proc.</u> 42:1721-1725.
36. Macieira-Coelho, A. (1981): In: <u>Aging: A Challenge to Science and Society, Vol. 1</u>, edited by Danon, D., Shock, N.W., and Marois, M., pp. 136-160. Oxford University Press, Oxford.
37. McKeehan, W.L. (1982): <u>Cell Biol. Internatl. Rep.</u> 6:635-650.
38. McKeehan, W.L., Hamilton, W.G., and Ham, R.C. (1976): <u>Proc. Natl. Acad. Sci. U.S.A.</u> 73:2023-2027.
39. Medina, D., Lane, H.W., and Tracey, C.M. (1983): <u>Cancer Res.</u> 43:2460s-2464s.
40. Packer, L. (1976): In: <u>Aging, Carcinogenesis and Radiation Biology: The Role of Nucleic Acid Addition Reactions</u>, edited by Smith, K.C., pp. 519-535.
41. Packer, L., and Fuehr, K. (1977): <u>Nature</u> 267:423-425.
42. Paynter, D.I. (1979): <u>Aust. J. Agric. Res.</u> 30:695-702.
43. Price, P.J., and Gregory, E.A. (1982): <u>In Vitro</u> 18:576-584.
44. Sakagami, H., and Yamada, M.-A. (1977): <u>Cell Struct. Function</u> 2:219-227.
45. Schneider, E.L., and Smith, J.R. (1981): <u>Internatl. Rev. Cytol.</u> 69:261-.
46. Shamberger, R.J. (1983): <u>Biochemistry of Selenium.</u> Plenum Press, New York.
47. Sies, H., Wendel, A., and Bors, W. (1982): In: <u>Metabolic Basis of Detoxication</u>, edited by , pp. 307-320. Academic Press, New York.

48. Simonian, M.H., Hornsby, P.J., Ill, C.R., O'Hare, M.J., and Gill, G.N. (1979): Endocrinology 105:99-108.
49. Simonian, M.H., White, M.L., and Gill, G.N. (1982): Endocrinology 111:919-927.
50. Simpson, E.R., and Boyd, G.S. (1971): Eur. J. Biochem. 22:489-499.
51. Tappel, A.L. (1980): In: Free Radicals in Biology, Vol. 4, edited by Pryor, W.A., pp. 1-50. Academic Press, New York.
52. Taylor, W.G., Camalier, R.F., and Sanford, K.K. (1978): J. Cell. Physiol. 95:33-40.
53. Thompson, K.G., Fraser, A.J., Harrop, B.M., and Kirk, J.A. (1980): Res. Vet. Sci. 28:321-324.
54. Voigt, J., and Autor, A.P. (1983): Fed. Proc. 42:806.
55. Walton, J. (1982): Mech. Ageing Develop. 19:217-244.
56. Williamson, J.R., and Cooper, R.H. (1980): FEBS Lett. Suppl. 117:K73-K85.
57. Zavala, C., Herner, G., and Fialkow, P.J. (1978): Exp. Cell Res. 117:137-144.

Effects of Antioxidants on Neuronal Lipofuscin Pigment

Kalidas Nandy

Geriatric Research, Education and Clinical Center, Edith Nourse Rogers Memorial Veterans Hospital, Bedford, Massachusetts 01730; and Department of Anatomy and Neurology, Boston University Medical School, Boston, Massachusetts, 02118

The term "lipofuscin" was used in the literature by Borst (4), although the pigment was first demonstrated by Hannover (21). The name was derived from a Greek word "lipo" (meaning fat) and a latin word "fuscus" (meaning brown). One of the consistent cytological changes associated with aging is the depositon of lipofuscin in the cytoplasm of the neurons, although it is found in other non-dividing cells such as myocardium and skeletal muscle. Various investigators demonstrated the pigment and used different names, such as chromolipid, age pigment, and ceroid (3, 11, 43, 53,). The genesis of lipofuscin has been investigated by a number of workers but the results are generally inconclusive (1, 5, 9, 15, 18, 19, 35, 64).

PROPERTIES OF LIPOFUSCIN

Lipofuscin pigment formation is a continuous process which starts early in life and increases progressively as a function of age. The distribution of the pigment in the neurons of CNS in various animals including man have been studied in different laboratories using histochemical, fluorescence, and ultrastructural methods. The neurons in the CNS exhibit a variable pattern of pigmentation in their cytoplasm. Lipofuscin pigment is generally not detected in the neurons of very young animals. At first the pigment is rather diffuse in its distribution and later gradually develops clumps, perinuclear or polar, as age advances. The percentages of nerve cells showing both heavy (clumpy) and moderate (diffuse) pigmentation among the Purkinje cells of cerebellum and pyramidal cells of cerebral cortex and hippocampus increase progressively with age.

The staining properties of the pigment also appear to vary in different ages. While the pigment in young animals is stained more easily with Sudan black B and PAS, it is more easily stained by Nile blue and ferricferricyanide methods in the old. A difference in the fluorescence properties is also noted in these two types of pigment. Upon activation at wave length of 3650 A° the pigment in younger animals gives mostly a green-yellow fluorescence with an emission spectrum of 4,000-4,600 A° and an intensity peak at 4,400 A°. The autofluorescence of the lipofuscin in older animals is mostly orange-yellow with an emission spectrum of 4,200 - 5,000 A° and a peak at 4,800A°. The pigment in the young animals and not in the old are partially dissolved by trypsin, lipase, and lipid solvents (35, 36).

ANTIOXIDANTS

The metabolic roles of various antioxidants in the animal tissues have been the subject of numerous investigations but the results are quite controversial (66). Most important of the antioxidants in animals is vitamin E which was discovered by Evans and Bishop (16). A number of diseases are associated with vitamin E deficiency in animals (27, 39) and these include testicular degeneration, muscular dystrophy and hepatic necrosis and encephalomalacia (20). The appearance of darkish pigment in the myometrium, sex glands and the brain has been linked to vitamin E deficiency in rats (12, 31). Various investigators demonstrated that vitamin E reacts with free radicals formed by the interaction of polyunsaturated fatty acids with molecular oxygen, and these, in turn, produce a chain reaction producing lipid peroxides. The fluorescent end product of the lipid peroxidation is lipofuscin age pigment (58, 60, 61). Several investigations also provide evidence of lipid peroxidation reactions in vivo. These studies demonstrated a direct correlation between lipid peroxidation products (lipofuscin and ceroid) and thiobarbituric acid reactant (malonaldehyde). According to the antioxidant theory Vitamin E stops chain reactions of lipid peroxidation by destroying free radicals formed by the interaction of polyunsaturated fatty acids and oxygen (22). In addition, structurally dissimilar lipid antioxidants have successfully replaced vitamin E in the animal diet (62). The shiff base product of malonaldehyde cross-linked with primary amino groups of proteins, nucleic acids and their bases or phospholipid has fluorescence characteristics similar to lipofuscin. Although the presence of malonaldehyde as an indicator of lipid peroxidation in tissues in vivo has not been conclusively established, the method has been employed quite frequently to correlate lipid peroxidation and vitamin E (2, 38, 57).

Numerous studies have indicated that lipofuscin formation in various tissues is significantly increased in animals on vitamin E deficient diet (8, 12, 30, 31, 32, 46, 55, 63). On the other hand, Porta et al (45) used 15% coconut oil or 15% safflower oil or combination of the two and 2mg% or 200mg% vitamin E as diet to study cytoplasmic volume of lipofuscin in

TABLE 1

Effects of vitamin E deficient and excess diets on neuronal lipofuscin in the pyramidal cells of hippocampus and frontal cortex and Purkinje cells of the cerebellum in C57BL/6 mice

Mos. on diet (Age of mice)	Control			Vitamin E Deficient Diet			Vitamin E excess		
	H.C.	F.C.	P.C.	H.C.	F.C.	P.C.	H.C.	F.C.	P.C.
3 mos. (6 mos.)	8.2 ±3.1	6.2 ±3.0	3.0 ±1.2	12.7 ±4.1	13.1 ±6.2	9.5 ±4.0	6.3 ±4.4	8.4 ±9.0	4.6 ±4.6
6 mos. (9 mos.)	20.6 ±6.0	16.3 ±4.2	7.1 ±2.6	36.5 ±10.2	30.1 ±9.3	16.7 ±6.7	16.6 ±6.1	15.1 ±6.9	5.2 ±3.1
12 mos. (15 mos.)	38.6 ±6.2	40.7 ±7.2	22.0 ±4.5	52.6 ±16.4	56.3 ±17.7	30.7 ±14.5	24.3 ±9.1	26.2 ±10.3	7.3 ±3.9
24 mos. (27 mos.)	47.4 ±10.4	55.9 ±9.2	32.2 ±7.2	74.0 ±26.4	76.9 ±29.3	43.5 ±16.2	39.2 ±12.7	37.0 ±11.1	16.7 ±7.3

Animals were 3 mos. old at the onset of the experiment
Mice were subjected to Vitamin E deficient and Vitamin E excess for 3, 6, 12, & 24 mos.
The figures represent the numbers of intersections overlying pigment granules using an ocular grid in a fluorescence microscope

H.C. = Hippocampal neurons; F.C. = Frontal cortical neurons; P.C. = Purkinje cells of cerebellum

various rat tissues with inconclusive results. The effects of vitamin E excess on lipofuscin accumulation has also been the subject of several investigations. Although the results are somewhat contradictory, most studies indicate that vitamin E retards pigment formation in vivo (33, 41, 42, 46). Csallany et al (7) demonstrated a significant difference in the lipofuscin between animals on vitamin E deficient diet and highly supplemented diet. Tappel (58) suggested that vitamin E may act as a free radical scavenger and thereby stopping membrane damaging lipid peroxidation reactions. Furthermore, several investigators suggested that Vitamin E stabilizes membranes in a physio-chemical manner and thereby prevents any deteriorative action. Although the majority of the studies demonstrated a reduction of the pigment by vitamin E excess, its physiological effects have not been determined.

Since certain diseases due to vitamin E deficiency responded well with selenium treatment, the possible role of the latter as an antioxidant has been investigated (6, 50, 65). Selenium has also been shown as a component of glutathione peroxidase (49). In the presence of reduced glutathione, glutathione peroxidase catalyzes the conversion of hydrogen peroxide to water (34) and lipid peroxides to corresponding alcohols (26). It has further been suggested that vitamin E acts by preventing the formation of lipid peroxides while glutathione peroxidase destroys any peroxides that are formed (24). Tappel (62) also demonstrated that the diet containing high levels of several antioxidants including selenite reduced the accumulation of lipofuscin.

The effects of vitamin E deficient diet and vitamin E excess diet on the lipofuscin formation in the pyramidal cells of hippocampus and frontal cortex and Purkinje cells of the cerebellum in C57BL/6 mice have recently been studied in our laboratory (37). Young (3 mos) female mice were subjected to vitamin E deficient diet (Nutritional Biochemical Co.) for a period of 24 mos. Animals were sacrificed after 3, 6, 12 and 24 mos. on diet and lipofuscin pigment was quantified morphometrically under a fluorescence microscope using an ocular grid. The control mice of the same age and sex were fed ad libitum on Purina mouse chow. A progressive increase of the pigment was observed in the above neurons with age in both control and dietary groups and a significantly increased pigment accumulation was noted in the neurons in the animals on diet for 5 mos or more compared to the controls (Table 1). Young female mice (3 mos) were also subjected to vitamin E excess diet (2%) for the same period of time. The effects of vitamin E excess on the neuronal lipofuscin was less pronounced than those of the effects of E deficient diet. A significant reduction of the pigment in the pyramidal cells of the hippocampus and frontal cortex was noted in animals on diets for 12

TABLE 2

Lipofuscin pigment and malonaldehyde in the retinal pigment epithelium and frontal cortex in mice of different ages.

Age in months	Retina		Frontal Cortex	
	LP	MDA	LP	MDA
3	37.25±25.90	2.3±0.2	66.29±41.22	3.4±0.4
12	423.31±53.81	4.7±0.6	675.77±73.31	6.9 ±.9
24	1002.37±123.75	12.6±0.8	1235.17±160.32	15.7±1.5

The values on malonaldehyde are expressed as µg/mg of tissue using Thiobarbituric acid assay. The values on lipofuscin represent the number of intersections in the grid overlying the pigment particles in RPE and the neurons in the frontal cortex. The differences between age groups (3 and 12 mos; 12 and 24 mos) are statistically significant (p<.05).

LP = Lipofuscin

MDA = Malonaldehyde

months or more (Table 1). A similar study on the effects of vitamin E deficient diet was carried by Lai et al (25) on Sprague-Dawley rats. Young (45 days) male rats were fed on this diet for 14 months and controls of the same age and sex were fed ad libitum Purina rat chow. The intracellular accumulation of lipofuscin was significantly increased in the pyramidal cells of hippocampus and frontal cortex in the dietary group. Behavioral tests for short-term and long-term memory using condition avoidance response, performance of delayed-alternate response and retention of one-trial learning of aversive experience were also carried out on these animals prior to sacrifice. The dietary animals exhibited a significant impairment in learning and increased deposition of neuronal lipofuscin and an accelerated aging in these animals was suggested (25). Freund et al (17) studied the effects of vitamin E supplemental diet on the alcohol induced learning deficit (using the conditon avoidance test) as well as brain lipofuscin content by chloroform-methanol extraction method. Although vitamin E supplemental diet reduced lipofuscin contents in all groups, no improvement of the ethanol induced learning deficit was noted in these animals.

One theory of cellular aging proposed initially by Harman (23) is the process of lipid peroxidation caused by free radicals produced by either endogenous or exogenous sources. An increased lipid peroxidation in aging mammals has been demonstrated in brain and liver (44, 51). Stege et al (54) enzymatically isolated hepatocytes to evaluate levels of lipid peroxidation in young (3 months), adult (12 months) and aged (25 months) Fisher-344 female rats. Old animals exhibited an increase in lipid peroxidation and more so in the animals on antioxidant-free diet. A comparative study was carried out on the levels of lipid peroxidation and the amount of lipofuscin in the neurons of frontal cortex and in pigment epithelial cells of the retina in mice of different ages (3, 12 and 24 mos). Malonaldehyde was measured as an index of lipid peroxidation using thiobarbituric acid assay and lipofuscin was quantitated by morphometric method using a fluorescence microscope. It was rather interesting to note that both lipid peroxidation and lipofuscin increased progressively with age in both retina and frontal cortex (Table 2).

Tappel (59) proposed a synergistic redox relationship between vitamin C and E. This observation was further confirmed by Packer et al (40) who found a free radical interaction occuring between vitamin E and vitamin C in vitro. It has also been shown that vitamin C in low doses may induce lipid peroxidation in brain homogenate by enhancing free radical formation. On the other hand, at a higher concentration it may act as free radical scavenger (10). Sulkin and Sulkin (56) studied the effects of the vitamin C deficient diet on the ganglion cells in guinea pig and found an increased lipofuscin by electron microscopy (55). On the other

TABLE 3

Effects of dietary restriction on neuronal lipofuscin pigment in young and old mice

Duration of diet (age in mos)	Dietary restriction		Control animals	
	H.C.	F.C.	H.C.	F.C.
Onset of exp. (3 mos)				
6 mos (9 mos)	10.6 ±3.6	10.2 ±3.8	5.2 ±3.2	4.6 ±3.5
24 mos (27 mos)	36.3 ±12.0	29.0 ±9.2	15.1 ±3.1	14.2 ±3.0
			52.5 ±13.2	46.7 ±10.11
Onset at 24 mos.				
3 mos (27 mos)	54.6 ±16.1	50.0 ±13.3	52.5 ±13.2	46.7 ±10.1
6 mos (30 mos)	51.9 ±16.1	59.3 ±17.3	62.7 ±15.7	56.7 ±14.7

Young (3 mos) were subjected to dietary restriction for 6 and 24 mos; old mice on diet for 3 and 6 mos. The figures represent the number of intersections overlying pigment granules in the cytoplasm of the neurons using an ocular grid in a fluorescence microscope.

H.C = Pyramidal neurons of hippacampus

F.C = Pyramidal neurons of frontal cortex

hand, no significant difference in the extractable fluorescence of brain homogenates was seen in guinea pig fed on ascorbic acid deficient diet (13).

EFFECTS OF DIETARY RESTRICTION

The effects of dietary restriction on the neuronal lipofuscin pigment in the CNS have been studied in our laboratory. Young (3 mos) and old (24 mos) female C57BL/6 mice were subjected to dietary restriction by feeding 2.5 gm of Purina mouse chow with supplemental multi-vitamins as needed. Young animals were kept on the diet for a period of 24 months and were sacrificed at 6, and 24 months time points after the beginning of the experiment. Control animals consisted of mice of the same age and sex and fed ad libitum with average daily intake being 5 gm of Purina Mouse Chow. A significant difference in the lipofuscin pigment in the pyramidal neurons of frontal cortex and hippocampus was found in animals on the diet for 6 months or longer (37). On the other hand, restricted diet did not produce any significant difference in the lipofuscin content in the brain of old animals (Table 3). The dietary restriction has also been shown to be most effective in extending lifespan in young animals and is less as these grow older (47, 48). The influence of dietary restriction on lipid fluorescent products on lipofuscin was examined in brain and heart in male swiss albino mice at 3, 5, 7 and 12 months of age (14). These authors observed that lipofuscin was significantly lower in all age groups in the brain and only at 12 months of age in the heart in the dietary group as compared to the controls. The study on the effects of protein deprivation during gestation of squirrel monkeys also showed an acceleration of age pigment formation by histochemical methods in the neurons of cerebellum, spinal cord, motor cortex of developing fetuses (28, 29, 52).

SUMMARY

One of the consistent cytological changes associated with aging is the deposition of lipofuscin age pigment in the cytoplasm of the neurons. The pigment is generally visualized by fluorescence, histochemical and electron microscopic methods. The quantitation of the pigment has been done morphometrically by light microscopy and biochemically following extraction of the fluorescent materials from the brain homogenates. Lipofuscin probably represents the end product of lipid peroxidation induced by free radicals. A number of antioxidants including vitamin E, glutathione peroxidase, selenium and vitamin C have been studied, and these act by scavenging free radicals and reducing lipid peroxidation of the cell membranes. The effects of the antioxidants on the lipofuscin pigment in the neurons have been studied by a number of investigators. It appears that vitamin E deficient diet is quite effective in enhancing lipid peroxidation and increasing lipofuscin. On the other hand, vitamin E excess diet was able to reduce the pigment over a longer period of time. The dietary restriction also reduced lipofuscin in the neurons and myocardium of experimental animals but the mechanism of this change is not clear. It appears that lipofusin pigment is a very reliable marker of neuronal aging and the antioxidants probably might play an important role in regulating the aging process in various tissues, particulary the neurons.

ACKNOWLEDGEMENTS

The work was supported by Veterans Administration Research Fund as well as National Institutes of Health Grant NS-11069. I also appreciate typing assistance of Mrs. Deborah Murchie in the preparation of the manuscript.

REFERENCES

1. Barka, T., and Anderson, P.J. (1963): Histochemistry: Theory, practice and bibliography. pp 190-193. Harper and Row, Inc. Hoeber Div., New York.
2. Bieri, J.G., and Anderson, A.A. (1960): Arch. Biochem., 90:105-110.
3. Bondareff, W. (1957): J. Gerontol., 12:364-369.
4. Borst, M. (1922): Pathologische Histologie. Vogel, Leipzig.
5. Brody, H. (1960): J. Gerontol., 15:258-261.
6. Bunyan, K., Green, J. and Diplock, A.T. (1963): Brit. J. Nutr., 17:117-123.
7. Csallany, A.S., Ayaz, K.L. and Su, L.C. (1977): J. Nutr., 107:1792-1799.
8. Dam, M. and Granados, H. (1945): Acta. Physiol. Scand., 10:162-171.
9. DeDuve, C. (1963): In: Lysosomes, edited by A.V.S. deReuck and M.P. Cameron, pp. 1-31. Little, Brown and Company, Boston.
10. Dormandy, T.L. (1978): Lancet, 1:647-650.
11. Duncan, D., Nall, D., Morales, R. (1960): J. Gerontol., 15:366.
12. Einarson, L. and Ringsted, A. (1938): Effect of Chronic Vitamin E deficiency on the nervous system and skeletal musculature in adult rats, Levin and Munksgaard, Copenhagen.
13. Ellery, P.M., Hughes, R.E. and Jones, E. (1979): Exp. Gerontol., 14:49-50.
14. Enesco, H.E. and Kruk, P. (1981): Exp. Geront., 16:357-361.
15. Essner, E. and Novikoff, A.B. (1960): J. Gerontol., 15:366.
16. Evans, H.M. and Bishop, K.S. (1923): J. Metab. Res., 3:233-316.
17. Freund, G. (1979): Life Sci., 24:145-152.

18. Friede, R.L. (1962): Acta. Neuropath., 2:113-125.
19. Goldfischer, S, Villarvede, H., Forschrim, R. (1966): J. Histochem. Cytochem., 14:641-652.
20. Green, J. and Bunyan, J. (1969): Nutr. Abstr. Rev., 39:321-345.
21. Hannover, A. (1842): Videnskappsselsk Math AFth (Copenhagen) 10:1.
22. Harman, D. (1968): J. Gerontol., 23:476-482.
23. Hartman, D. (1956): J. Gerontol., 11:298-300.
24. Hoekstra, W.G. (1975): Fed. Proc., 34:2083-2089.
25. Lal, H., Pogacar, S. Daly, P.R. and Puri, S.K. (1973) In: Neurobiological aspects of maturation and aging. edited by D. Ford, Elsevier, New York.
26. Little, C. and O'Brien, P.J. (1968): Biochem. Biophys. Res. Commun., 31:145-150.
27. Majaj, A.S. Dinning, J.S., Azzam, S.A. and Darby, W.J. (1963): Amer. J. Clin. Nutr., 12:374-379.
28. Manocha, S.L. and Sharma, S.P. (1977): Acta Histochem. (Jena), 58-219-231.
29. Manocha, S.L. and Sharma, S.P. (1978): Experienta (Basel), 34:377-379.
30. Martin, A.J.P. and Moore, T. (1936): Chem. and Ind., 55:236-238.
31. Martin, A.J.P. and Moore, T. (1939): J. Hyg. (Lond.) 32:643-650.
32. Mason, K.E. and Emmel, A.F. (1945): J. Anat. Rec., 92:33-59.
33. Miquel, J. and Johnson, J.E. (1975): Gerontologist, 15:25.
34. Mills, G.C. (1957): J. Biol. Chem., 229-189:197.
35. Nandy, K. (1971): Acta Neuropath., 19:25-32.
36. Nandy, K. (1978): Senile dementia: A biomedical approach, edited by K. Nandy, Elsevier North Holland, Inc. New York.
37. Nandy, K. (in preparation) Effects of dietary restriction on neuronal lipofuscin pigment in aging mice.
38. Noguchi, T., Cantor, A.H. and Milton, L. (1973): J. Nutr., 103:1052-1511.
39. Oski, F.A. and Barness, L.A. (1967): J. Pediat., 70:211-220.
40. Packer, J.E., Slater, T.F. and Wilson, R.L. (1979): Nature (Lond.) 278:737-738.
41. Packer, L. (1976): In. K.C. Smith (ed.) Aging Carcinogens and Radiation Biology, pp. 519-535 Plenum, New York.
42. Packer, L. and Smith J.R. (1974) Proc. Nat. Acad. Sci. (Wash) 71: 4763-4767.
43. Pearse, A.G.E. (1964): Histochemistry, theoretical and applied. Little, Brown and Co., Boston, pp. 661-675.
44. Player, T.J., Mills, D.J. and Horton, A.A. (1977): Biochem. Biophys, Res. Comm., 78:1397-1403.
45. Porta, E.A., Nitta, R., Kia, L. and Joun, N.S. (1979): Fed. Proc., 38:1341-1342.

46. Reddy, K., Fletcher, B., Tapel, A. and Tappel, A.L. (1973): J. Nutr., 103-908-915.
47. Ross, M.H. (1961): J. Nutr., 25:197-210.
48. Ross, M.H. (1972): Am. J. Clin. Nutr., 24:834-838.
49. Rotruck, J.T., Pope A.L., Ganter, H.E., Swanson, A.B., Hafeman, D.G. and Hoekstra, W.G. (1973): Science, 179: 588-590.
50. Schwarz, K. and Foltz, C.M. (1957): J. Amer. Chem. Soc., 79:3292-3293.
51. Sharma, O.P. (1977): Biochem. Biophys, Res. Commun., 78:469-474.
52. Sharma, O.P. and Namocha, S.L. (1977): Mech. Age. Develop., 6:1-14.
53. Samorajaski T, Keefe J.R., Ordy, J.M. (1964): J. Gerontol., 19:262-276.
54. Stege, T.E., Mischke, B.S. and Zipperer, W.C. (1982): Experimental Gerontol., 17:273-279.
55. Sulkin, N.M. and Srivanij, P. (1960): J. Gerontol., 15:2-9.
56. Sulkin, D.F. and Sulkin, N.M. (1967): Lab. Invest., 16:142-152.
57. Takeuchi, N., Fumiko, T., Katayama, Y. and Yamamura, T. (1976): Exp. Gerontol., 11:179-185.
58. Tappel, A.L. (1965): Fed. Proc., 24:73-78.
50. Tappel, A.L. (1968) Geriatrics, 23:97-105.
60. Tappel, A.L. (1970): Amer. J. Clin. Nutr., 23:1137-1139.
61. Tappel, A.L. (1972): Ann. N.Y. Acad. Sci., 203:12-28.
62. Tappel, A., Fletcher, B. and Deamer, D. (1973): J. Gerontol., 28:415-424.
63. Weglicki, W.B., Reichel, W. and Nair, P.O. (1968): J. Gerontol., 23:469-475.
64. Whitefore, R. and Getty, R. (1966): J. Gerontol. 21:31-44.
65. Zalkin, H., Tappel, A.L. and Jordan, J.P. (1960): Arch. Biochem. 91:117-122.
66. Zuckerman, B.M. and Geist, M. (1981) In: Age Pigments, edited by R.S. Sobal, pp 283-302. Elsevier/North-Holland, Biomedical Press.

Antioxidants and Longevity

Richard G. Cutler

Gerontology Research Center, National Institute on Aging, Baltimore City Hospital, Baltimore, Maryland 21224

ANTI-AGING RESEARCH: THE FUTURE FRONT OF BIOMEDICAL RESEARCH

Much evidence now indicates that humans have evolved the biological capacity to maintain optimum vigor and health (both physiologically and mentally) to an age of about 30-40 years. Data supporting this conclusion is based on the rapid decline in physiological and mental maximum performance capacity observed in humans and the increase in onset frequency of diseases after the age of 30-40 years (19,20,23,25). Particularly striking is the rapid increase in the onset frequency of a broad spectrum of diseases such as cancer after the age of 30-40 years (43).

This 30-40-year limit on the ability of humans to maintain optimum function and resistance to disease can be understood on a population genetics and evolutionary basis. A fundamental observation in this regard is that few animals living in their natural ecological niche ever live to a chronological age where they begin to suffer seriously from the disabling effects of senescence. For instance, the mean lifespan for human populations up to about 400-500 years ago was about 30-40 years. The recent increase of mean lifespan to the present level of about 70 years was largely the result of lowered environmental hazards, a decrease in infectious diseases and better nutrition. This increase in mean lifespan resulted in an increased percent of "aged" individuals in human populations and is therefore seen as an artifact of recent civilization. We therefore do not have to search for answers as to what might be the benefit of senescence and, indeed, there appears to be no evolutionary selective advantage for an animal to live in a declining health condition due to an aging process.

A major factor accounting for the recent increase of mean lifespan of humans was a decreased incidence of infectious diseases for the younger segment of the population achieved by improved sanitation. More recent developments have further decreased infectious diseases by the use of antibiotics and various vaccines, but these have had less effect on increasing mean lifespan. Thus, the increase of mean lifespan beyond the age of 30-40 years has been largely the result of a reduction in exogenous types of pathogens. Today, much research in the biomedical sciences whose aim is presently to increase the duration of human health still takes this approach, to eliminate and/or improve resistance against exogenous-source pathologies. Although reduction of exogenous-source pathologies was remarkably successful in the past to increase mean lifespan, it should be emphasized that none of these advances had significantly increased the normal duration of optimum health. That is, the normal rate of aging has remained unchanged, and only the random component of death has been significantly reduced (33). Thus, the increased survival past the age of 30-40 years meant in reality a life growing deeper into the problems and dysfunctions associated with senescence.

These findings demonstrate that the major cause of human suffering and disease today is mainly a result of endogenous rather than exogenous causes. The general dysfunctions that have an endogenous cause and which increase with increased chronological age are collectively called the diseases of aging. A key problem now is determining if these diseases are a result of the aging process or are not related to the aging process and are just age-dependent. There is reluctance by some investigators to treat the major age-dependent diseases now accounting for most human suffering as actively being caused by the aging process. Instead, many age-dependent dysfunctions such as cancer, arthritis, or senile dementia are said to be diseases caused by viruses or some other exogenous factor and not a result of the endogenous causative factors associated with aging. If these dysfunctions were caused by an aging process, then it is felt a cure would be much less probable and accordingly the project less likely to be successful.

Recent analysis of the the major causes of death in populations of developed countries have shown that the elimination of the major killers today will result in only a surprisingly small average increase in lifespan for the population at large (14,37,41,42,61,71). For example, the elimination of all forms of cancer would on the average only increase lifespan in the United States by about two years. Similarly, the elimination of all cardiovascular and renal problems would increase average lifespan in the United States by about 10 years. It is also important to note that these additional years of lifespan would be similar to those gained in the past. That is, with these types of advances in the biomedical sciences, one is able to live deeper into old age but still not able to have a longer and healthier lifespan (23,25).

The basis of this insignificant impact on increasing the average longevity of the population is the ubiquitous nature of the aging process. Aging affects essentially all physiological processes to approximately the same extent as one grows older.

Those individuals that die early in life (about 20-50 years) usually do so from some type of specific localized dysfunction. There is no evidence that these individuals die of an accelerated aging process. On the other hand, most individuals living 70 years or more have been able to age more uniformly with no particular physiological weakness. Thus, for these older individuals, the elimination of one specific age-related dysfunction simply results in uncovering another that is now more serious. These data must be taken seriously, for they indicate that the most serious health-related problem of today is not cancer or heart disease but the aging process. Furthermore, the problems related to the aging process cannot be solved by a piecemeal approach. Clearly, what is needed are means to reduce the general effects of aging itself.

It is also frequently stated that what we need is "more life in our years" rather than "more years of life". Actually, most people would probably like to have both a healthy and a longer lifespan. However, neither is likely to be achieved without slowing down the rate of aging. There is no evidence I am aware of that suggests the possibility of being able to significantly increase only the "life in our years" without also increasing the "total years of life".

The general importance of aging as being the most important underlying cause of human diseases and dysfunction is now slowly being recognized by the biomedical community (3,50). For example, we see growing numbers of scientists in the fields of cancer, heart disease, dementia and eye dysfunctions such as cataracts being drawn by necessity into the study of the aging process. However, the major difficulty in studying the aging process has been its great complexity in affecting all biological processes of an organism. Thus, the general reluctance many investigators have in studying age-related diseases and in hoping that their particular disease or dysfunction of interest is not caused by the aging process is understandable.

On the other hand, for many of those investigators that do accept aging as being important in underlying most of the diseases humans suffer from today, the general approach to increase the duration of human health has been to discover some key process or other factor such that its repair and/or addition might result in some degree of rejuvenation or slowing down the aging rate. The rationale of this approach is that aging is largely viewed as a wearing-out process. The hope is to find wearing-out points and/or missing factors and to replace these much like one would do for a wearing-out automobile. However, there is much evidence indicating that aging is remarkably uniform in nature and that there are not likely to be key wearing-out points. Thus, a piecemeal fixing up of a few aged components does not seem likely to be able to significantly extend healthy lifespan.

The prospect to be able to significantly decrease the rate of aging in a uniform manner has therefore had a long history of discouragement. In studying this problem, I came to some conclusions that made me more optimistic that something might be done in the near future to slow down the rate of aging. These conclusions were based on a new approach to the problem of anti-aging research that utilize a comparative and evolutionary biology of mammalian species having different lifespan potentials (15,23,25). The idea was to learn of the biochemical basis determining the different aging rates of mammalian species. In particular, I was interested in learning why human is naturally the longest-lived of all mammalian species and to determine if it was possible to then apply this knowledge to further extend the duration of healthy lifespan. In other words, if it would be possible to understand the biological processes that "naturally" account for determining longevity in humans, then it might be possible to further enhance these "same" processes to further reduce the human aging rate.

What I discovered in these studies was the possibility that, in spite of the vast complexity of the human aging process, the human aging rate appears to be governed by a relatively few biochemical processes largely separate from those processes causing aging. These processes have been called "longevity determinant processes". Thus, instead of finding out what goes wrong with an organism in terms of the aging process and then to try to repair this problem, the new strategy is to learn what processes naturally govern species' longevity and then to further enhance these same processes to further reduce aging rate.

Because of the vast complexity of the aging process, any significant extension of human maintenance of optimum health and the postponement of all the many age-related diseases and dysfunctions appears to depend on the existance of these hypothetical longevity determinant processes. This chapter reviews some of the major findings suggesting that longevity determinants do exist and that antioxidants may form one class of the longevity determinant processes.

BIOLOGICAL NATURE OF AGING

The aging process can theoretically be a result of either "passive" or "active" processes (69). The "active" hypothesis postulates that aging evolved for the good of the organism or species, and that it is an integral part of the genetic developmental program. The "active" hypothesis of aging states that, just as changes in gene expression resulted in the development of an organism, so does the continued change of gene expression past the age of sexual maturation result in its aging. This hypothesis suggests the existance of specific "aging genes"

(structural or regulatory in nature) for the sole purpose of
creating aging in the organism for its own good or for the good of
the species. If such genes did not exist, aging presumably would
not occur.

No such "aging genes" have been discovered, although there
is suggestion for "death hormones" or the genetic programmed
shut-down of proper cell function (23,25). However, some
arguments against the "active" gene hypothesis of aging are (a)
there is no evidence that aging evolved. Rather, it is more
consistant with evolutionary theory that longevity evolved, where
the problem has always been preservation of life, and this has led
to the evolution of anti-aging processes, (b) few animals living
in their natural ecological niche lived to an age where a
significant degree of senescence was reached, so there would be
little incentive in terms of evolutionary selective pressure to
evolve special genes to age an organism for the purpose of
eliminating it from the population, and (c) if excess lifespan was
a problem, termination of life would not appear likely have
evolved by a slow aging process that can extend for up to one-half
of an animal's lifespan. Instead, a genetic program of death
rather then senescence would have evolved where optimum health
would be maintained throughout most of the lifespan and then
suddenly at a certain age the organism would die. Examples of this
are seen in many annual plants and insects.

If aging is not "active" in nature, then this leaves us with
the "passive" hypothesis of aging. Here, aging is considered to be
caused by the by-products of the normal essential metabolic
processes of life. These by-products normally have no significant
detrimental effect on the organism unless it lives longer than it
was evolved to live. That is, if one takes an organism from its
natural ecological niche and places it in much safer conditions
(such as in captivity) then, as the organism grows older, the aging
process would be allowed to be manifested. Another way of
defining this "passive" hypothesis of aging is to state that aging
is "pleotropic" in nature (15,23).

If aging is the result of normal by-products of metabolism,
then longevity of different species must be a result of means to
reduce the aging effects of these by-products. Thus, an important
aspect in the evolution of life itself as well as to the evolution
of lifespan potential (anti-aging process) of various organisms
must have been the various types of trade-offs that occurred
between the beneficial and long-term toxic effects of metabolic
processes and the various protective, repair and detoxifying
processes that were possible (15). In this sense, I have argued
that a study of the biological basis of longevity in different
species is likely to represent an extremely fundamental biological
science. That is, an understanding of much of the biological
processes (morphological as well as biochemical) in an organism is
likely to have been shaped by a long evolutionary history of the

selection of optimum life functions as opposed to their long-term toxic age-producing side effects (15,19). If one does not consider anti-aging processes as an important life preservation strategy in the studies of the life sciences, then important aspects of the living process are likely to be missed or misunderstood.

But what might be the endogenous by-products of metabolism most important in causing aging and the corresponding anti-aging defense processes that exist to counter their aging effects? In dealing with the mammalian species, it appears likely that most of the endogenous processes causing aging are similar in all the species, regardless of their different lifespan potentials. Indeed, a reasonable explanation of the evolutionary origin of species is based on evidence indicating that most mammalian species share a common set of structural genes and that it is the regulation of these genes in terms of when they are to be turned on or off and for how long and how much that determines their different morphological features (73). If this is the case, then it is easy to understand why the qualitative nature of aging (disease and physiological dysfunctions) are remarkably similar and independent of species' lifespan. It would simply be because all species have "similar" metabolic processes giving rise to "similar" by-products and thus "similar" causes of aging.

This possibility greatly simplifies the problem since it is also likely that those processes that govern species' longevity would also be similar in different species. In other words, species' differences in longevity may only be a matter of timing and expression of a specific set of genes found in all mammalian species (58).

What might then be some of the major metabolic by-products that cause aging? I have been able to identify two potential classes that may be the major generators of aging (19,23,25). The first class is related to by-products of differentiation and development and I have called these "developmentally-linked biosenescent processes". I will not deal any further with these and refer the readers to other publications on this topic (22,58). The second class is related to by-products of energy metabolism, and I have called these "continuously-acting biosensenescent processes" (19).

Evidence that by-products of energy metabolism are an important cause of aging is that the product of specific metabolic rate (SMR) and maximum lifespan potential (MLSP) is a constant for many different mammalian species. This product has been called lifespan energy potential (LEP)(24,26,27,70). Thus, for many species, "aging rate" is proportional to "metabolic rate". This evidence is consistant with the concept that by-products of energy metabolism cause aging. There is now much experimental evidence supporting this concept, such as the relationship of metabolic

rate to aging rate in flies and to the effects of hibernation to
longevity (66). Other evidence is that it is relatively easy to
identify toxic by-products of metabolism that could conceivably be
important causative agents in aging (26). One potentially
important class is associated with oxygen metabolism, producing a
wide spectrum of free radicals, peroxides and aldehydes. Such
by-products have been shown to have the potential to cause
age-related diseases. In turn, there is known to be a number of
defense processes against the toxic effects of oxygen metabolism
that could be important as anti-aging or longevity determinants
(26,70). These are the antioxidants that will be discussed later
in this chapter.

An important question at this point is how toxic by-products
of oxygen metabolism could cause aging (26). The common view is
"wear and tear", where these toxins such as free radicals are
thought to cause damage to cells (or extracellular structures)
that accumulate and eventually result in cell death or other types
of impairment. The problem here is of course to identify what
processes in the cell might be most sensitive to these toxins in
terms of affecting proper cell function. Because there is little
evidence of a significant age-dependent accumulation of damage in
cells, I have been concentrating on the possibility that aging
results from the slow loss of the proper differentiated state of
cells (22,23,25,58). Thus, instead of aging being a result of a
wide spread accumulation of damaged cell components, I suggest
that the basic functions in a cell not related to its
differentiated state are not impaired. Instead, what does change
that has the most far-reaching biological effect is gene
expression. This is because the most sensitive target in the cell
to the toxic effects of by-products of oxygen metabolism are
predicted to be those processes determining the differentiated
state of cells. This prediction suggests that most of the
important age-related changes that are involved in causing aging
are those changes that affect the differentiated state of the
cell. These changes could be mutagenic or epigenetic in nature.
Aging, then, is seen as a slow reversal of the developmental
process that created the organism. I have called this process
dysdifferentiation, and there are a wide number of age-related
changes that can be reinterpreted as dysdifferentiation processes
(22,23).

The dysdifferentiative nature of aging suggests that the
processes most important in governing species lifespan are those
processes involved in stabilizing the proper differented state of
cells (22,23,26). Thus, the most serious effect of the toxic
by-products of oxygen metabolism is predicted to be their ability
to change the differentiated state of cells and, in turn, the
major biological effect of antioxidants are predicted to be
stabilizers of differentiation.

BIOLOGICAL NATURE OF LONGEVITY

The working hypothesis thus stated suggests that the biological nature of aging is dysdifferentiation. Cells are thought to be naturally unstable in their ability to maintain their proper state of differentiation, and by-products of development and energy metabolism accelerate their normal dysdifferentiation process. Important to this argument is the fact that extremely low levels of mutagens known to work by way of "active oxygen" species do indeed dysdifferentiate cells (26,29,30). In fact, this may be an important mechanism leading to the formation of cancer cells and would support the concept that cancer is a normal result of the dysdifferentiation process that produces less dramatic results in other cells.

Indeed, the expression of oncogenes in cancer cells and the genetic mechanisms leading to their expression may very well be the same processes causing aging. In turn, longevity of species must be determined at least in part by processes acting to stabilize differentiated cells against the toxic effects of active oxygen species. This would explain why longer-lived species have in proportion a lower endogenous onset frequency of cancer. Some of the stabilizing factors of differentiation may be the endogenous antioxidants, where it is known that antioxidants protect against the mutagenic effects of carcinogens (3). However, it should be kept in mind that many other stabilizing processes may be even more important than antioxidants.

If the genetic processes governing aging rate is less complex than the aging process per se, then just how many different genetic processes may be involved in determining longevity? In other words, what is the genetic and biochemical complexity of these hypothetical processes determining species' longevity? As noted before, the "qualitative" similarity of aging in different species, independent of their innate aging rates, suggests that the "qualitative" nature of the causes of aging and those determining longevity are also similar. Furthermore, it appears that longevity, a species' characteristic, is also largely determined by changes in the expression of regulatory genes governing a common set of structural genes just as other morphological features are determined. The question then is, just how many of these same types of regulatory genes governing the same type of anti-aging structural genes might be involved in governing species' longevity?

An answer to this question was sought by measuring the evolutionary rate of increase in longevity during the emergence of the primate species (16,18,21). Longevity was found to increase at a surprisingly rapid rate during late hominid evolution. A comparison of this rate to the maximum rate amino acid and base substitutions could occur suggests that no more than 0.5 % of the genes in a cell could be involved in determining species' longevity (16). This result strongly supports the existance of a specific class of genes governing species' longevity (the longevity determinant hypothesis) and also supports the concept that only a few key changes in regulatory genes were involved in the evolution of longevity in the primate species.

ANTIOXIDANTS AS A CLASS OF LONGEVITY DETERMINANT PROCESSES

How does one evaluate the possibility that antioxidants may play an important role in determining species' longevity? Clearly, free radicals, peroxides and aldehydes could potentially cause aging, but this is a trivial suggestion. Potential to cause aging does not prove causality, and it is relatively easy to draw up a long list of potential causes of aging. In addition, it is also clear that endogenous antioxidants are important defense processes, but this does not automatically imply their importance in governing the aging rate of an organism. Most processes important for maintaining proper cell function are likely to have little to do with governing lifespan.

It is possible that aging is a result of a steady age-dependent loss of proper protection against the toxic by-products of oxygen metabolism. That is, antioxidant levels could be at the same concentration in all species when they are in the prime of their life and then their aging rate from this point onward would be determined by the rate this protection is lost. This possibility appears unlikely, for (a) antioxidant levels appeared to have evolved to reduce the rate of aging (not to eliminate it) to the point that aging does not seriously affect the performance of an animal before it is normally killed by causes other than those related to aging. In other words, antioxidant levels are not likely to exist in excess of that required. This would argue that tissue levels of antioxidants would correlate to a species' lifespan potential, where the levels of antioxidants would be expected to be constant throughout most of the species' lifespan, and (b) a slow life-long decline in protection would imply an evolutionary selected genetic program of aging, which we have already reasonably ruled out.

In any case, a way to check the possible role antioxidants may play in determining species' lifespan is to determine if (a) tissue levels of antioxidants correlate to species' lifespan potential and (b) if an age-dependent loss of antioxidant levels does occur, whose rate of loss would be correlated with species' aging rate. Of course, I expect that little age-related change would exist except possibly for the last one-third of lifespan, where many normal body homeostasis processes begin to fail. In this case, this would not be a "cause" but an "effect" of aging.

In addition to seeking a correlation of tissue levels of antioxidants with lifespan potential, we also seek to determine if lifespan potential is correlated with the tissue level of an antioxidant per specific metabolic rate (SMR) of this tissue. That is, species' longevity would be expected to correlate to the amount of antioxidant protection a tissue has per amount of toxic free radicals produced (26,70). If we assume that primary free radical production rate is linearly related in a positive manner to metabolic rate, then this correlation appears reasonable (70). Thus, we seek to determine if:

$$\text{MLSP} \propto \frac{\text{tissue concentration of antioxidant}}{\text{specific metabolic rate}}$$

This is the same on rearranging the above equation as:

$$(\text{MLSP})(\text{SMR}) \propto \text{antioxidant concentration}$$

but we have already defined this product as being equal to a species' lifespan energy potential (LEP) value:

$$(\text{MLSP})(\text{SMR}) = \text{LEP}$$

Thus, one of the best tests to determine if an antioxidant is important in determining species' longevity may be to prove if its LEP value is proportional to tissue levels of antioxidants. Also, because the types of antioxidant are likely to be similar in all species, then the increased protection found is predicted to be a result of an increase of concentration of the same type of enzyme or product, not a different antioxidant that is more effective in the longer-lived species. Thus, the genetic change leading to an increase of antioxidant protection would be expected to be a regulatory change, not a modification of structural genes coding for the antioxidant.

Superoxide Dismutase

Superoxide dismutase (SOD) is one of the most important defensive enzymes against the toxic effects of oxygen metabolism (32). This enzyme removes the superoxide free radicals O_2^-. Any organism that uses oxygen (aerobic organism) cannot live without this enzyme or an equivalent type of protective mechanism. Tissue concentrations of SOD were measured in brain, liver and heart tissues of 12 primate and two rodent species (70). An excellent correlation was found between the ratio of SOD per SMR and MLSP, as shown in Figure 1 for liver tissue.

Thus, we find that a species' LEP value is proportional to the level of SOD it has in its tissues. This makes sense if the ratio of production of superoxide free radical per amount of oxygen consumed is a constant for the species investigated.

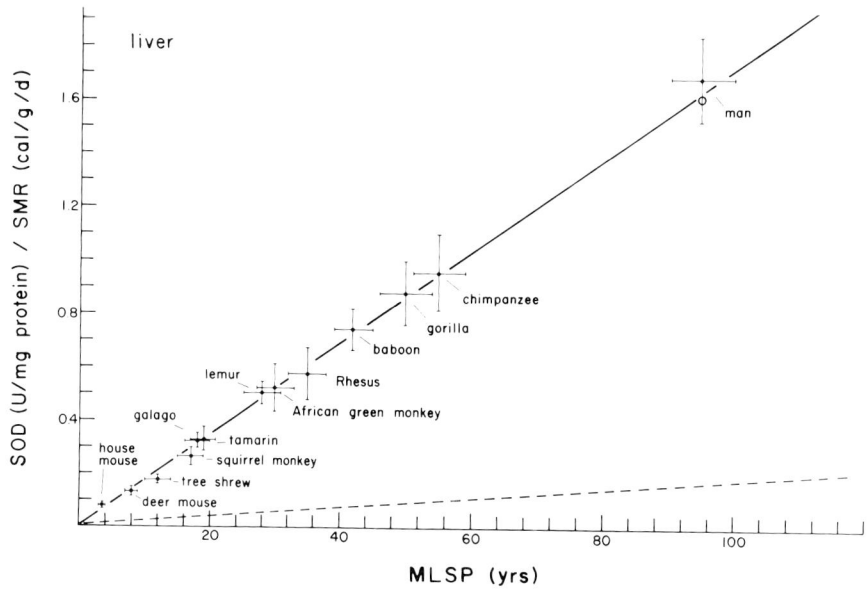

Fig. 1. Correlation of superoxide dismutase (SOD) activity per specific metabolic rate (SMR) against MLSP in liver of primate and rodent species. Correlation coefficient, r = 0.998 (70).

Because of the remarkable linearity of the correlation found for SOD per SMR with MLSP, this ratio may indeed be a constant, indicating that the amount of superoxide free radicals produced per amount of oxygen consumed is similar in mammalian species. Thus, the total amount of oxygen a tissue is able to utilize over a lifespan is directly proportional to the amount of protection offered by SOD that tissue has against the toxic by-products of oxygen metabolism.

The discovery that SOD tissue levels are proportional to species' LEP values suggests a biochemical basis of why humans are able to have such a large LEP value (815 kcal/g compared to about 210 kcal/g for most other organisms) (24). Moreover, it has been found that the SOD enzyme has a similar structure in a number of different mammalian species, so the high SOD levels found in human tissue is likely to be the result of more SOD enzyme, not the result of a better more efficient enzyme. Thus, the mechanism for increased LEP and consequently increased MLSP appears to involve a regulatory gene change, resulting in higher cellular concentrations of SOD.

Finally, it is known that the biological function of SOD is as an antioxidant. However, tissues contain many other compounds that may have antioxidant properties but may not be acting as an antioxidant in terms of their biological function. So a good test to determine if these compounds are antioxidants contributing to lifespan by affecting a species' LEP value is to see if a correlation exists in their concentration per SMR with a species' MLSP. Such a correlation is possible if either the antioxidant increased and/or the SMR decreased with increased MLSP.

More recent work has been done measuring the ratio of CU/Zn-SOD to Mn-SOD. This data is shown in Figure 2, where it is found that (1) primate species appear to have considerably higher levels of Mn-SOD compared to Cu/Zn-SOD than other species like the rodents, and (2) the ratio of Mn-SOD to total SOD increases with MLSP in brain but not liver in the primate species.

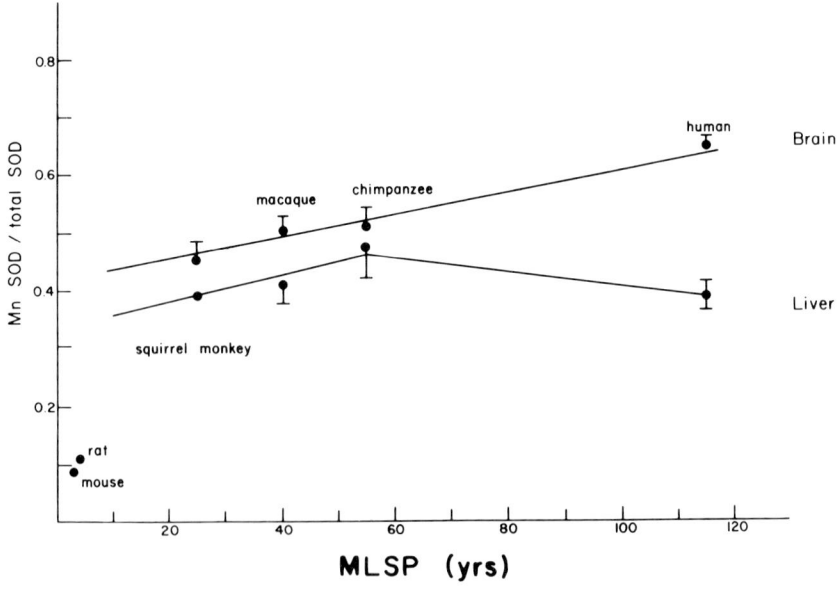

Fig. 2. Ratio of Mn-SOD to total SOD (Mn-SOD + Cu/Zn-SOD) in liver and brain tissues of primate species as a function of MLSP. Note comparison of ratio with rat and mouse also shown in this figure. (27).

Cu/Zn-SOD has been reported to decrease with age in the tissues of mice (60), but our studies have not been able to confirm that finding in both human and mouse (70).

The level of Cu/Zn-SOD in mouse liver tissue was also reported to vary according to the H-2 type of congenic mouse strain (56). This report was of considerable interest because H-2 congenic strain mice are known to have different lifespans, where cAMP and UV repair levels were also controlled by genes located in or near the H-2 region (23). Indeed, I found a good correlation of lifespan with higher levels of cAMP and UV repair in H-2 strain mice (see Table 1). Because cAMP levels were thought to be controlled by serum factors determined by the H-2 loci, we tried to determine if SOD might also be determined by serum factors (23). This experiment was based on forming parabiotic pairs of H-2 mice, each having a different level of SOD. However, on doing the controls for these experiments, I found that SOD did not vary with H-2 congenic strain type as reported, although differences were found in SOD in the different strains (see Table 2).

TABLE 1. Correlation between expression of potential longevity determinants and major histocompatibility genes in congenic mouse strains

Mouse strain	H-2 allele	Lifespan[a] (days)	SOD[b]	cAMP[c]	Ultraviolet DNA repair[d]	Bleomycin DNA repair[d]
B10Br/Sg	k	149 ± 1.1	23 ± 2	0.95 ± 0.2	2.0 ± 0.18	72
C57BL/10	b	155 ± 4.0	46 ± 2	1.29 ± 0.18	3.7 ± 0.28	120
B10A/2R	h2	-	46 ± 2	1.76 ± 0.32	-	-
B10A/4R	h4	-	48 ± 2	1.76 ± 0.32	-	-
A.BY	b	114 ± 3.5	45 ± 1	1.39 ± 0.16	-	-
A.CA	f	127 ± 3.0	48 ± 2	1.48 ± 0.13	-	-
C3H/HeDi	k	138 ± 0.7	18 ± 1	0.99 ± 0.28	1.7 ± 0.15	90
C3H.SW	b	150 ± 1.2	58 ± 1	1.25 ± 0.37	2.1 ± 0.23	125

[a] Tenth decile, male (65).
[b] Units SOD/mg protein in liver (56).
[c] Picomoles cAMP/mg liver wet weight (46).
[d] Expressed as amount of unscheduled DNA replication in treated lymphocytes at a given dose divided by amount of DNA replication of untreated cells (70).

TABLE 2. Superoxide dismutase levels in liver of congenic mouse strains[a]

Mouse strain	H-2 allele	10th decile lifespan (wks)	Whole cell (crude) units/mg protein		Cytosol (crude) units/mg protein		Cytosol (purified) units/mg protein	
			Cu/Zn[b]	Mn	Cu/Zn	Mn	Cu/Zn	Mn
C57Bl/10Sn	b	155	24.2 (46)	37.9	33.5	16.8	1297	42.7
B10.A/SgSn	a	154	20.0 (15)	16.3	22.3	22.3	1336	47.1
B10A(2R)/SgSn	h	—	32.8 (46)	21.8	23.4	19.4	—	—
B10A(5R)/SgSn	i	—	31.1 (18)	13.5	30.1	18.1	—	—
C3H.SW/Sn	b	150	21.5 (58)	13.1	29.5	43.0	1862	45.0
C3H/HeSn	k	138	29.0 (18)	13.5	20.9	17.1	1653	46.2
B10.AKM/Sn	m	139	34.4 (23)	3.46	36.6	3.89	1495	62.3
B10.D2/nSn	d	157	10.2 (15)	16.9	20.3	13.4	1508	49.9
A.BY/Sn	b	114	35.1 (45)	10.0	18.7	16.2	1003	37.8
B10.BR/SgSn	k	149	29.2 (23)	20.6	32.1	10.5	1102	26.6

[a] SOD determined by the xanthine/xanthine oxidase assay.
[b] Values in parentheses are those reported by Novak et al. (56).

Uric Acid

Uric acid is a by-product of purine metabolism and has usually been thought to be a waste product with no biological function (64,74). Recently, it has been found that uric acid is an excellent antioxidant, capable of protecting membranes from lipid peroxidation (4). An evaluation of urate as a potential longevity determinant was made by correlating tissue levels of uric acid with species' MLSP and LEP values. Figure 3 shows the correlation found of plasma urate levels with MLSP in primates and Fig. 4 for non-primate species. Fig. 5 shows that urate per SMR also correlates well with lifespan in brain tissues although uric acid is presumably not synthesized in brain (36).

These data strongly indicate that the unusually high tissue and plasma urate levels in human are playing an important role as an antioxidant by contributing to the human LEP value.

In addition to being a potentially important antioxidant, uric acid may also play a role as a nervous system stimulant. Uric acid has a similar structure to caffeine and other known neural stimulants. Indeed, it is known that men who have suffered from gout are frequently remarkably successful. It is also known that the probability to suffer from gout is related to the serum uric acid level, the higher the uric acid level, the higher the probability of suffering from the symptoms of gout.

Recent experiments to test the possible relationship of serum uric acid levels to intelligence, achievement and the need for achievement were made in persons 12 to 18 years of age. A highly significant correlation was found to each behavioral category (40). Longer-lived species in general appear to be more intelligent and more dependent on learned than instinctive behavior. Indeed, I have argued that the major driving force of mammalian evolution, and particularly in the primate species, is learned behavior capacity, coupled with longevity to make it possible (18,20,25). Thus, uric acid may have two important effects in primates, one as an antioxidant and the other as a neural stimulant--playing a role in enhancing learned behavior.

The biochemical basis of what determines plasma and tissue levels of uric acid is the tissue level of uricase, an enzyme that degrades uric acid, and the ability of the kidney to excrete uric acid from the blood. A study of these two regulatory mechanisms of uric acid concentration was made, where uricase and kidney ability to excrete uric acid were determined for primates species in relation to their MLSPs. It is found that only a few genetic regulatory alterations are required to account for the evolutionary increase of urate in primate species with increasing MLSP. This is consistent with the prediction of the working hypothesis. Furthermore, these studies indicate for the first time the point during primate evolution when endogenous synthesis capacity for vitamin C was lost, about 40-50 million years ago along the hominid descendant sequence (38).

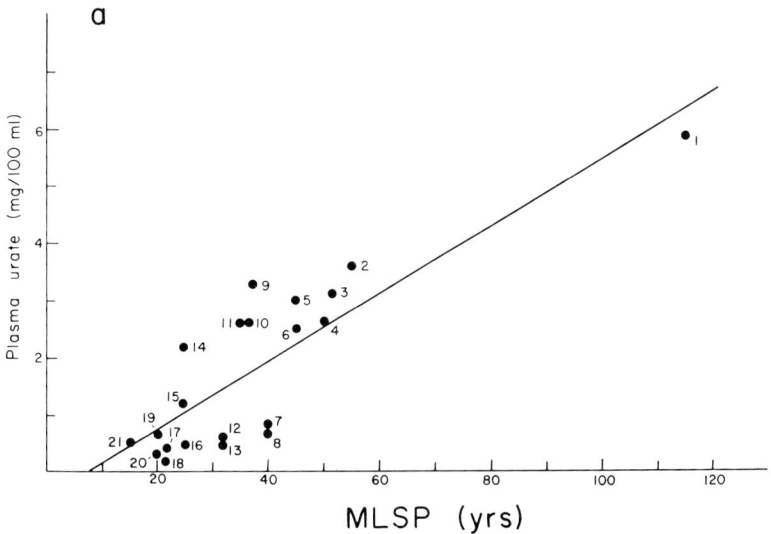

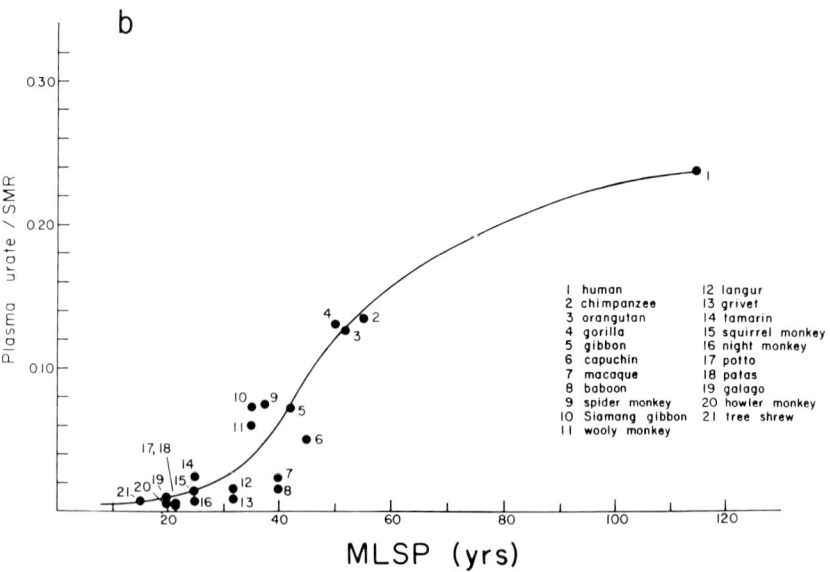

Fig. 3. Plasma urate levels and urate level per SMR in primates as a function of MLSP (27). Values taken from the literature. Correlation coefficient, r = 0.82. Identification of numbers is listed on b.

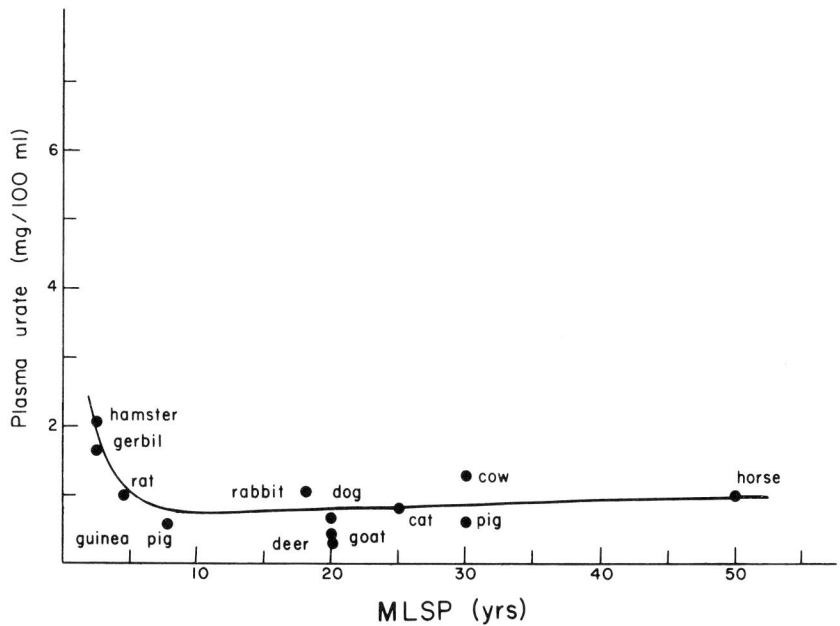

Fig. 4. Plasma urate levels in non-primates as a function of MLSP. Note that for these species, LEP values are approximately equal. (27).

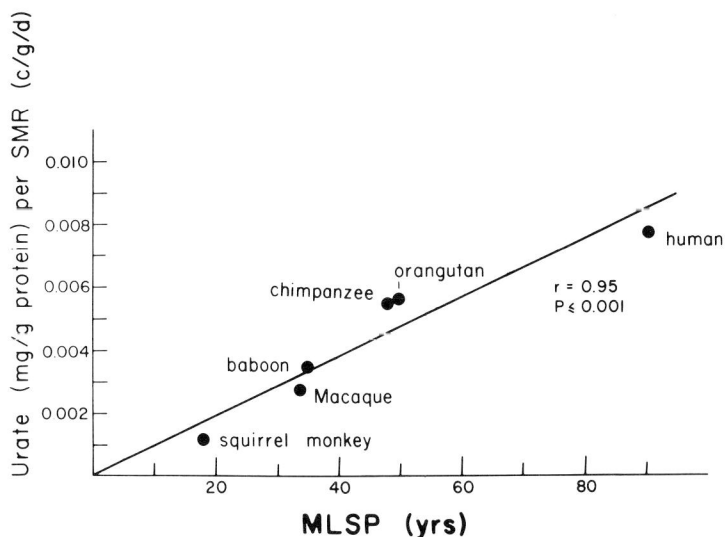

Fig. 5. Brain urate per SMR as a function of MLSP in primates. (27).

An important finding here is that the increase in uric acid levels with increased MLSP is related to a loss of uricase enzyme activity. Humans, having the highest tissue levels of urate, have no uricase. Thus, as for SOD, we find that all mammalian species have urate and that the mechanism of increase of urate is by a regulatory gene alteration. In this case, the increase of urate was a result of a decrease in an enzyme activity. This is a process that has a high probability of occurrence and therefore is likely to evolve at a high rate.

Carotenoids

Carotenoids are substances synthesized only in plants which serve to protect the plants from the free radicals generated during photosynthesis (54). Beta-carotene (one of the carotenoids) has been thought to be of value to humans and other species only as a precursor to vitamin A. However, recently beta-carotene has been found to have excellent antioxidant properties (44,45). Also, people with low tissue levels of beta-carotene are found to be unusually prone towards getting a number of different types of cancers (59). For this reason, there is considerable interest in pharmacological prevention of cancer by retinoids (49,67,68).

Serum carotenoid concentrations as a function of LSP are shown in Figure 6. Although a good correlation was found for the carotene, the correlation was not significant for vitamin A for species of MLSP about 30 years or so. Similar results were found against LEP value. This point is emphasized in Table 3, showing the ratio of carotene to vitamin A in the serum as a function of MLSP or LEP. Particularly striking is the unusually high levels of beta-carotene found in human serum.

Serum levels of carotene do not appear to change with age, as indicated by the literature for humans and for chimpanzee. Thus, aging does not appear to be a result of a loss of carotenoids. These results also suggest that the carotenoids may be more important in our diet than vitamin A by itself, where they may act both as an antioxidant and as a precursor to vitamin A.

The mechanism of tissue regulation of carotene is not fully understood, but the level of the enzyme carotenoid dioxygenase found in the intestine is known to play an important role (35,39). This enzyme cleaves the beta-carotene molecule in half, forming vitamin A. Although we have not yet tested the level of this enzyme in different species, it is known that the human level of carotenoid deoxygenase is unusually low. Thus, in general, an inverse correlation of beta-carotene tissue levels with carotenase activity might exist. If so, then (like SOD and urate) tissue levels of beta-carotene would be determined by regulatory genes, where an increase of carotene would be a result of a loss of carotenase activity.

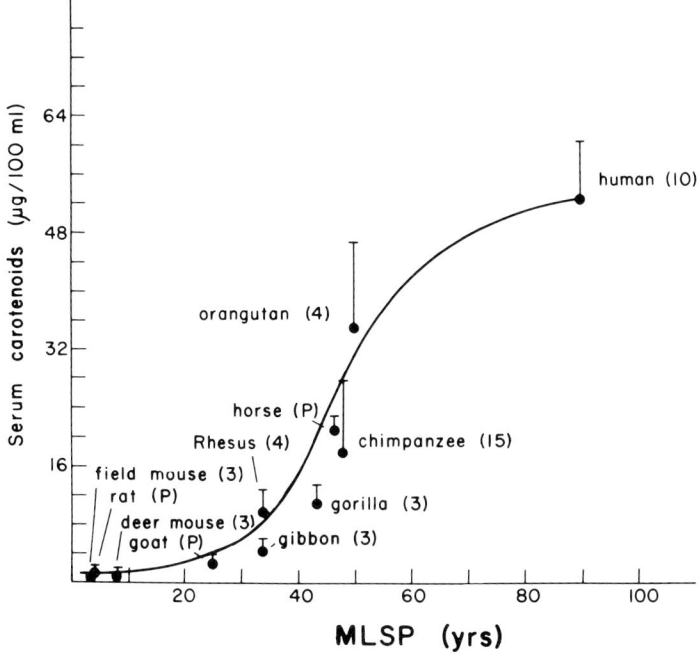

Fig. 6. Serum level of carotenoids as a function of MLSP (27).

TABLE 3. Serum Beta-Carotene/Vitamin A ratio as a function of MLSP and LEP

Species (common name)	LSP (yrs)	LEP (kcal/g)	BC	VA	BC/VA
human	90	815	52.0	25.5	2.0
orangutan	50	447	35.5	30.0	1.18
horse	46	235	22.0	9.4	2.34
chimpanzee	48	469	18.0	24.5	0.734
gorilla	43	309	11.0	18.8	0.585
Rhesus	34	517	10.0	11.5	0.869
gibbon	35	569	4.5	31.0	0.145
goat	25	277	3.0	18.2	0.164
deer mouse	8	440	1.0	17.0	0.0588
rat	4	115	2.5	16.2	0.154
field mouse	3.5	232	0.7	15.0	0.0466

BC/VA vs MLSP, $r = 0.741$, $n = 11$, $P \leq 0.01$
BC/VA vs LEP, $r = 0.342$, $n = 11$, $P \leq$ not significant
$52/0.7 = 74.2$, $2/0.0466 = 42.9$. (27).

Alpha-Tocopherol

Alpha-tocopherol or vitamin E is a well known tissue antioxidant (51), but how important vitamin E is to human health maintenance is still highly controversial (51). To determine the possible role vitamin E may have in determining human longevity, plasma levels of vitamin E for different species were obtained from the literature and are shown in Table 4 as a function of their MLSP and LEP values. An excellent positive correlation is found for the ratio of alpha-tocopherol level in plasma as a function of both the species' MLSP and LEP values.

TABLE 4. Plasma levels of Alpha-tocopherol in different species

Species (common name)	MLSP (yrs)	SMR (c/g/d)	LEP (kcal/g)	α-tocopherol (mg/100 ml)	α-tocopherol/SMR
Homo (human)	90	24.8	815	1.2	0.0483
Equus (horse)	46	13.9	233	0.25	0.0179
Cebus (capuchin)	42	52.2	804	0.50	0.00957
Papio (baboon)	35	30.9	394	0.73	0.0236
Macaque (Rhesus)	34	41.3	512	0.56	0.0135
Bos (cow)	30	15	164	0.40	0.0266
Sus (pig)	30	20	219	0.16	0.008
Aotus (night monkey)	20	72.7	530	0.53	0.00729
Ovis (sheep)	20	25.6	186	0.020	0.000781
Canis (dog)	20	35	255	0.41	0.0117
Rattus (rat)	4	104	152	0.31	0.00298
Mus (mouse)	3.5	182	232	0.75	0.00412

MLSP vs α-tocopherol, r = 0.554

LEP vs α-tocopherol, r = 0.661

MLSP vs $\dfrac{\alpha\text{-tocopherol}}{\text{SMR}}$, r = 0.864

Non-lifespan data taken from (2,6).

Alpha-tocopherol is seen to be a good potential longevity determinant, acting importantly in determining LEP values as an antioxidant. Nothing appears to be known of what regulates vitamin E levels in tissues.

Ascorbic Acid

Ascorbic acid (or vitamin C) has long been advocated to be essential for human health and longevity. Indeed, the inability of human to synthesize ascorbic acid, as is found in most other species, has been considered to be a serious human hereditary defect (9,48,55). Ascorbic acid is used in a number of metabolic processes including collagen synthesis. However, it is also a well known antioxidant. On evaluating ascorbic acid as a potential LDP by searching the literature for ascorbic acid levels in different tissues in mammalian species, I found to my surprise no significant positive correlation with species' MLSP or LEP values. Typical data are shown in Figure 7, where plasma ascorbic concentration is shown as a function of MLSP.

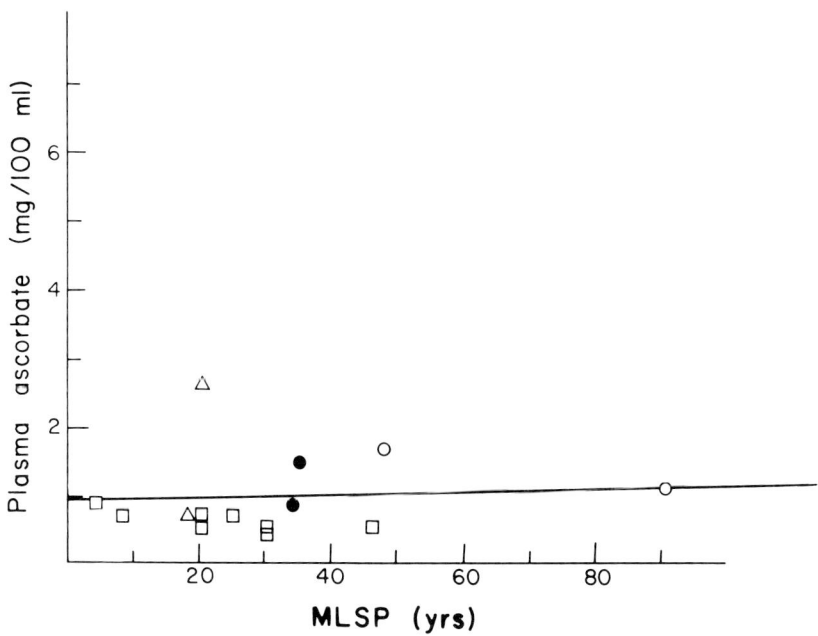

Fig. 7. Plasma ascorbate levels in primate and non-primate species. (O) Hominidae; (●) Old World monkeys; (△) New World monkeys; (□) non-primate mammals (27).

Human tissues do not have unusually high levels of ascorbate as compared to shorter-lived species. In fact, it appears that longer-lived species actually have lower concentrations of ascorbate in cerebrospinal fluid, adrenal gland and liver. For eye lens and whole brain tissue, species' levels of ascorbate appear to be about the same, irregardless of MLSP or LEP values. It is important to note, however, that ascorbate levels do decrease significantly with age in most tissues and this decrease may be an important effect of aging.

Thus, it appears that ascorbate may not have played an important role in determining the unusually long MLSP or high LEP value of human. Indeed, it appears that ascorbate may even be harmful, as indicated by some tissues in longer-lived species having lower ascorbate levels. Clearly, there is no indication that the inability of human to synthesize ascorbate is a disadvantage, and perhaps the loss of this synthesis was an advantage or at least had a neutral effect.

One interesting finding related to this possibility is that urate is very effective in removing Fe^{++} from tissues (personal communication with Paul Hochstein). In the presence of oxygen and Fe^{++}, ascorbate is transformed to the ascorbate free radical, which is highly toxic. Urate therefore may protect against the toxic effects of ascorbate and, by preventing ascorbate radical formation, urate may act to preserve the use of ascorbate for non-antioxidant-related functions such as in the synthesis of collagen. Thus, with higher urate levels, less ascorbate is needed.

Glutathione

Glutathione is thought to be one of the most important tissue antioxidants (52,53). To our surprise, when tissue levels of glutathione from the literature were compared to species' MLSP or LEP values, a negative correlation was found. These data are shown in Table 5. These data suggest that (a) glutathione is not a longevity determinant, (b) metabolic reactions involving glutathione may be toxic and/or a result of increase of LSP, and (c) detoxifying processes requiring high levels of glutathione were reduced during the evolution of longer MLSP. This prediction for a toxic effect of glutathione has recently found support by studies showing that glutathione and cysteine are positive in the Ames mutagenicity test (34).

Glutathione Peroxidase

Glutathione peroxidase has also been considered to be a major protective enzyme against the accumulation of peroxides (31). However, a survey of the literature on tissue levels of

TABLE 5. Glutathione concentration in tissues

Species	MLSP (yrs)	LEP (kcal/g)	Brain	Heart	Liver	Kidney	Spleen
				(relative values)			
human	90	815	12.8 ± 3.45 n = 5	11.3 ± 1.95 n = 4	13.0 ± 9.41 n = 4	7.42 ± 2.89 n = 4	6.2 n = 1
baboon	35	394	17.4 ± 2.68 n = 4	15.7 ± 4.59 n = 2	36.8 ± 13.0 n = 4	8.66 ± 6.35 n = 3	7.66 ± 6.78 n = 3
pig-tailed macaque	34	517	17.5 ± 1.41 n = 2	13.0 ± 0.981 n = 3	42.6 ± 2.46 n = 3	4.30 ± 1.13 n = 2	13.1 ± 4.38 n = 2
deer mouse	8	440	54.5 ± 2.1 n = 2	-	-	-	-
field mouse	3	232	49.0 ± 1.4 n = 2	-	-	-	-

n is number of different individuals measured.
MLSP vs GSH, r = -0.803 (brain); LEP vs GSH, r = -0.636. (27).

TABLE 6. Glutathione peroxidase activity in liver and brain

Species	n	MLSP (yrs)	LEP (kcal/g)	Liver[a]	Brain[a]
mouse	5	3	232	1140	23 ± 3
rat	5	4	152	153 ± 23	5
guinea pig	5	6	206	57	14 ± 4
rabbit	5	12	257	381	20 ± 5
dog	2	20	268	ND	3
cow	5	30	153	ND	5 ± 1

[a]Activity of cytosol fraction: mean ± SD. Nanomoles GSH oxidized/min/mg protein.
n is number of animals used for determination.
MLSP vs GPX: r = -0.315, liver; r = -0.551, brain.
 LEP vs GPX: r = 0.476, liver; r = 0.412, brain.
Non-lifespan data taken from (28).

glutatione peroxidase showed a decrease in concentration with increased MLSP. These results are shown in Table 6 for liver and brain. Similar results are found for blood (red blood cells) and liver (Se and non-Se activity). This is true both for the selenium and non-selenium-dependent types of glutathione peroxidase. Because glutathione peroxidase requires glutathione as a co-factor, then a decrease in glutathione level because of its toxicity might be expected to also result in a decrease of glutathione peroxidase in longer-lived species.

Catalase

Catalase is considered to be an important enzyme in removing hydrogen peroxide (8). However, few data on catalase concentrations in tissues could be found in the literature for different species to make an MLSP correlation. We therefore made a systematic survey of catalase concentrations in tissues of primate species as a function of their MLSP. Methods of assay described by Cohen et al. were used (13). Data is shown in Fig. 8, where a negative correlation was found for catalase activity with lifespan potential. This negative correlation may be related to the toxic effect of iron, which is a cofactor of catalase. Also, it is possible that other more effective antioxidants evolved to replace ascorbate, glutathione and glutatione peroxidase, which led to less need for catalase.

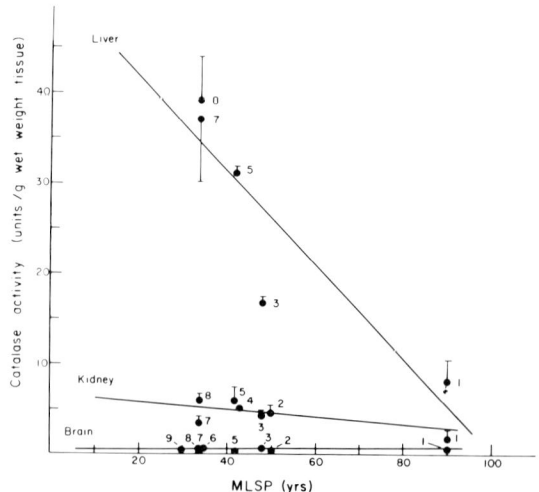

Fig. 8. Catalase activity as a function of MLSP in primate species. 1 human, 2 orangutan, 3 chimpanzee, 4 gorilla, 5 baboon, 6 lemur, 7 Rhesus, 8 pig tailed macaque, 9 African green monkey. Liver: $r = -0.895$, $P \leq 0.001$. Kidney: $r = -0.644$, $P \leq 0.001$ Brain: $r = 0.296$, P not significant (27).

The comparative antioxidant studies just reviewed cover the major known antioxidants. Other potential antioxidants that have shown a weaker but positive correlation with MLSP are DHEA, choline, glucose and cholesterol (25). Much of the data used were taken from the literature, coming from a number of different laboratories. More exact and detailed experiments need to be undertaken, particularly in the assay of antioxidants in regulatory tissues involving the neuroendocrine system (pituitary, hypothalamus and adrenal glands) and the immune system (T and B lymphocytes) to be able to establish confidently whether a given antioxidant may be important in determining species' longevity.

In this regard, a potentially important antioxidant is the catecholamines found in high concentrations in certain regions of the brain (11,13). These groups of compounds associated most frequently as neurotransmitters may also be important as antioxidants, where deficiencies may lead to free radical damage associated with Parkinson's and Alzheimer's disease.

RATE OF TISSUE AUTOXIDATION AS A FUNCTION OF AGE AND MLSP

There are probably many other important tissue antioxidants we do not yet know about. Thus, to fully test the hypothesis that human MLSP is determined in part by tissue levels of antioxidants, it is necessary to determine if indeed human tissues are more resistant to the toxic effects of free radicals and other active oxygen species as compared to shorter-lived species. That is, just because we have shown a correlation of a few specific antioxidants with MLSP or LEP, it does not prove that the net antioxidant capacity of longer-lived species is greater than the shorter-lived species. This is partially true in view of the antioxidants that appear to be less concentrated in the longer-lived species.

To meet this need, net tissue antioxidant protection was estimated by measuring the autoxidation rate of tissues. This was done by homogenizing a tissue in a neutral aqueous solution and then placing this homogenate in a shaking water bath at the normal body temperature of 37°C. Under these conditions, the tissue (in the presence of air) will autoxidize, which is the reaction of the complex tissue constituents with the oxygen in the air, forming a complex array of free radicals, aldehydes and peroxides. In effect, the tissue is allowed to go rancid. The extent of the autoxidation process is measured by determining the amount of malonaldehyde produced using the thiobarbituric acid assay (57). It is the same type of reaction that is thought to take place normally within a tissue (in vivo), but at a much slower rate. The rate of autoxidation of the tissue depends on the composition of the homogenate, how much unsaturated fatty acids are present and the net level of antioxidants (5,7). Thus, if human MLSP is truely related to innate ability of tissues to protect themselves from the toxic effects of oxygen, then we would expect to find that human tissues would be the most resistant to autoxidation. Figure 9 shows the rate of autoxidation of whole brain homogenate in air as a function of species' lifespan potential. A slightly better correlation is seen with species' LEP value. It was found that indeed human brain is extraordinarily resistant to autoxidation and that brain homogenates from shorter-lived species are in proportion more sensitive.

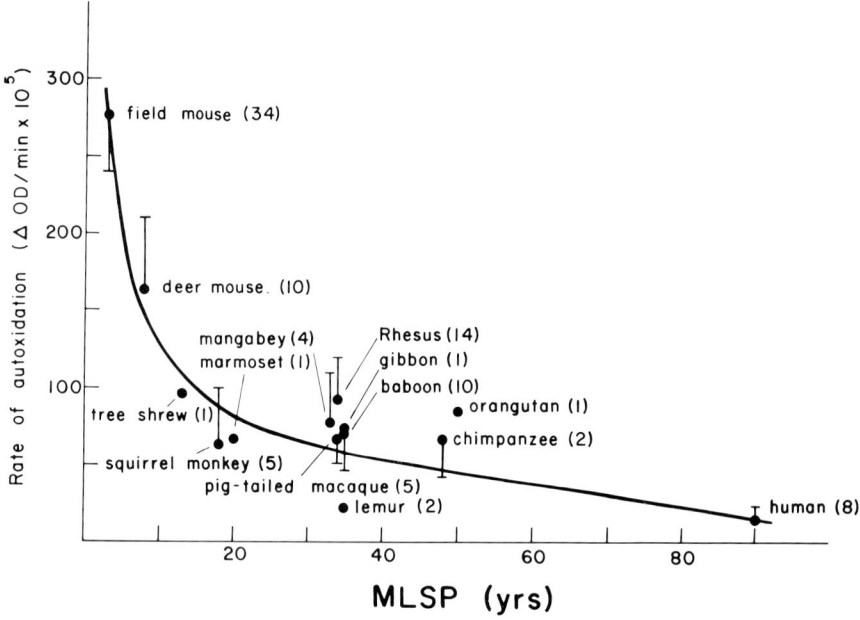

Fig. 9. Rate of autoxidation of whole brain homogenate in air from different mammalian species as a function of MLSP. Numbers in parentheses represent number of individuals used in each determination (27).

These data establish in my opinion the most convincing demonstration yet that human MLSP is determined at least in part by the innate resistance of the tissues to undergo autoxidation. For example, addition of alpha-tocopherol has been shown to slow down the rate of autoxidation. Thus, the resistance of the tissues to autoxidation is likely to be a result of higher levels of the same types of antioxidants found in all mammalian species. The LEP correlation with rate of autoxidation is likely then to reflect the relative antioxidant levels in the tissues. This conclusion is based on the findings of some preliminary experiments where the plateau levels of autoxidation of the tissues (the maximum amount of autoxidation obtainable) were found to be similar for all

primate species (data not shown). For Peromyscus and Mus, the
plateau levels are slightly higher. Thus, the composition of the
tissues, in terms of autoxidizable material, appears to be similar
in the different species. In addition, it is known that rate of
autoxidation is dependent on levels of antioxidants in the
tissues.

EVALUATION OF THE COMPENSATIONAL NATURE OF ENDOGENOUS ANTIOXIDANT CONCENTRATIONS IN TISSUES

One serious difficulty in the concept that antioxidants are
important as longevity determinants is that many are derived from
the diet and, accordingly one finds large differences in the
tissues of different individuals of the same species. For example,
we have found two-fold differences between human individuals for
urate, alpha-tocopherol and carotene, and yet there is no
evidence of associated differences in their aging rates. In fact,
the intrinsic aging rate of different human individuals appears to
be extraordinarily similar, with a variation less than that found
in height differences. Also, it seems improbable that such an
important characteristic as longevity would be subject to whatever
diet an individual happens to eat. Finally, the strongest argument
used against antioxidants playing an important role in determining
longevity has been the experimental results showing that
antioxidant supplementation to the diet does not significantly
increase either the mean nor the MLSP of normal long-lived
wild-type laboratory mammals (22-24,26).

Recent data emphasizing this conclusion was found, where
vitamin E supplementation to the diet failed to increase either
the mean or MLSP of mice even though the rate of accumulation of
lipofuscin pigment in tissues was decreased two-fold (26). If
antioxidants were truly important in determining longevity, one
would expect to be able to at least double or triple lifespan by
increasing the LEP of a mouse for example from 200 to 800 kcal/g,
as it is in human.

A possible answer to this dilemma has now been found, which
may prove important in all of our future studies concerning the
genetic and biochemical nature of longevity determinants. This is
that the net antioxidant levels of tissues is under compensatory
or homeostatic control (23,25,26). That is, the net protective
antioxidant level of a tissue is what probably determines aging
rate, and it is this net level not the level of individual
antioxidants that appears to be genetically regulated to determine
a species' MLSP. There are a number of different antioxidants with
overlapping capabilities, where one antioxidant can partially
compensate for another. For example, on feeding a rat a long-term
vitamin E-deficient diet, glutathione peroxidase, glutathione
reductase, and superoxide dismutase are found to dramatically
increase, thus compensating in part for the vitamin E-deficiency

and minimizing its effects (10). On the other hand,
supplementation of vitamin E in the diet depresses the levels of
other antioxidants, thereby seriously diminishing the net gain in
antioxidant protection the vitamin E could offer. Thus, over a
wide range of amounts and types of antioxidants taken in by food,
the net tissue antioxidant protection level may remain fairly
constant.

These results could explain the remarkable consistency of
aging rate found among different human individuals in spite of
large variations in dietary uptake of antioxidants or endogenous
synthesis levels of enzymatic antioxidants. These results also
explain the failure of antioxidant-supplemented diets to
significantly increase MLSP and inhibition of antioxidants to
shorten lifespan (1). That is, little effective net gain of
antioxidant protection in the tissue was ever achieved in normal
wild-type animals. Mutant strains of mice do appear, however, to
benefit from antioxidant supplementation, perhaps because they are
abnormally deficient in antioxidant synthesis, absorption, or in
the compensatory control of antioxidants. However, the results are
clear that normal long-lived mouse strains do not benefit from
simple dietary supplementation of antioxidants. Thus, antioxidant
supplementation by the diet does not appear to be the method of
choice to enhance net tissue antioxidant levels for normal healthy
persons. Instead, some means to intervene in the antioxidant
compensatory regulation mechanism is required.

SUMMARY

A positive correlation in the concentration of antioxidants
in tissues with species' lifespan potential was found, and this
data supports the hypothesis for the existance of specific
longevity determinants. These antioxidants are superoxide
dismutase (Mn and Cu/Zn), beta-carotene, alpha-tocopherol and
urate. On the other hand, a number of antioxidants were found to
have a negative correlation in their tissue levels with species'
lifespan potential. These are catalase, glutathione, and
glutathione peroxidase. In addition, a number of enzymes involved
in detoxification processes, such as the P-450, P-448 and
glutathione S-transferase enzymes, are also negatively correlated
with species' lifespan potential (25,26). Ascorbate showed no
correlation (negative or positive) with species' longevity. These
negative data can be taken to argue either against antioxidants
and/or detoxification processes as being important as determinants
of human longevity or that their decreased concentrations in
longer-lived species played an important causative role in the
evolution of increased longevity. Thus, an investigation into the
possible toxic effects of ascorbate, glutathione, glutathione
peroxidase, and related detoxification reactions appears warranted.
Also, it is important to measure the turnover rate of
antioxidants, since tissue level itself is only a measure of

potential protection, not actual utilization of this protection.

Most of the changes in the antioxidant levels (both the negative and positive correlations) are likely to be a result of changes in the levels of enzymes, thus supporting the prediction that changes in regulatory genes were the major genetic mechanisms in the evolution of longevity.

Lastly, it should be noted that recent studies on heat-shock proteins support the longevity determinant hypothesis (47,62). This work has indicated that organisms from bacteria to mammals increase the synthesis of a specific set of proteins when placed under a stress such as produced by heat, mutagens or free radicals (oxidative stress). These proteins are few in number (from 5-10) and similar in sequence in most species. Their increased synthesis is associated with the animal being less susceptible to a repeat of the initial stress or to other types of stress. The heat-shock proteins, which have also been called alarmones (47), therefore appear to be protective in nature. Although it is not known how these proteins protect the organism, it would not be surprising to find that they represent more than is implied in the alarmone concept, and instead represent an important new class of longevity determinants which act to stabilize the proper differentiated state of cells.

I thank Edith Cutler for laboratory assistance, drawing of the figures, and typing of the manuscript. Her assistance was made possible by support from the Paul Glenn Foundation for Medical Research.

REFERENCES

1. Allen, R.G., Farmer, K.J., and Sohal, R.S. (1983): Biochem. J., 216:503-506.
2. Altman, P., and Dittmer, D., editors (1961): Blood and Other Body Fluids, Fed. Am. Soc. Exp. Biol., Bethesda, MD.
3. Ames, B.N. (1983): Science, 221:1256-1264.
4. Ames, B.N., Cathcart, R., Schwiers, E., and Hochstein, P. (1981): Proc. Natl. Acad. Sci., USA, 78:6858-6862.
5. Barber, A.A., and Bernheim, F. (1967): In: Adv. Gerontol. Res., Vol. 2, edited by B.L. Strehler, pp. 355-403. Academic Press, New York.
6. Bernirschke, K., Garner, F.M. and Jones, T.C., editors (1978): Pathology of Laboratory Animals, Springer-Verlag, Basel.

7. Bieri, J.G., and Anderson, A.A. (1960): Arch. Biochem. Biophys. 90:105-110.
8. Chance, B., Sies, H., and Boveris, A. (1979): Physiol. Rev. 59:527-605.
9. Chatterjee, I.B. (1978): World Rev. Nutr. Diet. 30:69-87.
10. Chen, L.H., Thacker, R.R., and Chow, C.K. (1980): Nutr. Reports Intern. 22:873-881.
11. Cohen, G. (1982): In: Pathology of Oxygen, edited by A.P. Autor, pp. 115-126. Academic Press, New York.
12. Cohen, G., Dembiec, D., and Marcus, J. (1970): Anal. Biochem. 34:30-38.
13. Cohen, G. and Heikkila, R.E. (1977): In: Superoxide and Superoxide Dismutases, edited by A.M. Michelson, J.M. McCord and I. Fridovich, pp. 351-365. Academic Press, New York.
14. Cole, T.R. (1983): The Hastings Center Rept., 13:34-40.
15. Cutler, R.G. (1972): In: Adv. Geront. Res. Vol. 4, edited by B.L. Strehler, pp. 219-312. Academic Press, New York.
16. Cutler, R.G. (1975): Proc. Natl. Acad. Sci., USA, 72:4664-4668.
17. Cutler, R.G. (1975): Exp. Gerontol. 10:37-60.
18. Cutler, R.G. (1976): J. Human Evol. 5:169-202.
19. Cutler, R.G. (1976): In: Interdiscipl. Topics Gerontol., Vol. 9, edited by R.G. Cutler, pp. 83-133. Karger, Basel.
20. Cutler, R.G. (1978): In: The Biology of Aging, edited by J.A. Behnke, C.E. Finch, and G.B. Moment, pp. 311-368. Plenum Press, New York.
21. Cutler, R.G. (1979): Gerontology 25:69-86.
22. Cutler, R.G. (1982): In: The Aging Brain. Aging, Vol. 20, edited by E. Giacobini, G. Giacobini, G. Filogamo, and A. Vernadakis, pp. 1-18. Raven Press, New York.
23. Cutler, R.G. (1982): In: Testing the Theories of Aging, edited by R. Adelman and G. Roth, pp. 25-114. CRC Press, Boca Raton, Fla.
24. Cutler, R.G. (1983): Gerontology 29:113-120.
25. Cutler, R.G. (1984): In: Aging and Cell Structure, Vol. 2, edited by J.E. Johnson, Plenum Press, New York, in press.
26. Cutler, R.G. (1984): In: Free Radicals in Biology, Vol. 6, edited by W.A. Pryor, Academic Press, New York, in press.
27. Cutler, R.G. (1984): submitted for publication.
28. DeMarchena, O., Guarnieri, M., and McKhann, G. (1974): J. Neurochem. 22:773-776.
29. Fahmy, M.J., and Fahmy, O.G. (1983): Cancer Res. 43:801-807.
30. Fahmy, M.J., and Fahmy, O.G. (1983): Teterogenesis, Carcinogenesis and Mutagenesis, 3:27-39.
31. Flohe, L. (1982): In: Free Radicals in Biology, Vol. V., edited by W.A. Pryor, pp. 223-254. Academic Press, New York.
32. Fridovich, I. (1979): In: Oxygen Free Radicals and Tissue Damage, Ciba Found. Symp. 65, pp. 77-93. Excerpta Medica, Amsterdam.

33. Gavrilov, L.A., Gavrilova, N.S., and Nosov, V.N. (1983):
    Gerontology, 29:176-180.
34. Glatt, H., Protic-Sabjic, M., and Oesch, F. (1983): Science
    220:961-963.
35. Goodwin, T.W. (1954): Carotenoids. Their Comparative
    Biochemistry, Chem. Pub. Co., New York.
36. Hallgren, R., Niklasson, F., Terent, A., Akerblom, A., and
    Widerlov, E. (1983): Stroke, 14:382-388.
37. Handler, P. (1982): Regulatory Toxicol. & Pharmacol. 2:3-14.
38. Hersch, D., Cutler, R.G., and Ames, B.N., in preparation.
39. Isler, O., editor (1971): Carotenoids, Birkhauser Verlag,
    Basel.
41. Keehner, J.M. (1979): Dissert. Abstr. 1283-A.
41. Keyfitz, N. (1977): Demography, 14:411-418.
42. Keyfitz, N. (1978): Amer. J. Pub. Health, 68:954-956.
43. Kohn, R.R. (1978): Principles of Mammalian Aging,
    Prentice-Hall, Inc., Englewood Cliff, N.J.
44. Krinsky, N.I. (1982): In: The Science of Photomedicine,
    edited by J.D. Regan and J.A. Parrish, pp. 397-405.
    Plenum Press, New York.
45. Krinsky, N.I., and Deneke, S.M. (1982): J. Natl. Cancer
    Inst. 69:205-209.
46. Lafuse, W., Meruelo, D., and Edidin, M. (1979):
    Immunogenetics, 9:57-65.
47. Lee, P.C., Bochner, B.R., and Ames, B.N. (1983): Proc. Natl.
    Acad. Sci., USA, 80:7496-7500.
48. Lewin, S. (1976): Vitamin C: Its Molecular Biology and
    Medical Potential, Academic Press, New York.
49. Lotan, R. (1980): Biochim. Biophys. Acta 605:33-91.
50. Ludwig, F.C. (1980): Science, 209: editorial, Sept. 5.
51. Machlin, L.J. (1980): In: Vitamin E. A Comprehensive
    Treatise, edited by L.J. Machlin, pp. 637-645. Marcel Dekker,
    Inc., New York.
52. Meister, A. (1982): Biochem. Soc. Trans. 10:78-79.
53. Meister, A. (1983): Science, 220:471-477.
54. Moore, T. (1957): Vitamin A, Elsevier Pub., Co., Amsterdam.
55. Naito, H.K. (1979): In: Nutritional Elements and
    Clinical Biochemistry, edited by M.A. Brewster and
    H.K. Naito, pp. 69-115. Plenum Press, New York.
56. Novak, R., Bosze, Z., Matkovics, B., and Fachet, J. (1980):
    Science, 207:86-87.
57. Ohkawa, H., Ohishi, N., and Yagi, K. (1978): J. Lipid Res.
    19:1053-1057.
58. Ono, T., and Cutler, R.G. (1978): Proc. Natl. Acad. Sci.,
    USA, 75:4431-4435.
59. Peto, R., Doll, R., Buckley, J.D., and Sporn, M.B. (1981):
    Nature 290:201-208.
60. Reiss, U., and Gershon, D. (1976): Euro. J. Biochem.
    63:617-623.

61. Rosenwaike, I., Yaffe, N., and Sagi, P.C. (1980): Amer. J. Pub. Health, 70:1074-1080.
62. Schlesinger, M.J., Ashburner, M., and Tissieres, A., editors (1982): Heat Shock. From Bacteria to Man. Cold Spring Harbor Lab., New York.
63. Schneider, E.L., and Brody, J.A. (1983): New Eng. J. Med. 309:854-856.
64. Seegmiller, J.E. (1979): In: Contemporary Metabolism, Vol. 1, edited by E. Freinkel, pp. 1-85. Plenum Press, New York.
65. Smith, G.S., and Walford, R.L. (1977): Nature, 270:727-729.
66. Sohal, R.S., editor (1981): Age Pigments, Elsevier, Amsterdam.
67. Sporn, M.B. (1978): In: Carcinogenesis, Vol. 2, Mech. of Tumor Promotion and Cocarcinogenesis, edited by T.J. Slaga, A. Sivak, and R.K. Boutwell, pp. 545-551. Raven Press, New York.
68. Sporn, M.B., and Newton, D.L. (1979): Fed. Proc. 38:2528-2534.
69. Strehler, B.L. (1978): Time, Cells and Aging, Academic Press, New York.
70. Tolmasoff, J.M., Ono, T., and Cutler, R.G. (1980): Proc. Natl. Acad. Sci., USA, 77:2777-2781.
71. Tsai, S.P., Lee, E.S., and Hardy, R.J. (1978): Amer. J. Pub. Health, 68:966-971.
72. Walford, R.L., and Bergmann, K. (1979): Tissue Antigens, 14:336-342.
73. Wilson, A.C., Carlson, S.S., and White, T.J. (1977): Ann. Rev. Biochem. 46:573-637.
75. Wyngaarden, J.B., and Kelley, W.N. (1979): In: Contemporary Metabolism, Vol. 1, edited by N. Freinkel, pp. 1-130. Plenum Press, New York.

Clinical Laboratory Tests as Indicators of Free-Radical Reactions

T. L. Dormandy and D.G. Wickens

Department of Chemical Pathology, Whittington Hospital, London N19 5NF, United Kingdom

We all cherish the illusion that new diagnostic methods evolve in response to new clinical needs. The truth is usually the reverse : it is new clinical indications which are suddenly perceived when new diagnostic methods become available. Would diabetes mellitus exist as a clinical syndrome if the urine of diabetics did not taste sweet or if glucose were not easily measurable in blood? One may doubt it. (Of course the non-existance of many syndromes would be no great loss : diagnostic tests can create spurious as well as real entities.) Such philosophical ruminations are not without relevance to free-radical activity. That such activity is important and sometimes crucial in the pathogenesis of common disorders is no longer in doubt. If clinical chemistry evolved in response to such advances measurements relating to free radicals would be competing with electrolytes and liver function tests in the repertory of hospital service laboratories. They do not in fact because no such measurement exist. And because no such measurements exist no clinical syndromes exist either. Only during the last year or two have their been hopeful signs that the situation may be changing.

ESTABLISHED PROCEDURES

The TBA test

There have, of course, been many attempts to turn experimental procedures into clinical tests ; and, before turning to recent developments and without any attempt at completeness, a few of these may be briefly considered.

It will be recalled that when an oxygen free radical attacks a **polyunsaturated** lipid it triggers off not just one reaction but a whole chain (or "cascade" to use the trendy term) of secondary reactions. Indeed, in biological material - ie, in the presence of specific enzyme systems - it triggers off several branching and cross-reacting cascades. At any one moment, therefore, peroxidising biological systems may contain dozens or even hundreds of more-or-less unstable peroxidation products. A few of those can now be measured specifically though the measurements usually involve complex and sophisticated techniques and apparatus. But when the whole peroxidising mixture is boiled in an acid medium many of the unstable intermediates break down to a small, water-soluble molecule, malonyl-dialdehyde (MDA)(10,11). This too is intrinsically unstable ; but interacting with the complexing agent, thiobarbituric acid (TBA), it forms a stable and easily measurable pink complex. Under standardised conditions the colour can be interpreted as a measure of the initial lipid peroxidation. (to what extent MDA itself is present in mixtures before boiling is uncertain. It is a major product in peroxidising liver microsomes but not necessarily in all biological material.)

The TBA test is easy to perform and there can be no doubt of its value. Though often misinterpreted (eg. when used as a <u>specific</u> measure of prostaglandin synthesis), it is still the mainstay of free-radical biochemistry as pursued in research laboratories. The age pigments which accumulate in certain organs in man as well as in animals are largely composed of TBA-reactive polymers and complexes (18), probably the most telling single piece of indirect evidence linking free-radical damage to ageing. But whether the test is applicable to "routine" clinical chemistry is questionable. In its original form it does not detect MDA in fresh plasma, blood cells, inflammatory exudate or any other clinically available material in man, either normal or abnormal. This is not quite true of the various modifications developed in recent years (16). Using such modifications raised plasma MDA levels have been reported, for example in rheumatoid disease (17) and in pre-eclamptic toxaemia (13). These findings are of great interest in pointing to free-radical oxidation as a possible aetiological factor in these disorders. But in our own laboratory we have never found the modified TBA tests sufficiently specific or robust to regard them as potentially useful diagnostic procedures.

Indirect tests

A different philosophy underlies a wide range of available indirect measurements. Biological structures are highly protected against potential free-radical damage. The protective system is complex and multifactorial (8). It is far from being fully understood. It is nevertheless possible to measure its efficiency under standardised experimental conditions (19) ; and it is possible to measure selected mechanisms which may (or may not) be important in vivo. The TBA test itself can be used in this way. Fresh red blood cells contain no measurable MDA ; but they can be induced to peroxidise under oxidant stress (7). In many haemolytic states an increased susceptibility is reflected by increased MDA generation over a set period of time. This may be the only demonstrable biochemical abnormality to account for the shortened cell life-span (1).

Among specific protective mechanisms a number of "antioxidant" enzymes are readily measurable. A significantly decreased superoxide dismutase activity, for example, may be demonstrable in certain cells in certain diseases. Since the enzyme is known to scavange superoxide radicals it is reasonable to speculate that an enzyme deficiency may be associated in vivo with increased superoxide free-radical activity. But one cannot go beyond reasonable speculation. The point to appreciate is that in a complex defence system the efficiency of the whole is largely determined by the efficiency of the weakest link. An enzyme deficiency will be important only when it happens to be the rate-limiting step in antioxidant protection. A similar proviso applies to such well-established antioxidants as vitamin E. Many animal diseases are undoubtedly due to specific vitamin-E deficiency ; and in such cases the serum vitamin-E levels are a diagnostic guide. In man vitamin E is rarely the rate limiting antioxidant and serum levels are usually of limited help.

Conversely, certain factors are well-recognised as promoters rather than as inhibitors of free-radical generation ; and these too can provide indirect evidence. A raised level of non-protein-bound "catalytic" iron in cerebrospinal fluid in at least one type of neuronal lipofuscinosis supports the concept that oxygen free radicals and lipid peroxidation play an aetological role in this disorder (12). But the limitation that applies to antioxidant protection also applies to potential pro-oxidant mechanisms.

DIENE CONJUGATION

Turning to recent developments, diene conjugation (DC) has been used as an index of lipid peroxidation for many years (5,6) ; and it is necessary therefore to explain briefly why we regard it as a new opening. The term means a characteristic molecular configuration where two double bonds are separated by one single bond. None of the major native polyunsaturated fatty acids in man has this configuration ; but when these molecules are attacked by an oxygen free radical, one or more of the double bonds shift to become diene (or triene) conjugated. Linoleic acid, for example, the most abundant of the polyunsaturated fatty acids, is an 18-carbon chain with two double bonds in the 9 and 12 positions. When attacked by an oxygen free radical the double bonds and the peroxy free-radical and a number of subsequent products will be diene conjugated with the double bonds in the 9 and 11 or 10 and 12 positions. Such a bond sequence absorbs ultraviolet light at a specific wavelength (234nm), distinct from the absorption peak of the non-diene-conjugated parent compound. This is useful when measuring peroxidation in pure lipids and simple lipid emulsions (eg, when the presence of even low concentrations of peroxidation products might be critical). It is also, with certain limitations, a valuable measure in biological fluids. A lipid extract can readily be prepared from plasma, for example, and the height of the 234nm peak recorded in arbitrary absorption units. A number of preliminary clinical trials suggest that (taken in conjunction with other parameters) this gave some indication of the concentration of circulating free-radical-damaged lipids (15,20). There is also a significant correlation between this "total DC" and age. However, the limitations of the method could not be permanently ignored. Resolution between the large peak of the non-diene-conjugated parent lipids and the diene-conjugated derivatives was unsatisfactory. Many drugs and drug-metabolites interfered. And most importantly, the measurement gave no indication of what peroxidation products were being measured : indeed the very assumption that total DC in biological extracts could be regarded as the outcome of lipid peroxidation had to be tested. The third task in particular proved to be more difficult than had been anticipated.

A wide range of polyunsaturated lipids were exposed to free-radical oxidation by a variety of methods (eg, ultraviolet irradiation, chemical mixtures, activated polymorphonuclear leukocytes). The resultant material at various stages of peroxidation was fractionated by high-performance liquid chromatography (HPLC) and scanned at two wavelengths (210nm for unconjugated lipids and 234nm for the conjugated products). The series of extensive panoramas of peaks obtained in this way would, it was hoped, provide the reference maps for identifying biological free-radical products. Lipid extracts from plasma, bile, duodenal juice were also fractionated by HPLC and scanned (2,3). The main 234nm absorption peaks could readily be identified but they did not correspond (ie, co-chromatograph) with any of the 234nm peaks on any of the reference maps of pure lipid peroxidation. At this point the possibillity had to be envisaged that, cotradicting both expectation and circumstantial evidence, total DC in plasma and other clinical material was wholly unrelated to lipid peroxidation in vivo. The work of Cawood et al. (4) eventually provided the explanation. To generate in vitro the diene-conjugated products present in biological fluids polyunsaturated lipids have to be exposed to free-radical activity in the <u>presence of protein</u>. In the presence of albumin, for example, will then generate the main diene-conjugated peak in plasma. This product was further investigated and identified not as a peroxide or epoxide or other oxygen-containing derivative (as generated by peroxidising linoleic acid in the absence of protein) but as a non-oxygen containing diene-conjugated 9,11 isomer (14). Simple methods are now being developed for its measurement in the lipid classes and in lipoprotein fractions.

PROTEIN FLUORESCENCE

A second recent development holds out hope of clinical application. It has been suggested that intense preoccupation with the effects of free-radical oxidation of polyunsaturated lipids over the past twenty years may have engendered the assumption that free-radical biochemistry is essentially the biochemistry of lipid peroxidation (9). Wickens et al. (21-23) have shown over the past two or three years that proteins are no less susceptible to direct oxygen free-radical attack ; and the consequences if less "spectacular" than the cascades associated with lipid peroxidation, may be no less important. Apart from functional changes (eg, altered immunological behaviour), these effects are readily demonstrable and measurable. In particular, proteins exposed to free-radical activity rapidly lose their "native" fluorescence (eg, human gamma-globulin Ex 298nm, Em 336nm) and, more slowly, aquire characteristic new fluorescence properties (Ex 360nm, Em 454 nm). This allows the differential measurement of free-radical-damaged proteins in inflammatory exudates (eg, synovial effusions) and potentially in other biological fluids.

CONCLUSIONS

We have thus begun to aquire tools with which we can measure evidence of free-radical activity in plasma and other diagnostic material. Such evidence does not, of course, provide us with a ready-made interpretation of the underlying free-radical processes any more than a raised blood-sugar reveals the metabolic defect in diabetes. To understand these processes will require the use of experimental model systems. Conversely, to assess the diagnostic significance of the findings will require the clinical assessment of whole patients. What these measurements provide at present is a link between the experimental investigator and the clinician, a means perhaps for one to stimulate and to exploit the other.

REFERENCES

1. Barnes,A.J.,Gutteridge,J.M.C.,Stocks,J.,Friedman,M., and Dormandy,T.L.(1973):Brit Med J, 1:462-463.

2. Braganza,J.M.,Wickens,D.G.,Cawood,P., and Dormandy, T.L.(1983):Lancet,2:375-379.

3. Cawood,P.,Iversen,S.A., and Dormandy,T.L.(1984):In: Oxygen Radicals in Chemistry and Biology,edited by W.Bors,M.Saran,and D.Tait,in press,de Gruyter, Berlin.

4. Cawood,P.,Wickens,D.G.,Iversen,S.A.,Braganza,J.M., and Dormandy,T.L.(1983):Febs Lett,162:239-243.

5. Comporti,M.,Benedetti,A.,and Casini,A.(1974): Biochem Pharmacol,23:421-432.

6. Di Luzio,N.R.(1972):J Agr Food Chem,20:486-490.

7. Dormandy,T.L.(1971):Brit J Haem,20:457-461.

8. Dormandy,T.L.(1978):Lancet,1:647-650.

9. Dormandy,T.L.(1983):Lancet,2:1010-1014.

10. Esterbauer,H.and Slater,T.F.(1981):IRCS Med Sci,9:

749-750.

11. Gutteridge,J.M.C.,Stocks,J.and Dormandy,T.L. (1974):Analyt Chim Acta,70:107-111.

12. Gutteridge,J.M.C.,Rowley,D.A.,Halliwell,B.,and Westermarck,T.(1982):Lancet,2:459-460.

13. Ishihara,M.(1978):Clin Chim Acta,84:1-9.

14. Iversen,S.A.,Cawood,P.,Lawson,A.,Meddigan,M.,and Dormandy, T.L.(1983):Febs Lett,in press.

15. Lunec,J.,Halloran,S.P.,White,A.G.,and Dormandy,T.L. (1981):J Rheumatol,8:233-245.

16. Matsushita,S.,Asakawa,T.(1980):Lipids,15:137-140.

17. Muus,P.,Bonta,I.L.and den Oudsten,S.A.(1979): Prostaglandins & Med,2:63-65.

18. Siakotos,A.N.,and Munkres,K.D.(1982):In:Ceroid-lipofuscinosis (Batten's disease),edited by D. Armstrong,N.Koppang,and J.A.Rider,pp.167-183, Elsevier Biomedical Press,Amsterdam.

19. Stocks,J.,Gutteridge,J.M.C.,Sharp,R.J.,and Dormandy,T.L.(1974):Clin Sci Mol Med,47:215-222.

20. Wickens,D.G.,Wilkins,M.H.,Lunec,J.,Ball,G.,and Dormandy,T.L.(1981):Ann Clin Biochem,18:158-162.

21. Wickens,D.G.,Graff,T.L.,Lunec,J.,and Dormandy,T.L. (1981):Agents Actions,11:650-651.

22. Wickens,D.G.,Norden,A.G.,Lunec,J.,and Dormandy,T.L. (1983):Biochim Biophys Acta,742:607-616.

23. Wickens,D.G.,and Dormandy,T.L.(1983):Agents Actions,13:520-521.

Free Radicals in Molecular Biology, Aging, and Disease, edited by D. Armstrong et al. Raven Press, New York © 1984.

Electron Spin Resonance Studies of Cancer: Experimental Results and Conceptual Implications

Harold M. Swartz

University of Illinois College of Medicine at Urbana-Champaign, Urbana, Illinois 61801

In addition to my assigned task to summarize studies of cancer utilizing electron spin resonance, I also have chosen to cover some of the critical conceptual problems involved in attempting to carry out studies to test theories that link free radicals to major types of biological processes such as cancer or aging. I believe that failure to understand the limitations of such studies has retarded adequate investigations and wider understanding of the role of free radicals in a number of physiological and pathological processes. I will therefore begin this chapter with a discussion of these concepts and then review some of the major areas of study that have employed electron spin resonance (ESR) spectroscopy in the study of cancer.

PROBLEMS AND LIMITATIONS OF THEORIES THAT ATTEMPT TO LINK/DEMONSTRATE AN ESSENTIAL LINK BETWEEN FREE RADICALS AND MAJOR BIOLOGICAL PROCESSES

There are at least three specific concepts which seem to be sources of potential confusion: 1) failing to differentiate between reactions of free radicals and other redox reactions, 2) failing to determine the biological significance of observed free radical (or other redox) reactions and/or intermediates, 3) failing to relate the reactivity of free radicals to the postulated key damaging reactions and the location of these reactions. As a consequence several major problems have developed in discussions in this area.

The Terms Used To Describe Theories And Results Related To Free Radicals Often Are Imprecise and Misleading.

"Free Radical Pathology" and similar terms are potentially misleading and should be discarded. In the last 10-20 years an impressive body of literature has been developed that indicates that redox reactions, especially oxidations, may play significant roles in a variety of physiological and pathological processes (e.g., the series on free radicals in biology edited by Pryor, 19). Some of the key intermediates are free radicals but many others are not. Transition metals (e.g., iron and copper) and hydrogen peroxide clearly are involved in many of the processes that have been studied. There is moderately strong evidence of involvement of organic hydroperoxides and aldehydes in some significant processes. There is also some evidence that singlet oxygen is involved in some key processes, although this area remains controversial. If transition elements, hydrogen peroxide, organic hydroperoxides, aldehydes and, perhaps, singlet oxygen are involved it is inaccurate and misleading to label the process as due to free radicals. Sometimes the problems are only semantic but at other times this terminology leads to inappropriately planned experiments and erroneous interpretations of experimental results. For example, the therapeutic and mechanistic implications of the observed effect of additives are quite different if they are acting via reduction of disulfides versus acting as direct free radical scavengers. In the long run, more progress can be made on elucidating the role of free radicals and in controlling associated deleterious reactions associated with them, if there is a clear understanding of what reactions actually involve free radicals. Ultimately it is likely that we shall find that there are a number of physiological and pathological processes that involve oxidations as intrinsic parts of their basic mechanisms and that the key reactions can occur via a variety of oxidative mechanisms, only some of which involve free radicals. Perhaps a more general and more accurate term such as "deleterious oxidations" may become developed and accepted, which will signify the important and common underlying oxidative mechanisms for such processes.

<u>A second major problem is the frequent failure to differentiate between the occurrence of free radical intermediates (or other potentially deleterious oxidants) and their importance in the biological process under study.</u>

At the most extreme level this is seen in theories or discussions that attempt to establish unifying "free radical theories" of aging, cancer, or other pathological processes. In view of the already known complexities of such processes, it seems very unlikely that they can be explained entirely on the basis of free radical reactions. Attempts to argue for such an all-encompassing role eventually may inhibit an appreciation of how free radicals and other oxidative reactions do contribute to the process under study. When a more limited role of free radicals is understood, then experiments and observations that indicate no significant role of free radicals (or other oxidants) in some aspect of a pathological process will not detract from other observations where uncontrolled oxidative processes appear to be essential components. The adoption of a more conservative approach should also lead to more objective considerations by other scientists of interpretations of data that link oxidative processes to some aspects of the pathological process under study.
Even when authors may not attempt to attribute all of the cause

of aging, cancer, etc., to free radicals, they frequently fail to demonstrate a quantified link between the occurrence of free radicals and their essential involvement in a process. Our ability to detect free radicals more readily than some other reactive intermediates has probably contributed to overinterpretations of the role of free radicals. Perhaps the clearest example of the problem is the area of free radicals in chemical carcinogenesis. Free radical intermediates can be demonstrated for many (but certainly not all) chemical carcinogens. It is much more difficult, however, to prove that the identified free radical is an essential or even an important intermediate in the reactions that lead to carcinogenesis.

A third major problem is the frequent failure to relate the reactivity of free radicals to the postulated key reactions, including the location of these reactions.

The two extremes of high and low reactivity of free radicals each have led to some basic misunderstandings. With highly reactive free radicals such as the hydroxyl radical, it is difficult to infer from the occurrence of such reactive free radicals that they are important in specific key reactions, because they will react with virtually any organic molecule they encounter. Special circumstances are required for the hydroxyl radical or similarly reactive species to cause significant damage, because of the high concentration of organic molecules in and around the cell, and because they are usually generated at a specific site in the cell (e.g., microsomes or mitochondria) that is not the site at which the critical biological damage is postulated (e.g., at DNA or the plasma membrane). In many cases the strongly oxidizing species is generated in a Fenton-like reaction; this requires the production of both reduced metal ions (e.g., Fe^{++}) and hydrogen peroxide. One or both components of the Fenton-like reaction are likely to be produced at specific sites and thereby localize the production of the strongly oxidizing species.

There are at least three circumstances in which highly reactive radicals can be the key cause of biological damage:

1) If they are generated at random and in large numbers.

Under these circumstances, some may occur close enough to the target to cause the observed damage. This is the case for ionizing radiation:

(1) $H_2O + h\nu \rightarrow OH° + H°$

While more than 99% of the hydroxyl radicals generated by ionizing radiation have no significant long-lasting biological effect, those few generated in or very close to DNA do account for the key lesions in cell damage due to ionizing radiation.

2) *If they are generated at the site of the key damaging reactions.*

This appears to be the case for the metal complexes of glycopeptide antitumor agents such as bleomycin and tallysomycin that appear to be able to catalyze the generation of hydroxyl radicals that attack DNA (14) (Fig. 1).

THE MECHANISM OF BLEOMYCIN ACTION

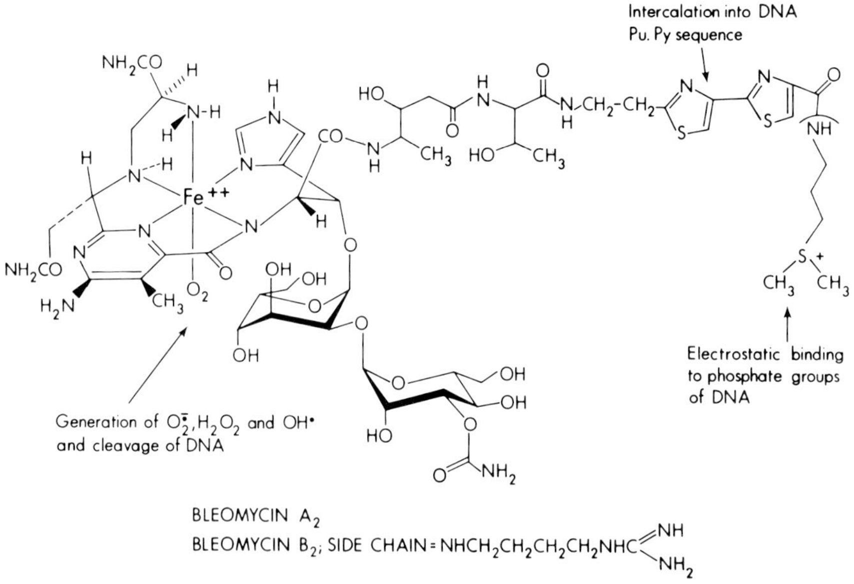

[Structural diagram of bleomycin showing Fe²⁺ complex, with annotations: "Intercalation into DNA Pu. Py sequence", "Electrostatic binding to phosphate groups of DNA", "Generation of O_2^-, H_2O_2 and OH• and cleavage of DNA"]

BLEOMYCIN A_2
BLEOMYCIN B_2; SIDE CHAIN = $NHCH_2CH_2CH_2CH_2NHC\underset{NH_2}{\overset{NH}{\diagup}}$

Figure 1 Possible Mechanism for Free Radical Mediated DNA Damage by an Antitumor Agent. The figure schematically indicates the several components that could facilitate such a mechanism: 1) binding to DNA via interacalation and/or electrostatic binding to phosphate groups; 2) complexing of a metal iron in an active reducing state (ferrous); and 3) potential generation of $O_2°$ and H_2O_2 with subsequent metal catalyzed generation of hydroxyl radicals (Figure from unpublished work of J. W Lown with permission).

In this illustrative example one part of the drug binds to the target and a second part of the molecule contains a complex metal ion (copper or iron) that upon reduction can cause a Fenton-like reaction in the presence of hydrogen peroxide, by the following reactions or their equivalent occurring:

(2) $Fe^{+++} + \text{Reducing Agent} \to Fe^{++} + \text{oxidized reducing agent}$

(3) $Fe^{++} + H_2O_2 \to Fe^{+++} + OH^\circ + OH^-$

Because of the location of the metal ion, The resulting flux of hydroxyl radicals has a reasonable probability of encountering the DNA before they can react with another cell constituent. As this case illustrates, it clearly requires very detailed understanding of the reactions and reactants involved in the production of strongly oxidizing species if one is to assess accurately the consequences of the production of these species.

3) <u>If the reactive radicals generate radicals of significant but lower reactivity.</u>

In order for damage to occur at sites distant from the site of generation of the strongly oxidizing species, the damaging species (or their precursors) must be able to diffuse a significant fraction of the dimensions of cells. While this is not possible for hydroxyl radicals, it is possible for some of the products of hydroxyl radicals. Again this indicates the need for much more detailed understanding of the nature and site of the production of strongly oxidizing species plus an understanding of the reaction products they produce. It is very likely that in some instances the chain of events triggered by the generation of strongly oxidizing species, such as hydroxyl radicals, will involve some nonradical intermediates and/or final reactants (4).

A complement of the concept that very reactive free radicals need special circumstances to cause significant biological damage is that very unreactive free radicals are also not likely to be very important biologically, except by indirect means. But in ESR studies of complex biological materials the free radicals one is most likely to observe are the very stable ones. More reactive radicals rapidly disappear (although in some favorable circumstances there may be a steady-state concentration of reactive free radicals that is sufficient for direct observation). In general, in complex biological systems, if you can see it easily, the free radical is likely to be unimportant!

The reaction of the generally unreactive superoxide anion radical with ferric ions or other metals, e.g.,

(4) $O_2^{\bar{\circ}} + Fe^{+++} \to O_2 + Fe^{++}$

may be a very important example of an exceptional circumstance in which relatively stable free radicals lead to important chemical reactions.

From these considerations it seems likely that species of intermediate reactivity are most likely to lead to significant biological damage. These may be able to diffuse over significant distances and then react, perhaps with some specificity, at critical targets. Very reactive radicals, if they are to cause damage directly, need to be generated at the critical sites. Very stable free radicals, while most readily observed, are unlikely to lead to significant biological damage except under special circumstances.

ELECTRON SPIN RESONANCE STUDIES OF CANCER AND FREE RADICAL REACTIONS WITH DNA

Many aspects of this subject have been dealt with fairly recently (e.g., 9,23,24) and I shall therefore limit my discussion to a description of some of the highlights. Similarly, the reader is referred to a recent review of ESR for biologically oriented scientists for details on the technique itself (26). The title ESR Studies of Cancer was chosen rather than Free Radical Studies of Cancer in order to emphasize that other ESR responsive species, especially paramagnetic transition elements and stable spin labels, are very important in ESR studies related to cancer. I have chosen to organize this section of the chapter by the consideration of some pertinent concepts and my estimation of their current validity. These concepts are considered in rough order (lowest to highest) of my evaluation of their proven pertinence and/or validity.

Intrinsic Role of Free Radicals in Cancer.

At this time the vast preponderance of the evidence indicates that free radicals do not have an intrinsic role in cancer. While the theories that postulated an essential role of free radicals in cancer were attractive inasmuch as they implied a potential means to prevent or control cancer, it now is clear that they are incorrect or, at best, vast oversimplifications. In retrospect, the theoretical bases of these theories were not substantial.

The lack of an essential role of free radicals in cancer does not imply that they have no importance in some aspects of cancer. As discussed later in this chapter, free radical intermediates are often involved in the mechanism of action of some carcinogens, in some anticancer agents and, perhaps, in some aspects of promotion. They also are important in many physiological and pathological processes. Thus the study of free radical reactions remains an important and useful avenue of studies of carcinogenesis as long as one does not attempt to force the results into a unifying theory of cancer or carcinogenesis. The same limitations apply even if one extends the theories to include other oxidative processes in addition to free radical reactions.

Free Radicals are Intrinsically Involved in Tumor Promotion.

Quite recently some interesting and potentially exciting results have been reported that indicate a potential role of superoxide anions in the mechanism of action of some tumor promoters (e.g., 3,6,11,27) (Fig. 2). The data indicate that tumor promoters such as phorbol myristate acetate: 1) can stimulate the production of superoxide anions by leukocytes; 2) can cause strand breaks in DNA; 3) have their effects on DNA inhibited, at least in part, by superoxide dismutase or catalase.

The data are consistent with a hypothesis that the production of superoxide can lead to a process in which a reactive species is generated that is able to reach and damage DNA and which includes the generation of hydrogen peroxide as an intermediate. A possible description of the process, based on present knowledge, is:

(5) $O_2^{\circ -} + Fe^{+++}_{(complexed)} \rightarrow O_2 + Fe^{++}_{(complexed)}$

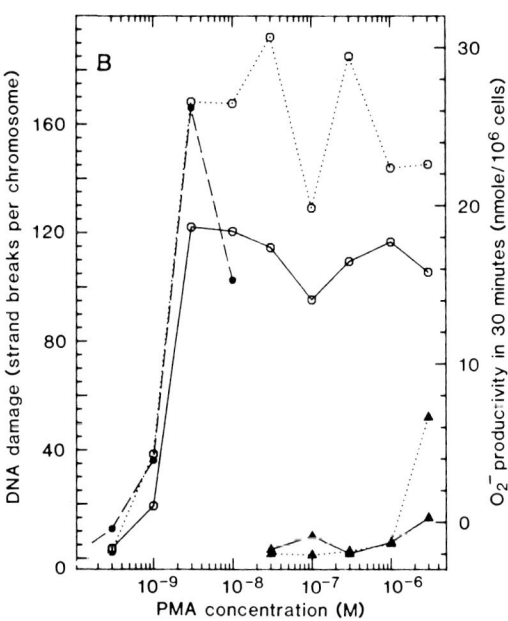

Figure 2 Some Evidence for a Possible Role of Superoxide Production and DNA Strand Breakage in the Mechanism of Action of Tumor Promoters. Using the tumor promoter phorbol myristate acetate (PMA) to stimulate human leukocytes, Birnboim (3) measured DNA breaks (open circles) (in two experiments) and production of superoxide anions (closed circles) as a function the concentration of PMA (with permission, (3)).

(6) $O_2^{-\circ} + HO_2 \xrightarrow{+H^+} H_2O_2 + O_2$

(7) $H_2O_2 + Fe^{++}\text{(complexed)} \rightarrow OH^\circ + OH^- + Fe^{+++}\text{(complexed)}$

(8) $OH^\circ + XH \rightarrow X^\circ + H_2O$

(9) $X^\circ + DNA \rightarrow$ DNA Strand Break

(10) DNA Strand Break $\rightarrow$ Promotion

I have written the above simplified scheme to provide the reader with insights into both the complexities of the process and on how many arbitrary conclusions are required at this time to extend the experimental observations into a plausible mechanism. Even if the above scheme turns out to be correct, it clearly is an oversimplification that neglects many important factors. More likely, it contains one or more major errors.

My conclusions on free radicals and promotion include: a key role of free radicals in this process is not yet proven; if free radicals are involved they may be an alternate rather than an essential pathway; other mechanisms of promotion are likely to occur that do not involve this or other free radical-linked pathways.

Role of Oxygen Radicals in Cancer.

The term "oxygen radicals" is not used consistently, but often encompasses small (and not so small) oxygen containing radicals (the superoxide anion, $O_2^{-\circ}$; the peroxy radicals, HO_2°; the hydroxyl radical, OH°; organic peroxyl radicals, RO_2°; and alkoxy radicals RO°); and related nonradical species (hydrogen peroxide, H_2O_2; organic hydroperoxides, ROOH; ozone, O_3; and singlet oxygen $^1O_2$). With such a diverse (but often inter-related) group of reactive compounds it is not surprising that many reactions associated with carcinogenesis, anticancer drugs, etc., involve some of this group. But the same arguments sketched out in previous sections of this chapter apply to "oxygen radicals"—there is no cohesive and defensible theory that links them intrinsically to cancer. Again, in particular instances some of these are involved and in other instances there is no evidence of their involvement.

Role of Quinone-Hydroquinone Couples.

These systems are attracting increasing attention and study in biological systems and are becoming increasingly recognized as key components of many of the systems. The quinone-hydroquine equilibria usually involve a semiquinone intermediate that is a free radical:

(11) [Quinone] + [Hydroquinone] → 2 [Semiquinone] + 2H⁺

Quinone   Hydroquinone   Semiquinone

The quinone is a reducing agent, the hydroquinone is an oxidizing agent. The intermediate semiquinone free radicals will have a range of activities depending on the nature of the substituent groups of the parent compounds. The semiquinones can be fairly stable; if they react, the most common reactions are likely to be abstraction of an electron or disproportionation to form the parent quinone and hydroquinone. The reaction to form the semiquinone can be stabilized and driven to the right in the presence of metal ions that can complex to it:

(12) [semiquinone] + M^{n+} → [semiquinone-metal complex]

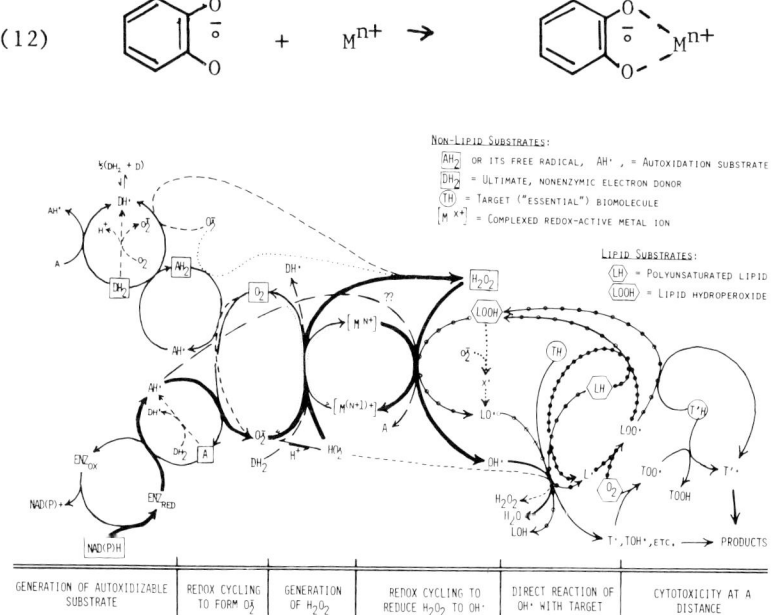

Figure 3 Diagram of Interactions Leading From Redox Couple (DH_2 and D) to Changes in the Eventual Target (T). The redox couple or autoxidizable substrate may be a quinone-hydroquinone system (from Borg & Schaich (4), with permission).

Recently Szent-Györgi and co-workers have postulated a fundamental role of quinones in cell regulation and have also proposed that they may be effective anticancer agents. They have published preliminary data on the generation of semiquinones when quinones and/or hydroquinones are added to cells, the efficacy of these agents as antitumor agents, and the possible role of ascorbic acid in these effects (18). Borg and co-workers (4) have developed a complex and comprehensive scheme showing possible interactions between oxidations, lipids, and eventual biological damage in which redox couples such as quinone-hydroquinones can generate the actual cytotoxic agents (Fig. 3).

The figure provides a good overview of some of the complexities of free radical reactions in cells and of some of the interactions with nonradical species. Similarly, the work by Pethig et al. (18) indicates the possible relationship of quinone-hydroquinones with ascorbic acid as does the work of Borg et al. (5).

Quinone-Hydroquinone chemistry may also be involved in the action of many metabolic intermediates (e.g., ubiquinones) and drugs. The quinone antibiotics such as adriamycin are important examples of the latter (2). In complex biological systems adriamycin readily becomes converted to a semiquinone (Fig. 4) which may be a source of some of the toxic side effects of this drug, especially the toxic effects on the heart.

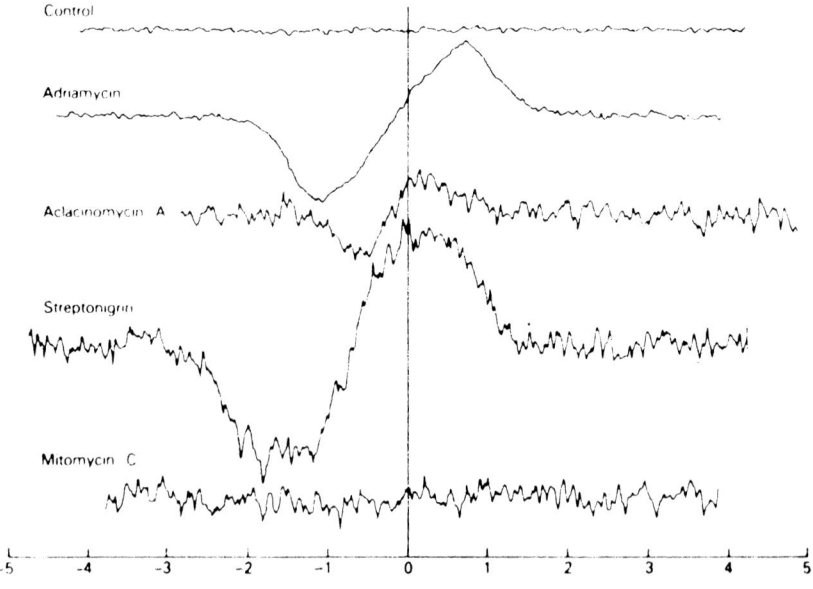

Figure 4 Electron Spin Resonance Spectra of Quinone Antibiotics Generated in Mixtures with Microsomes (from Bachur et al. (2), with permission).

It is less likely that the semiquinone is involved in the general cytotoxic properties of adriamycin which are the basis of its therapeutic use, although it is an attractive possibility that after adriamycin binds to DNA that it then could provide a primary or secondary source of free radicals at a critical site. Analogs of the monomers of melanin are an example of another quinone based anticancer drug, one with a more direct link of its free radical intermediates to the mode of action (20).

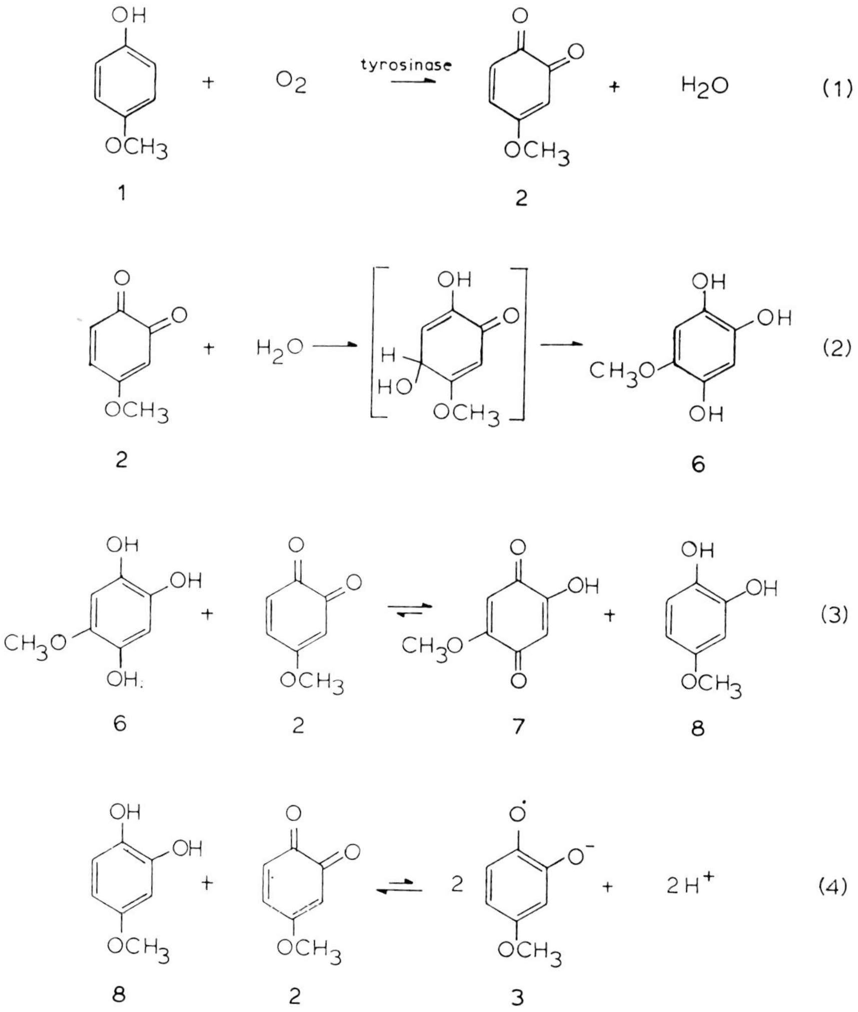

Figure 5 Some of the Early Steps and Intermediates Formed by Reaction of Substrates with the Enzyme Tyrosinase (from Nilges et al., 16, with permission).

As described in detail by Sealy and Hyde (22) elsewhere in this book, the biological pigment melanin is formed by polymerization of monomers such as dihydroxyphenylalanine via free radical intermediates and is, itself, a free radical due to quinone-hydroquinone equlibria of the monomers and their derivatives. The same enzyme (tyrosinase) that catalyses the early steps of melanin formation has been used to convert related monomers into toxic forms that selectively attack malignant tumors derived from melanocytes (melanomas) by converting these drugs into free radicals and/or other toxic intermediates. Although the concept is straightforward and the approach has led to some promising therapeutic results, the detailed chemistry turns out to be illustrative of the complexity of these types of reactions (Fig. 5).

In order to optimize the clinical efficacy of this approach we shall probably need to learn the details of the factors that result in different yields of products and of the toxicity, lifetimes, etc., of the products.

These brief examples are intended to illustrate the widespread occurrence and biological pertinence of quinone-hydroquinone reactions, and the complexity of such systems. While quinone-hydroquinone systems are important to understanding aspects of cancer in specific instances, they are unlikely to provide systematic insights into the process of cancer or its treatment.

Role of Ascorbyl Radicals in Cancer.

Our understanding of the role of ascorbic acid in cancer is similar to our understanding of its role in normal metabolism: pervasive, apparently important and without fundamental knowledge of the details of its role. The subject of the role of ascorbic acid in free radical reactions and cancer has been reviewed recently (7,23,25). The conclusions included: the ascorbyl radical itself is rather unreactive; its significance may lie in part as an intermediate in the ascorbic acid--dehydroascorbic acid equilibrium; and its occurrence may reflect the redox state of the biological system and the generation of free radicals in that system. Some of the early findings that malignant tissues had "increased numbers of free radicals" turned out to reflect in part the fact that lyophilization of complex biological materials generates free radicals and/or precursors that become free radicals when exposed to oxygen (24). The actual ESR signal observed turned out to be the immobilized ascorbyl radical. Similarly the ESR signal seen in irradiated wet tissues turned out to be the ascorbyl radical in aqueous solution (10). Borg et al. demonstrated that redox system with free radical intermediates (such as quinone-hydroquinone systems), in the presence of ascorbic acid, often have ascorbyl radicals as the only observable species (5). Very recently, Pethig et al. have reported studies in which mixtures of substituted quinone systems with ascorbic acid resulted in cytotoxic species (18).

Some generalizations of these and other studies involving the free radical chemistry of ascorbic acid and cancer include:

1) the redox properties of the ascorbic acid system result in the potential for a wide range of chemical and biochemical reactions.

2) the occurrence of detectable amounts of the ascorbyl radical reflects this nonspecific reactivity and the fact that the ascorbyl radical is relatively stable and hence the most detectable part of a series of reactions of free radicals.

3) if we understand more fully the factors that affect the amounts of ascorbyl radicals we see, we may be able to use this as an overall measure of the complex redox states within cells and tissues. Recent studies by Lohman et al. (13) on ascorbyl radicals in lyophilized blood of patients with leukemia may represent this type of use of the ascorbyl radical.

Role of Free Radicals in Chemical Carcinogenesis.

A number of years ago it was postulated that carcinogens act by free radical mechanisms (24). This speculation was based in part on the knowledge that ionizing radiation generates large numbers of free radicals and is a potent carcinogen. It was buttressed by some early studies of chemical carcinogens that indicated that some of these formed free radicals. Subsequently an enormous amount of knowledge has developed on chemical carcinogenesis and it now is clear that while some carcinogens have free radical intermediates, other potent carcinogens do not (e.g., 8,12). Of those chemical carcinogens that do have free radical intermediates, only for some is a free radical the ultimate carcinogen. A number of other carcinogens have free radical intermediates but they may occur only in a minor pathway or involve reactions not directly connected to carcinogenesis. Reactions of free radical intermediates with oxygen may contribute to their effects.

DNA--Radical Reactions.

The ultimate target in carcinogenesis must involve DNA if the progeny of cells are to be malignant. Although there are plausible means whereby damage at a site distant from nuclear DNA may eventually result in permanent alterations of DNA, in the vast majority of known cases the critical changes are induced directly in DNA.

Free radical reactions with DNA are clearly demonstrable for radiation damage. Virtually all molecular level events caused immediately by ionizing radiation are irrelevant to cell death, mutation, or carcinogenesis unless they involve DNA. The biologically significant effects of radiation usually do involve

formation of very reactive free radicals either directly in the
DNA as the result of electronic excitation and ionization from
radiation-induced fast electrons or indirectly from radicals
generated very close to the DNA (usually hydroxyl radicals,
generated in water adjacent to DNA). The chemistry of most of
the early events involved in radical reactions induced in DNA by
ionizing radiation is well worked out and has recently been
summarized by Scholes (21). Radical reactions can occur
directly with nucleic acid bases and with the sugar-phosphate
chains. These reactions appear to account for most or all of
the biologically important damage by radiation but there is
incomplete knowledge as to which of the reactions of free
radicals with DNA are most important.

There are reasonably strong data indicating that free radical
derivatives of some carcinogens react with DNA. There also is
good direct ESR evidence for the interaction of some antitumor
agents such as bleomycin with DNA (15) (Fig. 6).

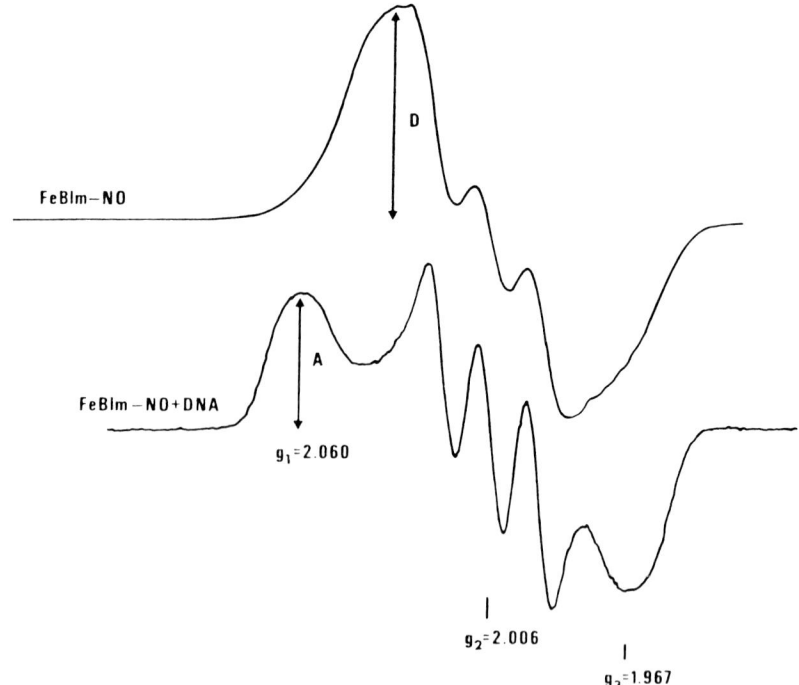

Figure 6 Evidence by Electron Spin Resonance Spectroscopy for
the Interaction of Metal-Bleomycin Complexes with DNA. The
spectra are from frozen solutions (-196°C) of the nitrous oxide
derivative of iron bleomycin in the presence or absence of calf
thymus DNA (from Antholine and Petering (1) with permission).

The interaction of bleomycin with DNA is particularly interesting because the metal complex of bleomycin (which is, most probably, the in vivo form of the drug) may be a free radical–metal complex with the unpaired electron distributed in both the metal ion and the bleomycin, and the complex probably generates superoxide and its products. As noted earlier, this provides a mechanism for localizing free radical production at the presumed site of significant reactions.

There are data that suggest that other drugs may react or bind to DNA and generate reactive species. Recently Pathak and Joshi (17) have presented evidence that the psoralens have an additional mode of action that involves generation of both singlet oxygen and superoxide anions. Because psoralens bind to DNA, this appears to be a case where reactive products may be produced at the critical target. (This presumes that the superoxide is converted to a more reactive species by a series of reactions such as those indicated earlier in this chapter.) The authors suggest that the free radicals and associated strongly oxidizing species that are produced contribute to the carcinogenic and mutagenic effects of these chemicals.

In summary, there are many direct and indirect lines of evidence indicating that free radicals and other strong oxidants can react with DNA and that such reactions contribute to significant biological effects such as carcinogenesis, mutagenesis, and cytotoxicity. Most of the free radical reactions require the generation of the free radicals at or near the DNA. There is no evidence that free radical reactions with DNA are an obligatory mechanism of carcinogenesis, mutagenesis, or cell killing.

Electron Spin Resonance as an Experimental Tool in Studies of Cancer.

The data cited in the previous parts of this chapter were obtained, to a great extent, by ESR techniques. These techniques have been and in the foreseeable future will continue to be essential for adequate studies of free radicals. In addition there is a powerful array of experimental approachs that employ ESR spectroscopy and artificial free radicals--spin labels. Spin labels provide very effective tools for measuring, among other things: 1) motion in regions such as membranes or cytoplasm, 2) polarity and changes in partition of probe molecules, 3) localization and environment of metal ion sites, 4) oxygen concentration, and 5) amount and sites of reduction within cells.

There are a number of new developments occurring in ESR spectroscopy that indicate this will be an even more valuable technique in the future. Areas of particular promise for studies of cancer include development of spectrometers with enhanced capabilities for resolving very fast events, development of capabilities for carrying out ESR studies in living animals, and development of spectrometers that are capable of obtaining data from very small samples, perhaps eventually as small as a single cell.

SUMMARY

In summary, I have attempted in this chapter to make it more feasible to study effectively those aspects of free radical reactions that are important to cancer (and aging) by summarizing some aspects of this field and by pointing out some of the misleading and/or erroneous concepts regarding free radicals. Free radical reactions are involved directly or indirectly in significant aspects of most processes related to cancer. Their role is often part of a more general role of deleterious oxidative reactions which can proceed by both free radical and nonfree radical pathways. In addition, the use of electron spin resonance has been and will continue to be an important experimental tool in studies related to cancer. It enables one directly to study free radicals and paramagnetic trace elements, and to utilize spin labels as probes.

Acknowledgements

We gratefully acknowledge support by the National Foundation of Cancer Research, Washington, DC.

REFERENCES

1. Antholine, W. E. and Petering, D. H. (1979): Biochem. Biophys. Res. Commun., 91:528-533.

2. Bachur, N. R., Gordon, S. L., and Gee, M. V. (1978): Cancer Res., 38:1745-1750.

3. Birnboim, H. C. (1982): Science, 215:1247-1249.

4. Borg, D. C. and Schaich, K. M. (1983): In: Oxy Radicals and their Scavenger Systems, Vol I, edited by G. Cohen and R. A. Greenwald, pp. 122-130. Elsevier Science Publishing Co., Inc.

5. Borg, D. C., Schaich, K. M., Elmore, J. J., Jr., and Bell, J. A. (1978): Photochem. and Photobiol., 28:887-907.

6. Copeland, E. S. (1983): Cancer Res., 43:5631-5637.

7. Dodd, N.J.F., and Swartz, H. M. Brit. J. Cancer, (in press).

8. Floyd, R. A. (1982): In: Free Radicals and Cancer, edited by R. A. Floyd, pp. 361-396. Marcel Dekker, Inc., New York.

9. Floyd, R. A., editor. (1982): Free Radicals and Cancer, Marcel Dekker, Inc., New York.

10. Floyd, R. A., Brondson, A., and Commoner, B. (1973): Ann. N.Y. Acad. Sci., 222, 1077-1086.

11. Goldstein, B. D., Witz, G., Amoruso, M., Stone, D. S., and Troll, W. (1981): Cancer Lett., 11:257-262.

12. Lesko, S. A., Ts'o, P.O.P., Yang, S. U., and Zheng, R. (1982): In: Free Radicals, Lipid Peroxidation, and Cancer, edited by D.C.H. McBrien and T. F. Slater, pp. 401-420. Academic Press, New York.

13. Lohmann, W., Bensch, K. G., Sapper, H., Pleyer, A., Schreiber, J., Kang, S. O., Löffler, H., Pralle, H., Schwemmle, K., and Filler, R. D. (1982): In: Free Radicals, Lipid Peroxidation, and Cancer, edited by D.C.H. McBrien and T. F. Slater, pp.55-73, Academic Press, New York.

14. Lown, J. W., Joshua, A. V., and Chen O. (1982): In: Free Radicals, Lipid Peroxidation, and Cancer, edited by D.C.H. McBrien and T. F. Slater, pp. 305-328, Academic Press, New York.

15. Nagata, C., Kodama, M., Ioki, Y., and Kimura, T. (1982): In: Free Radicals and Cancer, edited by R. A. Floyd, pp. 1-62, Marcel Dekker, New York.

16. Nilges, M. J., Swartz, H. M., and Riley, P. A., J. Biol. Chem. (in press).

17. Pathak, M. A., and Joshi, P. C. (1983): J. Inves. Derm., 80:665-745.

18. Pethig, R., Gascoyne, P. R., McLaughlin, J. A., Szent-Györgyi, A. (1983): Proc. Natl. Acad. Sci., 88:129-132.

19. Pryor, W., editor. (1976, 1976, 1977, 1980, 1982): Free Radicals in Biology, Vols I-V, Academic Press, New York.

20. Riley, P. A., Morgan, B.D.G., O'Neill, T., Dewey, D. L., and Galpine, A. R. (1982): In: Free Radicals, Lipid Peroxidation, and Cancer, edited by D.C.H. McBrien and T. F. Slater, pp. 421-438, Academic Press, New York.

21. Scholes, G. (1983): Br. J. Radiol, 56:221-231.

22. Sealy, R., Hyde, J., this book.

23. Swartz, H. M. (1982): In: Free Radicals, Lipid Peroxidation, and Cancer, edited by D.C.M. McBrien and T. F. Slater, pp. 5-26, Academic Press, New York.

24. Swartz, H. M. (1972): In: Advances in Cancer Research, edited by G. Klein and S. Weinhouse, pp. 227-252, Academic Press, New York.

25. Swartz, H. M. and Dodd, N.J.F. (1982): In: Oxygen and Oxy-Radicals in Chemistry and Biology, edited by M.A.J. Rodgers and E. L. Powers, pp. 161-168, Academic Press, New York.

26. Swartz, H. M. and Swartz, S. M. (1983): In: Methods of Biochemical Analysis, edited by David Glick, pp. 207-323, John Wiley & Sons, New York.

27. Witz, G., Goldstein, B. D., Amoruso, M., Stone, D. S., and Troll, W. (1980): Bioch. Biophys. Res. Comm., 97:883-888.

Free Radicals in Molecular Biology, Aging, and Disease, edited by D. Armstrong et al. Raven Press, New York © 1984.

Free Radicals, Lipid Peroxidation, and Cancer

T. F. Slater, K. H. Cheeseman, and Karen Proudfoot

Department of Biochemistry, Brunel University, Uxbridge, Middlesex UB8 3PH, United Kingdom

Environmental hazards that may result in cancer are radiation, chemicals and some viruses; only the first two of such causes of cancer will be discussed in this article: comprehensive reviews on oncogenic viruses that may be consulted are by Klein[20] and Phillips[24]. The exposure to radiation of different types and energies has been known for a long time to be associated with increased incidences of cancer of various kinds. For example, the increased incidence of leukemia and other cancers in survivors of the Japanese atomic bomb explosions[10]; the increased incidence of cancer in early pioneers working with radio-activity or with radiotherapy[33]; and the increased incidence of skin cancer in Caucasians exposed to strong sunlight[6].

That chemical substances may produce cancer can be traced to the observation of Sir Percival Pott that young boys working as chimney sweeps had an increased incidence of cancer of the scrotum[25]. A large number of chemicals are now known to be carcinogenic in one or more animal species; for review see Searle[30].

Many chemical carcinogens require an enzymic conversion to a reactive form in order to exert their full carcinogenic potential[22], this process is known as metabolic activation and often involves the NADPH-cytochrome P_{450} electron transport system. Although the latter system is mainly concentrated in the endoplasmic reticulum of liver cells, activities of the component NADPH-cytochrome P_{450} reductase and cytochrome P_{450} can be detected in the endoplasmic reticulum of many other tissues[3]. Within a particular tissue, liver[17] or lung[31] for example, the cytochrome P_{450} (and presumably the NADPH-cytochrome P_{450} reductase) can be distributed heterogenously between cells of different types, or between cells of the same type but different location. The metabolic activation of a toxic substance (T) to a more toxic intermediate (T*) may involve electron transfer reactions from either the NADPH-flavoprotein or from cytochrome P_{450} as indicated in Figure 1; such reactions are abbreviated in the remainder of this article as P_{450}-mediated reactions.

The metabolic activation of carcinogens by such P_{450}-mediated reactions is known to produce a variety of activated species: carbonium ions, epoxides, epoxide-dihydrodiols, and free radicals. Examples of substances known to be activated to free radical intermediates are given in Table 1: it is important to recognise that in some of these examples the free radical intermediate may not be the major metabolically activated intermediate or the most toxic intermediate in relation to cancer.

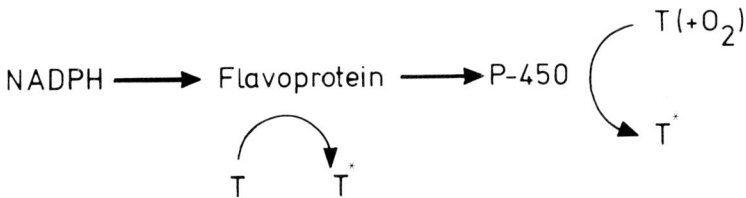

FIG.1. The NADPH-cytochrome P_{450} electron transport chain. Electron donation to a toxic agent (T), which may be a chemical carcinogen, to produce an activated intermediate T* is shown from the flavoprotein (FP) or P_{450}. T* can be either the simple electron adduct ($T^{\bullet -}$), or some subsequently rearranged or transformed product (such as $CCl_3^{\bullet}$ from $CCl_4^{\bullet -}$).

TABLE 1. Compounds that give free radical intermediates on metabolic activation

Halogenoalkanes :	Nitro-compounds
CCl_4	Aromatic amines
$CHCl_3$	Nitrosoamines
$CBrCl_3$	Quinones
dibromoethane	Polycyclic hydrocarbons
halothane	Polyunsaturated fatty acids

Free radical intermediates appear to be of major significance in some types of chemically-induced cancer[15,21]; as well as in radiation-induced cancer. It is of importance, therefore to understand more of the damaging events that reactive free radicals can produce in cells, and of the kinetic aspects of such events. In consequence, a brief survey will be given of methods used to detect free radicals; of procedures used to study the kinetics of free radical reactions in solution, and of the major ways in which reactive free radicals can produce cell injury.

FREE RADICALS AND TISSUE INJURY

Detection of free radicals

The direct and unequivocal method for detecting and studying free radicals is electron spin resonance (esr) spectroscopy[7]. The method is relatively insensitive, however, and many free radicals formed by metabolic activation in biological systems are present in concentrations too low for direct esr study. An unusually strong esr signal has been reported[4] in samples of intact frozen human cervix, and in frozen powders prepared from that intact material; the signal was much decreased in samples of cancer of the cervix.

Whole tissue and tissue fractions often give complex esr spectra; most studies are done on frozen samples to decrease the disturbing effects of aqueous water. Studies with lyophilised material may give artefactual signals that complicate interpretation. A useful review on the use of esr with normal tissues and cancer samples is by Swartz[39].

An indirect esr method that is useful for detecting and identifying free radical intermediates occurring in low concentrations is spin trapping. In this procedure the spin trap interacts with the free radical, which may be present in very low concentration but is continuously produced, to yield a relatively stable nitroxy-radical adduct (Figure 2).

A review of spin trapping techniques in biochemistry is by Janzen[18]; illustrations concerning metabolically activated intermediates are given in references[2,28].

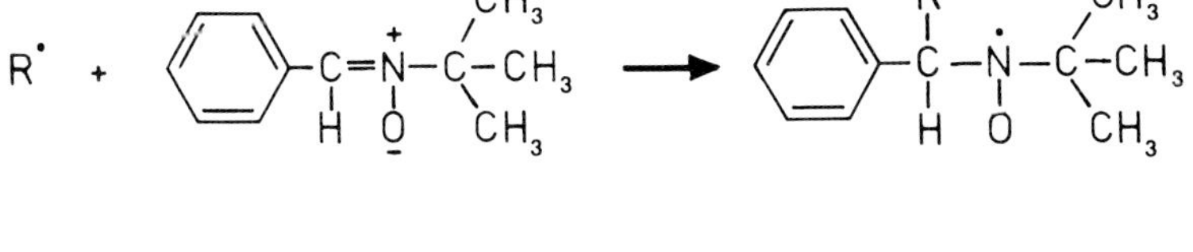

FIG.2. The interaction of a reactive free radical, R•, with the spin trap phenyl-t-butyl-nitrone to yield the nitroxy-radical adduct.

Kinetic studies on free radical reactions in solution

Classical chemical techniques for the study of free radical reactions in solution can be found described in many text-books and reviews such as that by Walling[42]. Rate constants and the spectra of transient intermediates can be obtained by fast reaction techniques such as flash photolysis and pulse radiolysis[5].

When the pulse radiolysis technique was applied to a study of the chemical reactivity of the trichloromethyl free radical, which is produced *in vivo* by the metabolic activation of CCl_4, it was found that this free radical species is relatively unreactive; however, it interacts rapidly with oxygen to form a peroxy-radical of much higher reactivity[23]. Other examples are known where the presence of oxygen can greatly increase the reactivity of a transient intermediate[16]. Under conditions *in vivo*, therefore, the local concentration of oxygen around the locus of formation of free radical intermediates formed by metabolic activation can be of crucial importance in modifying the chemical reactivity of the free radical.

In assessing the contribution that techniques such as pulse radiolysis can make to free radical interactions *in vivo*, it is important to recognise that the data obtained with homogeneous phases in pulse radiolysis can give absolute rate constants that may be considerably different to the apparent rate constants that apply to the heterogeneous phases of biological systems.

Damaging reactions of free radicals

Reactive free radicals formed by metabolic activation in the endoplasmic reticulum, or by irradiation throughout the cell, can cause metabolic disturbances and damage to membrane structure through a variety of ways[34]. Figure 3 is a diagrammatic representation of major types of free radical mediated damage.

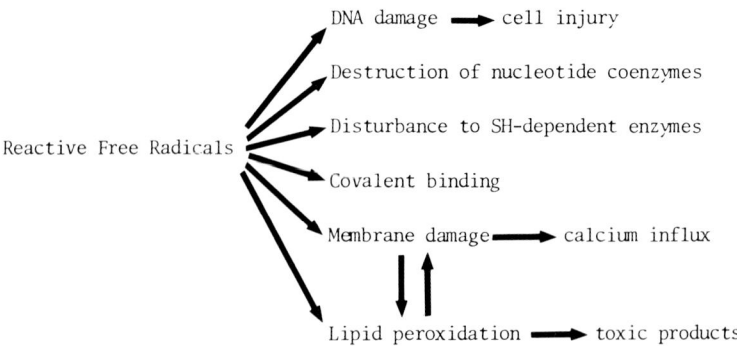

FIG.3. Major types of damage that reactive free radicals can produce within cells.

Although damage to DNA is indicated in Figure 3 it is important to remember that highly reactive free radicals are essentially trapped in the immediate vicinity of their site of formation as a consequence of their rapid interaction with neighbouring molecules. In consequence, their radius of diffusion is often very small in cellular terms. Reactive free radicals formed in the endoplasmic reticulum are unlikely to diffuse far enough to react with DNA in the nucleus. It has been postulated[35] therefore, that metabolically activated free radicals must have an intermediate chemical reactivity in order to interact directly with DNA: highly reactive free radicals cannot diffuse far enough; free radicals of very low reactivity, whilst able to diffuse will not be reactive enough to produce significantly important covalent adducts. This argument will not apply to free radicals formed close to DNA as occurs after irradiation.

Figure 3 indicates other ways in which major disturbances to the cell may occur through free radical reactions: changes in the thiol-content of proteins that may adversely affect their enzyme activity, and destruction of nucleotides through interactions close to the site of metabolic activation. An example of the latter reaction has been reported to follow metabolism of CCl_4 when the liver's content of NADPH rapidly decreases[38]. Reactive free radicals can also cause cellular malfunction by covalent binding and by stimulating lipid peroxidation. The latter situation is attended by a number of important biological consequences and is discussed separately below.

Lipid Peroxidation

Lipid peroxidation is a free radical mediated chain reaction that results in the oxidative degradation of unsaturated lipids, especially polyunsaturated fatty acids such as $C_{20:4}$ and $C_{22:6}$, which occur in the phospholipids of most biological membrane assemblies. The essential reactions that occur in lipid peroxidation have been reviewed by Pryor in this volume[27].

Many methods have been used for measuring lipid peroxidation in biological samples; the most often used method is that based on the thiobarbituric acid reaction with malonaldehyde and malonaldehyde-like substances. Each method has its characteristic advantages and disadvantages; an overview of these methods is by Slater[37].

Various model systems are used to study lipid peroxidation in biological materials. With liver microsomes, for example, peroxidation can be stimulated by (i) γ-irradiation; (ii) uv-irradiation; (iii) cumene hydroperoxide; (iv) $\pm Fe^{2+}$; (v) NADPH; (vi) NADPH + ADP/Fe^{2+}; (vii) NADPH + CCl_4; (viii) cysteine + Fe^{2+}. With such systems some marked differences have been observed in the behaviour of microsomes prepared from different species[19,26]. Some data for rabbit, rat, mouse and guinea-pig are shown in Figure 4.

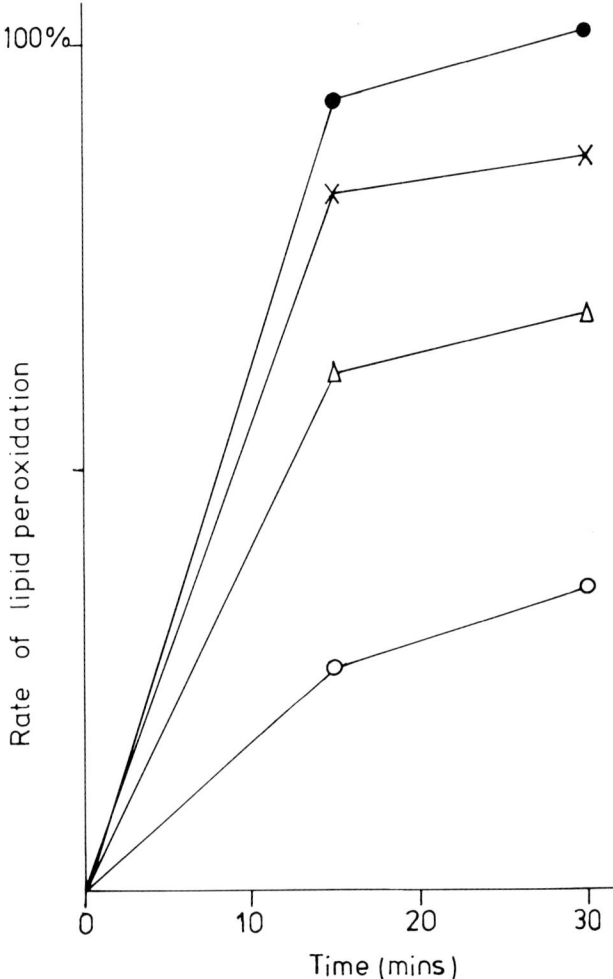

FIG.4. Relative rates of lipid peroxidation in liver microsomes prepared from rat (●), rabbit (o), mouse (x) and guinea-pig (Δ). The value obtained for rat microsomes at 30 min incubation has been set at 100%. Lipid peroxidation was stimulated by the addition of NADPH/ADP/Fe^{2+}, and measured by the thiobarbituric acid reaction.

Lipid peroxidation of biomembranes and of polyunsaturated fatty acids produces a variety of products (Table 2) some of which have very high biological activity.

TABLE 2. Products of lipid peroxidation and lipoxygenase

Leukotrienes [a]	Alkanals
Lipid hydroperoxides	Alkenals [c]
Hydroxy fatty acids	4-hydroxy-Alkenals [c]
Epoxy-fatty acids [b]	Ketones
	Alkanes

[a] reference 29, [b] reference 32, [c] reference 13

The aldehydic products of peroxidising microsomes are a complex family of substances that have been separated[14] by high performance liquid chromatography (hplc) and identified by gas chromatography and mass spectrometry (gc-ms). Among this family of aldehydic products are a number of 4-hydroxy-alk-2-en-1-als that have important biological reactivities (Table 3).

TABLE 3. Structural formulae of some 4-hydroxy-alkenals formed by lipid peroxidation of microsomal suspensions

$OHC-CH_2-CHO$	MALONALDEHYDE
$CH_3(CH_2)_3CHO$	PENTANAL
$CH_3(CH_2)_4-CH=CH-CHO$	OCT-2-EN-1-AL
$CH_3(CH_2)_3\underset{OH}{CH}-CH=CH-CHO$	4-OH-OCTENAL
$C_5H_{11}-\underset{OH}{CH}-CH=CH-CHO$	4-OH-NONENAL

In particular, the 4-hydroxy-alkenals react readily with thiol-groups, and inhibit DNA-synthesis at relatively low concentrations. (Table 4).

Since these products of lipid peroxidation have an inhibitory effect on DNA-synthesis, and on cell division and tumour growth, it is of interest to consider the regulation of lipid peroxidation in cancer cells.

TABLE 4. The effects of 4-hydroxy-pentenal on DNA-synthesis in tumour cells [a,b]

Tumour	Pre-incubn. with HPE (min)	HPE μM	Incubn. with ^{3}H-T	% of control
TLX-5	60	100	20	37%
	60	200	20	4%
	-	100	20	65%
	-	200	20	40%
Walker-256	60	100	20	20%
	60	200	20	5%
	-	100	20	85%
	-	200	20	45%
Gardner	30	40	30	85%
	30	80	30	35%
	30	160	30	10%
	30	320	30	8%
Ehrlich	30	40	30	85%
	30	80	30	65%
	30	160	30	25%
	30	320	30	6%
Sarcoma-180	30	40	30	81%
	30	80	30	42%
	30	160	30	8%

[a] Unpublished results of J.T.Nodes, P.J.Conroy and T.F.Slater

[b] DNA-synthesis measured by ^{3}H-thymidine incorporation

Lipid Peroxidation in Cancer

In general, mitochondrial or microsomal suspensions prepared from cancer cells peroxidise slowly[1,11,8,40,41] although exceptions have been reported. With liver tumours, for example, it has been found[1] that tumours resulting from aflatoxin or ethionine administrations have much slower rates of microsomal peroxidation than normal liver microsomes.

A much publicized difficulty in such comparisons is the problem of finding an appropriate normal cell population with which to compare the cancer sample. However, since most if not all of the liver tumours that are passaged *in vivo* show a much reduced peroxidation compared to normal liver then it seems reasonable to conclude that for liver tumours, at least, lipid peroxidation is generally depressed.

There are a number of possible reasons for such a decreased rate of lipid peroxidation. Firstly, the content of polyunsaturated fatty acid in the tumour membrane suspension may be reduced. Secondly, the concentration of the P_{450}-system, which often acts as the agent for free radical formation and, hence, for the initiation of lipid peroxidation, may be decreased. Thirdly, the local concentration of free radical scavengers may be increased so that the chain reaction of lipid peroxidation is inhibited. Results consistent with each of these possibilities have been published; a summary of our results[12] with the Novikoff hepatoma is given in Table 5.

TABLE 5. Lipid Peroxidation, Lipid Content and Antioxidant: Lipid Ratios in Novikoff Hepatoma Cells and Normal Rat Hepatocytes [a]

	Control liver	Tumour cells
Total Lipid (g/100 g)	2.5	1.0
Cholesterol/Total Lipid	9.6	18.7
Fatty Acids (% total fatty acids)		
16:0	18.4	13.8
18:0	17.7	17.0
18:1	11.5	20.0
18:2	19.4	29.5
20:4	15.5	9.6
22:6	6.5	1.9
Cytochrome P_{450} (% control)	100	Not detectable
NADPH-cytochrome c reductase (% control)	100	10
NADPH ADP-Fe^{2+} lipid peroxidation (% control)	100	9
CCl_4-stimulated lipid peroxidation (% control)	100	Not detectable
Total antioxidant/Total Lipid	2.6	6.9

[a] For details see reference 12

It can be seen from Table 5 that there is a decrease in total lipid, an increase in cholesterol, a decrease in polyunsaturated fatty acids, a substantial decrease in the P_{450}-system, and an increase in total antioxidant in the Novikoff cells (and microsomes) compared to normal hepatocytes (or normal liver microsomal suspensions). Of course, the Novikoff hepatoma is a maximum deviation hepatoma with many enzymic disturbances[43] compared to normal liver; even with that proviso, however, the changes reported in Table 5 are substantial. It will be of considerable interest and importance to see how a group of less deviated hepatomas, with a range of growth rates, behave in respect of changes in lipid peroxidation and antioxidant content.

Lipid Peroxidation and Cell Division

The above discussion, directed principally at the liver, has reported that lipid peroxidation of biological membrane suspensions, such as microsome suspensions, produces substances that can inhibit DNA-synthesis and cell division: in particular, the group of 4-hydroxy-alkenals are very active in these respects. Also, that cancer cells, particularly liver cancer cells, have a much reduced lipid peroxidation. Restricting the following discussion to liver cells, we can then ask the following questions: is the decreased lipid peroxidation in liver cancer one of the contributory features that relieve cell division from physiological constraints ?; or is the decreased lipid peroxidation in cancer a secondary phenomenon that is of little significance to the failure of normal control mechanisms for cell division ? Although many attempts have been made to answer such questions, for example by following changes in lipid peroxidation in normal liver regeneration[9], or during normal liver circadian rhythms[36], or during the development of chemically-induced primary liver tumours[8], no clear answers have yet emerged. Probably, an unequivocal answer will have to await definite evidence for the occurrence and significance in normal liver cells of a "physiological" lipid peroxidation: that, in turn, must await the development of cytochemical methods of sufficient sensitivity that can be used with single cells.

Acknowledgements: We are grateful to the National Foundation for Cancer Research, and to the Cancer Research Campaign for generous financial support and encouragement.

REFERENCES

1. Ahmed, S.M. and Slater, T.F. (1981): In: Recent Advances in Lipid Peroxidation and Tissue Injury, edited by T.F.Slater and A.Garner, pp.177-194. Dept. of Biochemistry, Brunel University, Uxbridge, U.K.

2. Albano, E., Lott, K.A.K., Slater, T.F., Stier, A., Symons, M.C.R. and Tomasi, A. (1982): Biochem.J., 204: 593-603.

3. Benedetto, C., Dianzani, M.U., Ahmed, M., Cheeseman, K.H., Connelly, C. and Slater, T.F. (1981): Biochim.biophys.Acta, 677: 363-372.

4. Benedetto, C., Bocci, A., Dianzani, M.U., Ghiringhello, B., Slater, T.F., Tomasi, A. and Vannini, V. (1981): Cancer Research, 41: 2936-2942.

5. Bensasson, R.V., Land, E.J. and Truscott, T.F., editors (1983) Flash Photolysis and Pulse Radiolysis. Pergamon Press, Oxford.

6. Blum, H.F. (1964): Photodynamic action and diseases caused by Light. Hafner, New York.

7. Borg, D.C. (1976): In: Free Radicals in Biology, edited by W.A.Pryor, volume 1, pp.69-147. Academic Press, New York.

8. Burlakova, E.B. (1975): Russian Chemical Reviews, 44: 871-880.

9. Burlakova, E.B., Molochkina, E.M. and Pal'mina, N.P. (1980): In: Advances in Enzyme Regulation, edited by G.Weber, pp. 163-179. Pergamon Press, Oxford.

10. Cairns, J. (1978): Cancer, Science and Society. W.H.Freeman, San Francisco.

11. Cheeseman, K.H. (1982): Carbon Tetrachloride Metabolism and Lipid Peroxidation in Rat Liver Microsomes; Ph.D. thesis. Brunel University, Uxbridge, U.K.

12. Cheeseman, K.H., Burton, G., Ingold, K.U., Webb, Ann C.and Slater, T.F. (1984). In preparation.

13. Esterbauer, H. (1982): In: Free Radicals, Lipid Peroxidation and Cancer, edited by D.C.H.McBrien and T.F.Slater,pp.101-122. Academic Press, London.

14. Esterbauer, H., Cheeseman, K.H., Dianzani, M.U., Poli, G. and Slater, T.F. (1982): Biochem.J., 208: 129-140.

15. Floyd, R.A., editor (1982): Free Radicals and Cancer. Marcel Dekker, New York.

16. Forni, L.G., Packer, J.E., Slater, T.F. and Willson, R.L. (1983): Chem.Biol.Interactions, 45: 171-177.

17. Gooding, P.E., Chayen, J., Sawyer, B. and Slater, T.F. (1978): Chem.Biol.Interactions, 20: 299-310.

18. Janzen, E.G. (1980): In: Free Radicals in Biology, edited by W.A.Pryor, volume 4, pp. 115-154. Academic Press, New York.

19. Jordan, R.A. and Schenkman, J.B. (1982): Biochem.Pharmacol., 31: 1393-1400.

20. Klein, G., editor (1983): Advances in Viral Oncology, volumes 1 and 3. Raven Press, New York.

21. McBrien, D.C.H. and Slater, T.F., editors (1982): <u>Free Radicals, Lipid Peroxidation and Cancer</u>. Academic Press, London.
22. Miller, J.A. and Miller, E.C. (1984): In: <u>Chemical Carcinogenesis</u>, Part A, edited by P.O.P.Ts'o and <u>J.A.Di Paolo</u>, pp. 61-85. Marcel Dekker, New York.
23. Packer, J.E., Slater, T.F. and Willson, R.L. (1978): <u>Life Sciences</u>, 23: 2617-2620.
24. Phillips, L.A., editor (1983): <u>Viruses associated with Human Cancer</u>. Marcel Dekker, New York.
25. Pott, P. (1775): <u>Chirurgical Observations Relative to the Cataract, the Polypus of the Nose, the Cancer of the Scrotum, the Different Kinds of Ruptures, and the Mortification of the Toes and Feet</u>. Hawes, Clarke and Collins, London.
26. Proudfoot, K. and Slater, T.F. (1984). In preparation.
27. Pryor, W.A. (1984): Free Radicals in Autoxidation and in Aging. In: <u>Free Radicals in Molecular Biology, Aging, and Disease</u>, edited by D. Armstrong et al., pp. 13-41. Raven Press, New York.
28. Rosen, G.M. and Rauckman, E.J. (1981): <u>Proc.Natl.Acad.Sci. U.S.A.</u>, 78: 7346-7349.
29. Samuelsson, B. (1983): <u>Science</u>, 220: 568-575.
30. Searle, C.E., editor (1976): <u>Chemical Carcinogens</u>. ACS Monograph No.173, American Chemical Society, Washington, D.C.
31. Serabjit-Singh, C.J., Wolf, C.R., Philpot, R.M. and Plopper, C.G. (1980): <u>Science</u>, 207: 1469.
32. Sevanian, A., Mead, J.F. and Stein, R.A. (1979): <u>Lipids</u>, 14: 634-643.
33. Shimkin, M.B. (1977): <u>Contrary to Nature</u>. U.S.Department of Health, Education and Welfare, Publication No. (NIM) 79-720, U.S.Government Printing Office, Washington, D.C.
34. Slater, T.F. (1972): <u>Free Radical Mechanisms in Tissue Injury</u>. Pion Ltd., London.
35. Slater, T.F. (1976): <u>Panminerva Medica</u>, 18: 381-390.
36. Slater, T.F. (1978): In: <u>Biochemical Mechanisms of Liver Injury</u>, edited by T.F.Slater, pp. 1-44. Academic Press, London.
37. Slater, T.F.(1984): In: <u>Methods in Enzymology</u>, edited by L. Packer, 105: 283-293. Academic Press, New York.
38. Slater, T.F., Sträuli, U.D. and Sawyer, B.C. (1964): <u>Biochem. J.</u>, 93: 260-266.
39. Swartz, H.M. (1982): In: <u>Free Radicals, Lipid Peroxidation and Cancer</u>, edited by D.C.H.McBrien and T.F.Slater, pp. 5-20. Academic Press, London.

40. Ugazio, G., Gabriel, L. and Burdino, E. (1968): Boll.Soc. Ital.Biol.Sper., 44: 30-53.
41. Utsumi, K., Yamamoto, G. and Inaba, K. (1965): Biochim. biophys.Acta, 105: 368-371.
42. Walling, Ch. (1957): Free Radicals in Solution. John Wiley and Sons, New York.
43. Weber, G. (1983): Cancer Research, 43: 3466-3492.

Free Radicals in Molecular Biology, Aging,
and Disease, edited by D. Armstrong et al.
Raven Press, New York © 1984.

Oxy-Radical Production in Alloxan-Induced Diabetes: An Example of an *In Vivo* Metal-Catalyzed Haber-Weiss Reaction

Gerald Cohen

Department of Neurology, Mount Sinai School of Medicine of the City University of New York, New York, New York 10029

Alloxan, dialuric acid, and some related compounds induce tissue pathology via mechanisms that show strong dependence on generated free radicals. Alloxan and dialurate form a redox pair: Dialurate, a compound with a polyphenolic structure, autoxidizes rapidly in the presence of oxygen to yield alloxan, a quinoidal compound (equation 1). In turn, tissue reducing agents, ascorbate in particular, can reduce alloxan back to dialurate (equation 2) to permit another oxidative cycle to be initiated:

dialurate + oxygen ---> alloxan + hydrogen peroxide   (eq. 1)

alloxan + ascorbate ---> dialurate + dehydroascorbate   (eq. 2)

The aerobic oxidation and redox cycling reactions generate reactive oxygen species, namely, hydrogen peroxide, the superoxide radical, and the hydroxyl radical. Strong experimental evidence implicates these reactive forms of oxygen in the tissue pathology. Reactive semiquinone radicals may also be generated as intermediates between alloxan and dialurate. The monodehydroascorbate radical intermediate, on the other hand, is not itself very reactive.

Alloxan and dialurate induce highly localized tissue damage. The target cells are the insulin-producing beta cells of the pancreas. Destruction of pancreatic beta cells results in a diabetic state. Alloxan-induced diabetes is a popular animal model for study of the diabetic state. The specificity of damage appears to depend upon a specific accumulation of alloxan/dialurate in the beta cells of the pancreas. Other organs and tissues appear unaffected by injection of diabetogenic doses of alloxan or dialurate into rats or mice. An exception is seen in vitamin E deficiency, where dialurate induces hemolysis of red blood cells.

This article concerns the relationship of the mechanism for alloxan toxicity to the so-called "Haber-Weiss reaction" (equation 3):

superoxide + HOOH ---> .OH + oxygen + hydroxide ion   (eq. 3)

where HOOH is hydrogen peroxide and .OH is the hydroxyl radical. Published data provide strong evidence that the mechanism for alloxan-induced toxicity to the pancreatic beta cells in vivo is dependent upon an iron-catalyzed Haber-Weiss reaction. These data are briefly reviewed.

THE "HABER-WEISS REACTION"

When Haber and Weiss (8) first introduced their concept of free radical intermediates in the iron-catalyzed decomposition of hydrogen peroxide, two reactions were suggested (equations 4 and 5):

.OH + HOOH ----> HOO. + water   (eq. 4)

HOO. + HOOH ----> .OH + oxygen + water   (eq. 5)

where HOO. is the perhydroxyl radical (the protonated form of superoxide). The perhydroxyl radical, rather than the superoxide radical, was formed under the strongly acid conditions used to study the decomposition of hydrogen peroxide. If formed under biological conditions, at neutral pH, the perhydroxyl radical would exist in its ionized form, i.e, the superoxide radical (pK = 4.8). Equation 3 is the equivalent of equation 5 at neutral pH.

In the Haber-Weiss cycle (equations 4 and 5), the hydroxyl radical reacts with hydrogen peroxide via hydrogen abstraction to form water and yield a new radical, namely, the protonated superoxide radical. The protonated superoxide radical reacts with a second molecule of peroxide to form the hydroxyl radical, a molecule of water, and molecular oxygen. This cycle was proposed to explain unusual aspects of the kinetics of the iron-catalyzed decomposition of peroxide. The publication (8) represents a milestone in chemistry in that it introduced the concept of reactive free radical intermediates into reaction mechanisms. The novelty of the concept led to many years of debate and experimentation. Today, equation 4 has been fully validated. But equation 5 is considered to be too slow to represent an effective reaction mechanism.

In recent years, equation 3 was once again invoked by biochemists (e.g., 1) to explain phenomena in which both superoxide and hydrogen peroxide were involved, and in which a strong oxidizing species, ostensibly .OH, was generated. These studies evolved from the fundamental finding by McCord and Fridovich (5,14) of the superoxide dismutase (SOD) activity of certain copper-zinc proteins that are broadly distributed in mammalian organisms. It was subsequently found that many forms of coupled oxidative reactions or tissue damage seen during oxidation of a variety of substrates (e.g., the xanthine oxidase reaction, or redox cycling of quinoidal drugs) could be blocked by adding either catalase (scavenger of hydrogen peroxide) or superoxide dismutase (scavenger of superoxide). Hence, it was reasoned that the hydroxyl radical had been produced via the Haber-Weiss reaction (equation 3).

Recent reevaluation of equation 3 by competent inorganic and radiation chemists showed once again that the rate was far too slow to be important in biological phenomena. However, studies by investigators (e.g., 13) who desired an explanation for apparent Haber-Weiss phenomena, provided experimental evidence for a set of two equations in which iron serves as a catalyst for equation 3. This has become known as "the iron-catalyzed Haber-Weiss reaction" and is shown in equations 6 and 7 below:

ferric ion + superoxide ----> ferrous ion + oxygen   (eq. 6)

ferrous ion + HOOH ---> ferric ion + .OH + hydroxide ion   (eq. 7)

In equation 6, superoxide acts a reductant, and ferrous ions and molecular oxygen are products. In equation 7, ferrous ions are oxidized by hydrogen peroxide and the hydroxyl radical is a product. Equation 7 is the Fenton reaction. (In the Haber-Weiss experiments, the Fenton reaction was the source of .OH to initiate the Haber-Weiss cycle). When equations 6 and 7 are summed, the ferrous and ferric ions drop out, and equation 3 remains:

superoxide + HOOH ---> .OH + oxygen + hydroxide ion   (eq. 3)

Because the Fenton reaction is the source of the highly-reactive hydroxyl radical, the sequence shown in equations 6 and 7 can be looked on as a "superoxide-driven Fenton reaction".

The concept of the "Haber-Weiss reaction" or the "metal-catalyzed Haber-Weiss reaction" remains today a hotly-debated topic. The main point of this article is to summarize the experimental evidence in favor of a metal-catalyzed Haber-Weiss reaction as the mechanism for the cytotoxic effects of alloxan/dialurate on the pancreas.

ROLE OF CHELATES IN THE "HABER-WEISS REACTION"

The equations shown above depict ferrous and ferric ions as components in the reaction. However, certain iron chelates can also participate, whereas other iron chelates are not effective.

The prime example of a reactive iron chelate is iron-EDTA (where EDTA is ethylenediaminetetraacetate). This ferric chelate is readily reduced by superoxide (equation 6); it also engages in rapid Fenton chemistry (equation 7). Other chelates, such as iron-ADP appear to show similar reactivity. On the other hand, iron-DTPA (where DTPA is diethylenetriaminepentaacetate) is a prime example of a non-reactive (or poorly reactive) iron chelate. Although ferrous-DTPA reacts vigorously with hydrogen peroxide in a Fenton-type reaction (2), the ferric chelate is not readily reduced by superoxide. Hence, once iron is in the ferric form, recycling is blocked. Other poorly reactive chelates in the iron-catalyzed Haber-Weiss reaction are those with o-phenanthroline or desferrioxamine; poor reactivity of these chelates can also be attributed to weak Fenton-type reactions.

In biological systems, iron may be complexed or chelated with ADP, amino acids, or a variety of other ligands. Addition of DTPA to in vitro biological systems, or injection of DTPA into experimental animals, provides a valuable probe for the operation of a metal-catalyzed Haber-Weiss reaction.

IN VITRO STUDIES

Simple in vitro studies by Cohen, Heikkila and MacNamee (3) with buffered solutions of dialurate showed formation of superoxide and hydroxyl free radicals, and hydrogen peroxide. When dialuric acid is added to a neutral buffered solution (e.g., Kreb's-Ringer phosphate at pH 7.4), oxygen is rapidly consumed. The reaction is conveniently followed with an oxygen electrode. If catalase is added at the termination of the autoxidation reaction, approximately one-half of the consumed oxygen reappears in solution. This result is exactly as expected from the dismutation (decomposition) of two moles of hydrogen peroxide by catalase to yield a mole of water and a mole of molecular oxygen. Hence, it is clear that hydrogen peroxide accumulates during the autoxidation reaction.

The superoxide radical is an expected intermediate for a one electron transfer from dialurate to oxygen. A simple technique to detect superoxide is to follow the reduction of cytochrome c at

550 nm with a spectrophotometer. The superoxide radical acts as a reducing agent towards ferricytochrome c (analogous to equation 6). During the autoxidation of dialurate, ferricytochrome c is reduced (3). However, dialurate is itself a mild reducing agent. Therefore, a distinction must be drawn between the reducing properties of generated superoxide and those of dialurate. This is easily achieved by adding SOD. Inhibition of the reduction of cytochrome c by SOD confirms that the reductant is superoxide (3).

The ability of ascorbate to recycle alloxan and, thereby, continue the accumulation of hydrogen peroxide, is also readily followed with an oxygen electrode (11). Alloxan by itself does not significantly consume oxygen. However, when ascorbate is added, a vigorous consumption of oxygen takes place. The rate is very much faster than the slow consumption of oxygen seen with ascorbate alone.

Formation of a .OH-like species is also readily detected in vitro (3). Methional reacts with .OH and generates ethylene, an unsaturated hydrocarbon gas:

.OH + methional ---> ethylene + other products   (eq. 8)

The ethylene gas is readily detected and measured by gas chromatography. When dialuric acid is added to methional at or near neutral pH, the formation of ethylene ensues (Table 1). Table 1 also presents results with 6-hydroxydopamine, a neurotoxin with similar molecular and chemical properties as dialurate. If catalase is added, the yield of ethylene is suppressed; the same is true for addition of SOD. Hence, the results indicate the operation of a Haber-Weiss reaction in vitro.

TABLE 1. Production of hydroxyl radicals from autoxidizing dialurate

Addition	nMoles Ethylene
0.1 mM dialurate	1.53
0.1 mM 6-hydroxydopamine	2.79

The buffer was 0.2 M acetate at pH 6.4. Methional was present at 1 mM. Ethylene was measured by gas chromatography at 40 min. Data are from Cohen, Heikkila & MacNamee (3).

Conclusions: Dialurate at neutral pH generates hydrogen peroxide, superoxide radical and hydroxyl radical from molecular oxygen. Recycling of alloxan by ascorbate amplifies the yield. The formation of the hydroxyl radical shows dependence upon both hydrogen peroxide and superoxide (Cf., the Haber-Weiss reaction).

STUDIES WITH ISOLATED PANCREATIC ISLETS EX VIVO

Studies by Fischer and Hamburger (4) demonstrated that the cytotoxic action of alloxan on isolated islets of Langerhans shows the characteristics of a metal-catalyzed Haber-Weiss reaction.

Pancreatic islets were isolated by collagenase digestion and then they were incubated with alloxan. Subsequently, the islets were challenged with glucose, and the secretion of insulin was measured. The following observations were made (Table 2):

TABLE 2. <u>Alloxan toxicity to isolated pancreatic islets: Protection by catalase, SOD and DTPA.</u>

Addition	% Inhibition of Insulin Release
Alloxan	62 %
Alloxan + SOD	7 %
Alloxan	82 %
Alloxan + catalase	29 %
Alloxan	66 %
Alloxan + DTPA	10 %

Experiments were performed with 0.15 mg alloxan/ml. Exposure time was 5 min. Data are from Fischer & Hamburger (4).

Exposure to alloxan inhibits the ability of the pancreatic beta cells to release insulin in response to provocation with glucose (Table 2). This result signifies damage to the beta cells by alloxan. SOD protects, as does catalase. These results are in keeping with a Haber-Weiss reaction. The iron-chelating agent, DTPA, also protects. This result is in keeping with the inability of iron-DTPA to participate in a metal-catalyzed Haber-Weiss reaction. In these experiments, heat-inactivated superoxide dismutase and catalase were ineffective as protective agents. In the experiment with DTPA, protection by DTPA was overcome by adding excess ferrous sulfate (i.e., above that chelated by the DTPA). Similar results have been reported by Grankvist et al (7). However, Grankvist et al also reported protection of other measured parameters in pancreatic islets by .OH-scavengers in vitro, whereas Fischer & Hamburger did not obtain protection with n-butanol or thiourea.

Experiments by Heikkila and Cohen (10) with rat brain slices showed damage by added dialurate/alloxan, with strong potentiation by ascorbate. Hence, the recycling of alloxan by ascorbate is indicated.

Conclusions: Damage by alloxan to isolated pancreatic islets shows characteristics of a metal-catalyzed Haber-Weiss reaction: Protection is provided by catalase, SOD, and DTPA. Ascorbate can potentiate damage to cellular systems.

STUDIES IN VIVO

Hydrogen peroxide can be detected in vivo after administration of alloxan. Experiments by Heikkila & Cohen (11) showed formation of peroxide in red blood cells. The method for detecting peroxide was based on the inhibition of endogenous catalase by 3-amino-1,2,4-triazole. The mechanism for enzyme inhibition has been established: It requires the physical presence of hydrogen peroxide. The reaction pathway passes through compound I, which is an intermediate formed from the reaction between catalase and hydrogen peroxide. When alloxan is injected into mice, red blood cell catalase is inhibited. No effect is seen with either alloxan alone, or aminotriazole alone. These results establish a reduction of alloxan (to dialurate), ostensibly by ascorbate, with subsequent autoxidation to generate hydrogen peroxide. It should be borne in mind that red blood cells are targets for hemolysis by alloxan or dialurate in vitamin E-deficiency.

Heikkila et al (12) tested .OH-scavenging agents in vivo and observed protection against alloxan-induced diabetes. When alloxan is injected into mice, the insulin-producing beta cells of the pancreas are destroyed; a diabetic state develops, which is characterized by chronic elevation in blood glucose. The study of blood glucose levels is a simple and convenient laboratory measurement. Hydroxyl radical scavengers, such as primary aliphatic alcohols and thiourea, provide strong protection (Table 3).

Table 3. Alloxan-Induced Diabetes: Protection by .OH-Scavengers in Vivo.

Group	Blood Glucose (mg/100 ml)
Control	120
Alloxan	445
Ethanol + Alloxan	107
n-Butanol + Alloxan	141
Thiourea + Alloxan	147

Alloxan was given i.v. at a dose of 50 mg/kg. Scavengers were given 30-120 min earlier at the following doses: ethanol (4 g/kg), n-butanol (0.8 g/kg), thiourea (3 g/kg). Blood glucose was measured at 72 hours. Data are from Heikkila et al. (12).

When alloxan is administered, the blood glucose is elevated to 3- to 4-fold its normal level by 72 hours. Prior administration of ethanol, n-butanol, or thiourea is strongly protective. Histologic examination of the pancreas confirms the protection of the beta cells. A particular advantage to these scavengers is that they can be given in high concentration without harm to the mice.

Experiments by Grankvist, Marklund & Taljedal (6) showed that injected SOD is protective in vivo.

Table 4. <u>Alloxan-Induced Diabetes: Protection by SOD in Vivo.</u>

Group	Blood Glucose (mM)
Control	4.8
Alloxan	20.4
SOD + Alloxan	9.2
Inactivated SOD + Alloxan	21.1

Alloxan was given at 50 mg/kg. The SOD was coupled to polyethylene glycol to enhance its half-life. Inactivation of SOD was achieved by treatment with hydrogen peroxide. Blood glucose was measured at 72 hours. Data are from Grankvist, Marklund & Taljedal (6).

The diabetic state provoked in mice by injection of alloxan is prevented by administration of SOD (i.v.) 12 hours earlier (Table 4). SOD that has been 99.2% inactivated, is without protective action.

It seems unlikely that SOD protects by penetration into the beta cells of the pancreas. Therefore, a direct action of alloxan/dialurate on or in the cellular membrane seems probable. The protective effect of SOD, without a similar action by SOD protein that has lost its enzymatic activity, indicates a role for superoxide in the in vivo toxic mechanism.

Experiments by Heikkila and Cabbat (9) showed a role for metals in the in vivo mechanism of cytotoxicity (Table 5):

TABLE 5. <u>Alloxan-Induced Diabetes: Protection by DTPA in Vivo.</u>

Group	Blood Glucose (mg %)
Alloxan	384
DTPA + Alloxan	157
EDTA + Alloxan	394

Alloxan was given at 75 mg/kg. DTPA or EDTA (250 mg/kg) was given one hour earlier. Blood glucose was measured at 48 hours. Data are from Heikkila & Cabbat (9).

DTPA given one hour before alloxan blocks the development of a diabetic state (Table 5). However, EDTA is not protective. These results are in accord with the properties of DTPA and EDTA vis-a-vis the iron-catalyzed Haber-Weiss reaction: DTPA effectively blocks equation 3, whereas EDTA does not. Therefore, the presence of an endogenous metal catalyst is indicated.

Conclusions: Damage by alloxan to pancreatic beta cells in vivo shows the characteristics of a metal-catalyzed Haber-Weiss reaction: Superoxide dismutase, DTPA, and .OH-scavengers are all effective protective agents. Cellular hydrogen peroxide generated by alloxan/dialurate has been detected in vivo.

CONCLUSIONS

* Experimentally-induced diabetes, provoked by the alloxan/dialurate redox couplet, shows the characteristics of the "metal-catalyzed Haber-Weiss reaction" as the mechanism for destruction of pancreatic beta cells.

* In vitro, dialurate reacts rapidly with oxygen to generate superoxide and hydrogen peroxide. The hydroxyl radical is also generated, by a reaction mechanism that is blocked by catalase and SOD.

* Isolated pancreatic islets are damaged by alloxan/dialurate in vitro. The islets are protected by added catalase or superoxide dismutase. Protection is also provided by DTPA.

* In vivo, damage to pancreatic beta cells and the development of a diabetic state are blocked by SOD, hydroxyl radical-scavenging agents, and DTPA.

* Protection in vivo by SOD points to the cell membrane as a site of damage. Protection in vivo by DTPA indicates the presence of endogenous iron complexes or chelates that can catalyze the Haber-Weiss reaction.

* From a biologic point of view, the "Haber-Weiss reaction", catalyzed by endogenous **iron-chelates** (or other metal chelates), remains a viable explanation for cellular toxicity, and a reasonable experimental guide for cellular reaction mechanisms.

REFERENCES

(1) Beauchamp, C. & Fridovich, I. (1970): J. Biol. Chem. 245: 4641-4646.
(2) Cohen, G. & Sinet, P.M. (1982): FEBS Lett. 138: 258-260.

(3) Cohen, G., Heikkila, R.E. and MacNamee, D. (1974): J. Biol. Chem. 249: 2447-2452.
(4) Fischer, L.J. and Hamburger, S.A. (1980): Diabetes 29: 213-216.
(5) Fridovich, I., (1978): Science 201: 875-880.
(6) Grankvist, K., Marklund, S. and Taljedal, I.-B. (1981) Nature 294: 158-160.
(7) Grankvist, K., Marklund, S., Sehlin, J. and Taljedal, I.-B. (1979): Biochem. J. 182: 17-25.
(8) Haber, F. and Weiss, J. (1934): Proc. Roy. Soc. Ser. A 147: 332-351.
(9) Heikkila, R.E. and Cabbat, F. (1982): Experientia 38: 378-379.
(10) Heikkila, R.E. and Cohen, G. (1972): Molec. Pharmacol. 8: 241-248.
(11) Heikkila, R.E. and Cohen, G. (1975): Ann. N. Y. Acad. Sci. 258: 221-230.
(12) Heikkila, R.E., Winston, B., Cohen, G., and Barden, H. (1976): Biochem. Pharmacol. 25: 1085-1092.
(13) McCord, J.M. and Day, E.D. Jr. (1978) FEBS Lett. 86: 139-142.
(14) McCord, J.M. and Fridovich, I. (1969): J. Biol. Chem. 244: 6049-6055.

Free Radicals in Molecular Biology, Aging, and Disease, edited by D. Armstrong et al. Raven Press, New York © 1984.

Free Radicals and Damage to Ocular Tissues

Rex D. Wiegand, *Jule G. Jose, Laurence M. Rapp, and Robert E. Anderson

*Cullen Eye Institute, Baylor College of Medicine, Houston, Texas 77030; *College of Optometry, University of Houston, Houston, Texas 77002*

Over the past decade a large body of literature has accrued indicating that free radicals, peroxides (and their enzymic regulation), antioxidants and auto-oxidation may play a role in disease and aging of the eye. Specifically, free radical mechanisms have been implicated in cataractogenesis (23,114,145), ocular inflammation (27,131), ceroid-lipofuscinosis (9), drug-induced retinopathy (90), retrolental fibroplasia (76,89), photic retinopathy (72,87,108), and ocular siderosis (73,119). Although the agent in each of the insults may be different, cell destruction and death apparently proceed through pathways common to lipid or protein oxidation.

The aims of this chapter are to present an overview of the role of free radicals in damage to ocular tissue. We will describe briefly basic ocular anatomy with particular attention to the structural organization of the lens and the retina. We will then review the evidence for free radical involvement in the pathogenesis of cataract and retinal degeneration. Lastly, we will present studies from our laboratory on two experimental conditions that result in retinal degenerations: irradiation and intravitreal injection of ferrous ion. Results of these studies suggest that these two very different stresses mediate their destructive effects via lipid peroxidation.

CELLULAR ORGANIZATION OF THE LENS

To understand how oxidizing agents might induce cataracts,[1] it is necessary to appreciate the unique anatomical structure of this organ.

---

[1] For purposes of this discussion, we shall use "cataract" to mean any change in normal lens transparency.

The relevant features are presented in Figure 1, which depicts a cross-sectional view of the lens. The lens is encased in a collagenous basement membrane called the capsule. This structure is considered to be inert metabolically and is probably not a significant target of oxidative insult. Located only on the anterior face of the lens just under the capsule is a single layer of epithelial cells which carry out the bulk of the metabolic activity in the lens. Although all of these cells have the potential to replicate, the highest mitotic rate lies in a ring near the periphery of the lens (128) that has consequently been designated the "germinative zone." As the cells divide, they remain in a monolayer. Consequently, cell division pushes the cells peripherally toward the equator.

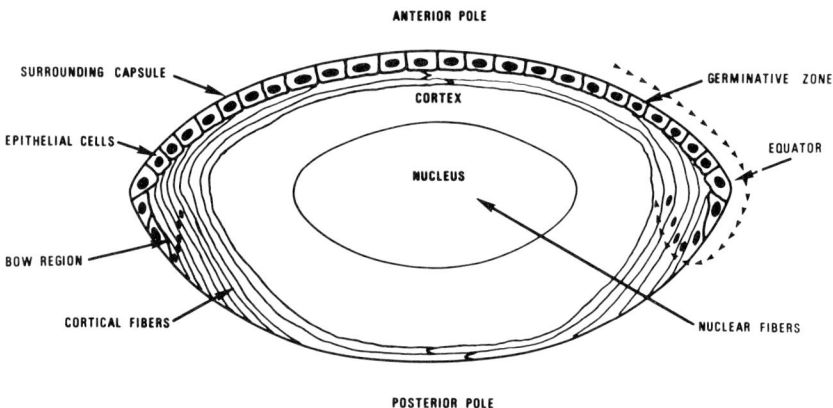

FIG. 1. Schematic representation of the mammalian lens in cross section. Arrows indicate direction of cell migration from the epithelium to the cortex. (From Jose, Ref. 82, with permission).

At the equator, the cells differentiate and elongate as they start to form the fibers that will make up the bulk of the lens. Just prior to differentiation, the cells lose the capacity to replicate. As differentiation proceeds, more of the metabolic machinery is lost and the fiber finally grows long enough to extend from a point near the anterior of the lens to another point about halfway around the lens near its posterior. Fibers grow from all around the periphery of the lens, with their ends meeting along a series of irregularly branched lines called the lens sutures. As the outer epithelial cells elongate, they entomb the older fibers deeper within the lens (54). These older fibers are never lost from the lens, but rather are compacted within the centermost region of the lens (called the nucleus). Those in the very center of the lens are in fact formed during embryogenesis. The lens is characterized by growth instead of turnover (54).

Lens clarity is dependent upon proper differentiation and placement of epithelial cells (91,171) and the nature of proteins within them. Protein synthetic activity is very high both in the lens epithelium and in the peripheral region of the lens which is called the lens cortex. There is no clear anatomical distinction between the cortex and the nucleus. Probably very little metabolic activity and no protein synthesis occurs in the lens nucleus (54).

In addition to protein synthesis, the differentiating epithelial cell must also synthesize lipids. Their chemical make-up would suggest that these membranes are fairly rigid (32) since they contain relatively high levels of cholesterol, sphingomyelin, and saturated fatty acids, and low levels of polyunsaturated fatty acids. However, no examination of lipids from only epithelial cells has been reported. As the fibers are compacted deeper and deeper into the core of the lens, interactions between the fiber membranes and proteins increase so that the protein becomes increasingly difficult to separate from the membrane (102).

CELLULAR ORGANIZATION OF THE RETINA

The vertebrate retina is a neuroepithelial tissue which lines the interior of the posterior globe and is comprised of six basic neural cell types: photoreceptor cells (rods and cones), bipolar cells, horizontal cells, amacrine cells, interplexiform cells, and ganglion cells. Each basic cell type may exhibit various degrees of heterogeneity with regard to morphology, relative position in the retina, and synaptic interconnections with other retinal cells (see Fig. 2). Histologically, the cells of the retina are organized in discrete layers: 1) the retinal pigment epithelium (RPE); 2) the photoreceptor outer segment layer (POS); 3) the "outer limiting membrane" (OLM) (a network of Muller cell apical processes which contact the photoreceptor cell "inner segments"); 4) the outer nuclear layer (ONL) (composed of photoreceptor cell nuclei); 5) the outer plexiform layer (OPL) (the region of synaptic contact between photoreceptor, bipolar, and horizontal cells); 6) the inner nuclear layer (INL) (perikarya of bipolar, horizontal, and amacrine cells); 7) the inner plexiform layer (IPL) (the level of synaptic contact between bipolar, amacrine, and ganglion cells); 8) the ganglion cell layer (GCL); 9) the nerve fiber layer (NFL) (collective axonal processes of the ganglion cells which ultimately exit the eye and form the optic nerve); and 10) the "inner limiting membrane" (ILM) (the network of Muller cell basal processes which forms the retinal-vitreal interface). In contrast to the neural cells, the Muller cells (glia) span the retina, extending from the vitreous interface to the photoreceptor cell layer.

The exchange of nutrients and gasses between the retina and the blood is accomplished in mammals via two independent circulatory systems (52,104). The choroidal system (located between the sclera and the RPE) consists of a network of fenestrated capillaries, separated from the basal surface of the RPE by a thin endothelial lining and Bruch's membrane (BM, Fig. 2). This blood supply serves the outer retinal layers,

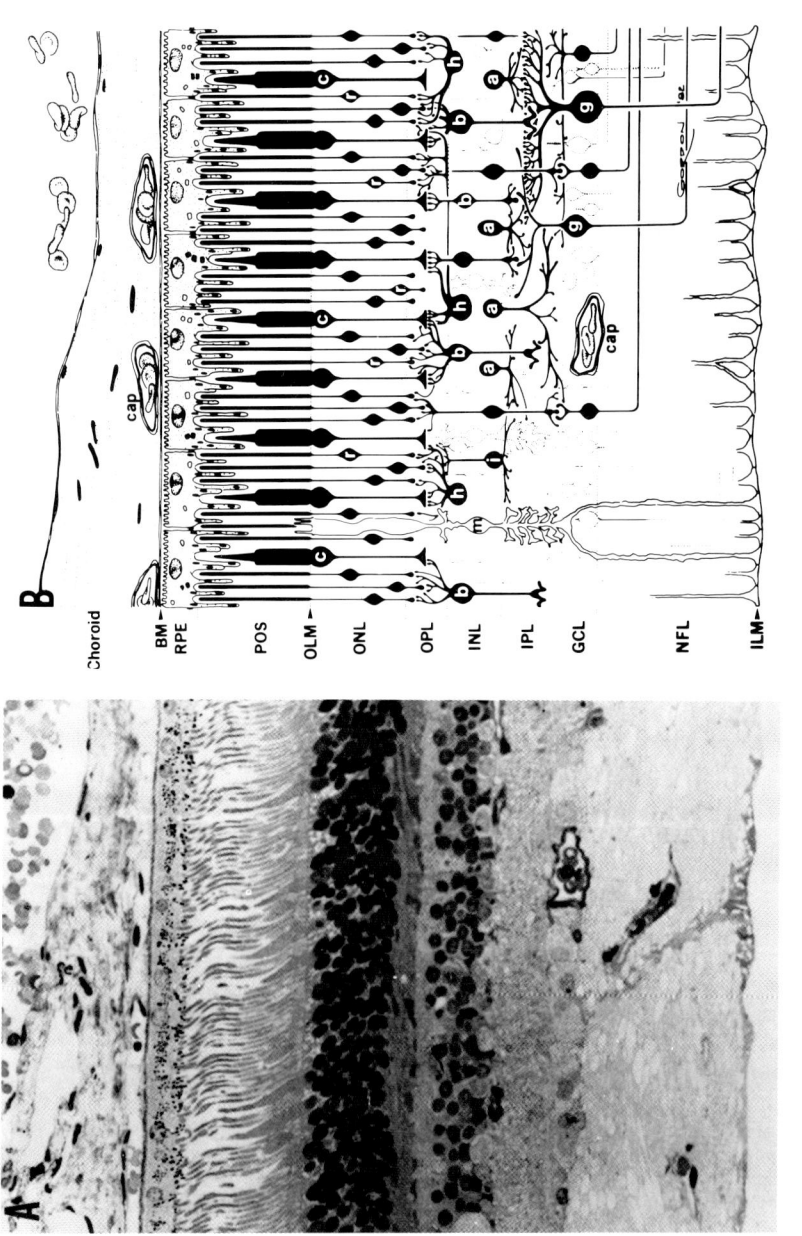

FIG. 2. (A.) Histological cross-section of a human retina. (B.) Schematic diagram depicting the cellular layers and representative cell types of the retina and their spatial arrangement. a, amacrine cell; b, bipolar cell; c, cone; g, ganglion cell; h, horizontal cell; i, interplexiform cell; r, rod; cap, capillary. See text for other abbreviations and details. (From Fliesler and Anderson, Ref. 51, with permission).

which are otherwise avascular. The exchange of various constituents between the outer retina and the choroidal circulation is mediated by the RPE (173). The photoreceptor outer segments are deeply invaginated into the extensive apical villous processes of the RPE cells (see Fig. 2b and Fig. 3), whose basal surfaces face the choroid. The direct flow of blood constituents to the outer retina via the extracellular space surrounding the RPE cells is prevented by the tight junctions on the apical lateral borders of the RPE cells. The RPE also participates in the daily maintenance and turnover of photoreceptor outer segments (163,164).

A second blood supply enters the retina through the optic disc and serves as a source of nutrients for all components of the inner retina (35). In contrast to the choroidal capillaries, the inner retina capillaries are not fenestrated and the passive exchange of material is therefore restricted to small molecules, ions, and gases.

Light enters the eye and must traverse the cornea, aqueous humor, lens, vitreous humor, and the various retinal layers before impinging upon the photoreceptor cells. The absorption of a photon by a single molecule of visual pigment in a photoreceptor cell is sufficient to alter the membrane conductance (by processes yet to be elucidated), resulting in hyperpolarization and a transient decrease in the photoreceptor dark current (141). The visual information encoded in this transient conductance change is transmitted sequentially from photoreceptors to bipolar cells to ganglion cells, whereupon the information is conveyed via the optic nerve to the brain. This linear information transfer may be modified to varying degrees by lateral inputs from horizontal and amacrine cells.

STRUCTURAL ORGANIZATION OF PHOTORECEPTOR CELLS

There are two structurally and functionally distinct photoreceptor cell types: the rods and the cones (31). Rods are elongate and cylindrical in shape, whereas cones are shorter than rods and in most instances possess a conical-shaped outer segment. Functionally, the rods are dim light receptors (i.e. they mediate scotopic vision), whereas the cones mediate color vision and have an operating range at relatively higher light intensities (I.e. photopic vision). With few exceptions, the number of rods is much greater than the number of cones in a given retina, and in certain species (i.e. the rat and various other nocturnal animals) only very few cones are found. Since much of our current understanding of photoreceptor cell metabolism and physiology has been derived from studies of rods, our discussion of photoreceptor cell degeneration will emphasize the rod.

A schematic diagram that depicts the typical features of a rod is shown in Figure 3. On the basis of morphology and function, the rod may be divided into four separate parts: 1) the outer segment (ROS); 2) the inner segment (RIS); 3) the nuclear region (N); and 4) the synaptic pedicle (SP). The ROS is a highly-specialized appendage composed of a rigidly-ordered stack of 500-2000 flattened membranous sacks (discs) enclosed by the plasma membrane of the cell. The majority of the discs are "free floating"; i.e. they are not in physical contact either with each other or

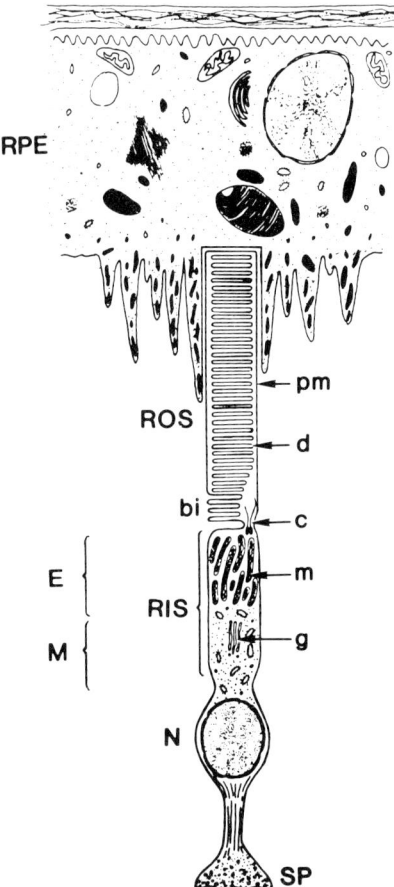

FIG. 3. Schematic diagram of a rod cell and its spatial relationship to the retinal pigment epithelium (RPE). See text for abbreviations and details. (From Fliesler and Anderson, Ref. 51, with permission).

with the plasma membrane (31), nor are they electrically coupled to the plasma membrane (126). Unlike the ROS, the entire cone outer segment is formed by repeated invaginations of the plasma membrane, resulting in a contiguous stack of parallel membrane lamallae (as opposed to free discs). The rat ROS is about 1.5-2 μm in diameter, 20-40 μm in length, and contains about 600-1000 discs, which is essentially the dimension of human ROS (74). The frog is about 5-7 μm in diameter, 35-50 μm in length and contains about 1700-2000 discs (124).

The outer segment is connected to the inner segment via a thin cytoplasmic bridge containing a cilium through which exchange of material takes place between the two cellular compartments. The RIS has two regions: the ellipsoid (E), the portion of the RIS closest to the outer segment, which contains numerous densely-packed mitochondria (m), and the myoid (M) which houses the network of smooth and rough endoplasmic reticulum and the Golgi apparatus (g). The presence of these organelles in the inner segment indicates that the RIS is the cellular compartment where the bulk of the metabolism takes place, in contrast to the organelle-devoid ROS. The nucleus of the cell lies just below the myoid region. An axonal process connects the RIS and nuclear regions with the synaptic pedicle of the cell.

Photoreceptor cells are very active metabolically. In mammals, 10% of the outer segment membranes are renewed daily (163,164). Distal tips of both rod and cone outer segments are shed each day and phagocytized and digested by the retinal pigment epithelium. Membrane constituents destined for the outer segments are synthesized on the endoplasmic reticulum of the inner segment and transported as vesicles to the base of the outer segment, where they fuse with the plasma membrane in the region of the connecting cilium. This results in an evagination of the membrane (see base of Fig. 3). Basal discs of rods, but not cones, pinch off eventually to become free-floating discs. The process of renewal of rod and cone outer segments provides a mechanism for the systematic replacement of membrane constituents.

EVIDENCE FOR FREE RADICAL DAMAGE IN THE LENS

There are several lines of evidence that free radicals and oxidizing agents play a role in development of lens opacities. First, agents that either directly or indirectly induce free radicals can also induce cataract formation. These include, for example, x-rays (47,112,117), ultraviolet (UV) (171), hyperbaric oxygen (130), and chemicals (113). Second, free radicals have been detected in the lens (154,155). Third, oxidative changes that have been shown to occur concomitant with the development of browning of the lens can be produced experimentally through the use of photosensitizers that act via free radical mechanisms or through the generation of H_2O_2 (63) or singlet oxygen (61,79). Fourth, epidemiological studies have demonstrated a correlation between sunlight exposure and cataract development (70). Fifth, agents that protect against free radical or other oxidative damage can prevent or reduce the severity of some experimental cataracts; conversely, agents that inhibit enzymatic defenses against oxidative damage enhance cataract formation (17,19,20). Finally, the lens is, by its nature, continually exposed to UV light, which can promote free radical formation (96) and alteration in lens function (92).

Potential targets of free radical and oxidative damage in the lens are protein, lipids, and DNA. It is widely accepted by those working in the lens that oxidative changes are frequently found in proteins from cataractous lenses (11,45,55). Certainly this seems to be the case when browning (brunescence) is present in the lens nucleus. Recently, lipid peroxidation has been suggested to be an initiating factor in the development of lens opacities (20,148). Several laboratories have begun studies in this area and on the role that antioxidants might play in protecting the lens from opacification. Another target of free radical and oxidative damage in the lens that has received very little attention is DNA (80). This may be due to the very low concentration of DNA in the lens, making it difficult to examine its role in normal lens function. Reports of oxidative damage to each of these lens components will be discussed.

Oxidation of Proteins

Protein comprises the bulk of the lens mass. Consequently, extensive research has been directed towards examining differences in the proteins extracted from normal and cataractous lenses. One hypothesis of cataractogenesis is that lens proteins denature and aggregate, becoming large insoluble masses that scatter light, resulting in lens opacities (15). Oxidation and also insolubilization have been suggested to be initiated by free radicals and other oxidants generated either in the lens or in the aqueous humor (97,110,134). The first report that lens proteins might become oxidized in cataracts was by Dische et al. (44) who found a decrease in free sulfhydral groups (SH) and an increase in disulfide bonds in cataractous lens proteins. However, although the amount of insoluble proteins increased with age in the lens, no increase in the amount of insoluble proteins in cataractous lenses was observed compared to normals of the same age (44). Testa et al. (140) also found no increase in the insoluble proteins, but confirmed the finding of decreased SH groups in cataractous lenses. More recent reports (11,12,46) also support the conclusion that protein insolubilization is not a feature of human cortical cataract formation.

There are two reports that insoluble proteins increase with cataract formation in animals: rabbit lenses exposed to x-rays (57) and mice exposed to a "black lamp" (171,172). Since both x-rays and the "black lamp" used could induce free radical formation, it is important to consider these reports carefully from the perspective of this review. When the data are examined carefully, it is noted that the absolute amount of insoluble protein increased only slightly, whereas the amount of soluble protein decreased dramatically in both types of cataracts. Also, although Dische et al. (44) confirmed that soluble proteins decreased in x-ray exposed animals, they found no increase in insoluble proteins. In fact, one group of their animals showed a decrease in insoluble proteins. Therefore, although x-rays and UV may generate oxidative damage in lens proteins, it is unlikely that any resulting insolubilization plays a major role in opacification of the lens cortex. These studies suggest that the loss of soluble proteins may be more important in cataractogenesis than an increase in insoluble lenticular proteins.

Pirie (111) also found no increase in the amount of insoluble protein in cataractous compared to normal human lenses, with one notable exception. In those lenses where there was a significant amount of brown color in the nucleus (brunescent cataract), there was a larger amount of insoluble protein compared to age-matched controls. Pirie (111) noted an increase in fluorescence from these brown nuclear proteins. Development of this fluorescence and the insoluble brown material has been associated with UV and free radical exposure (23,98,169,170). Lerman (97) has suggested that these fluorescent compounds could also be photo-activated to induce free radicals and protein cross-linking in the lens nucleus (23). However, there remains considerable controversy as to the nature, cause, and significance of the fluorescent molecules that develop in the lens with age (11). Recently, it has been reported that a fluorescent species (excitation peak, 407 nm; emission peak, 496 nm) develops in the lens of the mouse as it ages whether the animal is kept in the dark or under fluorescent light (165). Also, the mouse lens does not turn brown with age, nor has it been made to do so experimentally. Thus, at least for this animal, neither browning nor radiant energy exposure can be associated with development of fluorescent chromophores. Although the role of fluorescence and lens browning remains speculative, the latter has been found associated with insoluble proteins of human lenses in several independent studies (10,11,14,111,168), and these insoluble proteins show oxidative changes.

Many of the changes observed in nuclear proteins from human lenses can be produced photo-oxidatively in vitro. For example, it has been found that proteins will cross-link in the presence of photosensitizers plus radiant energy (25,61). Even in the absence of photosensitizers, prolonged exposure to high energy UV radiation can also induce cross-linking in solutions of lens proteins at very low concentrations (42). However, similar crosslinks could not be induced by irradiating either the whole lens or even in protein solutions at concentrations resembling those in the intact lens (42). Therefore, the significance of this cross-linking remains to be established for the in vivo situation.

Dillon et al. (42) also found that many amino acids, especially tryptophan, were also oxidized during irradiation. Many years previously, Pirie (110) had determined that sunlight turns lens proteins a brown color, especially in the presence of tryptophan or its photo-oxidation products. Such results suggest that tryptophan plays a major role in photo-sensitized reactions in the lens. There remains one major conflict with this suggestion: irradiation of lens proteins in vitro induces considerable breakdown of tryptophan, but the amount of tryptophan is not decreased in the cataractous lens (25,66). In spite of this one inconsistency, tryptophan is still considered the most likely photosensitizer in the lens. For example, free radical species have been detected in the lens by electron spin resonance, and these have been attributed to tryptophan (94,155). Also, tryptophan breakdown products, including kynurenine derivatives (69), beta carbolines (41), and anthranilic acid (142), have all been detected in cataractous lenses. Zigman et al. (169) found that photo-oxidation of free tryptophan produces pigmented or fluorescent species that bind to lens crystallins in vivo.

We are left with a paradox regarding the role of protein oxidation in cataract formation. On the one hand, there is a decrease in SH groups and in the soluble proteins in cortical cataracts, and protein oxidation, crosslinking, and insolubilization can be induced under certain conditions where free radicals are generated. On the other hand, there is no evidence in humans that lenses with cortical cataracts contain more insoluble protein than age-matched controls. Thus, if oxidative changes in proteins are responsible for opacities in the lens cortex, the effect must be mediated in the absence of significant changes in absolute amounts of insoluble proteins. However, protein oxidation and insolubilization are clearly associated with browning of the lens nucleus.

Oxidation of Lipids

The epithelial cells and the fibers of the lens are each individually surrounded by a lipid bilayer membrane. These lipids comprise three to five percent of the wet weight of the lens (5,24,50), and between 2% (123) and 10% (28) of the fatty acids are polyunsaturated. Despite the low concentration of polyunsaturates, their existence has spawned considerable interest in the possibility that lipid peroxidation could represent an initiating event in the development of senile cataracts (60,167). It has been suggested that superoxide or H_2O_2 generated photochemically in the aqueous humor or lens initiate this process (148). The process may be potentiated by endogenous or exogenous photosensitizers in the eye (166). The aldehydes produced during lipid peroxidation are thought to react with lens proteins, resulting in lipid-protein and protein-protein Schiff's base conjugates. Such cross-linking could inactivate enzymes as well as generate high molecular weight nuclear aggregates.

One of the breakdown products of lipid peroxidation, malondialdehyde (MDA), measured as a thiobarbituric acid (TBA) reactive species, has been found to be significantly elevated in cataractous lenses (20) or in lenses incubated under fluorescent light (148). In the latter experiment, lenses incubated overnight produced a sixfold increase in MDA over controls incubated in the dark. MDA formation was found to be partially inhibitable by concomitant incubation with superoxide dismutase (SOD) and was completely blocked by incubation with ascorbate (148).

MDA is elevated in human cataractous lenses (20) in inherited cataracts in the mouse (21), and in cataracts induced experimentally by x-rays, galactose, 3-aminotriazole (a catalase inhibitor) (18,19), or adriamycin (22). TBA reactive species were also increased in lenses incubated in vitro with H_2O_2 or 3-aminotriazole, and their formation could be prevented by SOD and catalase as well as by hydroxyl radical or singlet oxygen scavengers (19). Similarly, development of 3-aminotriazole cataracts was arrested in rabbits by vitamin E given parenterally (22). This also was true for the inherited cataract that occurs spontaneously in the Emory mouse (147). Vitamin E also inhibited in vitro photoperoxidation of lens lipids (146).

Lens epithelial cell membrane function has been shown to be disrupted by photodynamic treatment. The combination of riboflavin plus light damages the cation pump (144) of the lens epithelium. Riboflavin appears to act through a hydrogen peroxide intermediate, as its effects are completely inhibited by catalase (79,144). The effect is also partially inhibited either by SOD or ascorbate (144). Rose Bengal plus light also inhibits carrier-mediated transport in the lens, although the reactive species may be singlet oxygen (79). Other oxidants such as H_2O_2 also disrupt normal lens permeability (37,53), an effect that may be due to membrane alteration or to ATPase inhibition (56). Production of H_2O_2 has been demonstrated to occur in the aqueous humor in the presence of riboflavin plus light (110). H_2O_2 may also be elevated in the aqueous humor of humans (134) and experimental animals with cataract

(17). In the latter study, 3-aminotriazole was found to be a potent cataractogenic agent.

The mechanism of cataract production resulting from lipid peroxidation is not clear. Assuming, however, that its role is tenable, one consequence of peroxidation could be crosslinking of epithelial plasma membrane proteins, which could lead to decreased activities of the metabolic pump necessary for maintaining lens clarity. In the fiber layers, peroxidation and cross-linking could interrupt the orderly packing of the fibers. However, the role of lipid peroxidation in cataractogenesis remains to be established definitively.

Oxidative Damage to DNA

The epithelium contains all the DNA in the lens. For most larger species of animals, the epithelium contains only about 500,000 cells (127). However, these few cells are the precursors of the fibers and synthesize all the lipids and proteins that comprise the bulk of the lens. Since lens clarity is a function of the unique properties of the fiber proteins, any change in genomic expression may have a profound effect on lens clarity.

Although indirect, there is considerable evidence to suggest that DNA is a target of free radical damage in the lens (80). Cataracts can be induced by many oxidants that have been shown to induce DNA damage in other cell systems. For example, hyperbaric oxygen, which is cataractogenic, also induces pyknosis and chromatin clumping in lens epithelial cells. The latter was suggested to result from irreparable double-strand breaks in lens DNA (130).

X-rays, which induce cataract formation (47,112), are potent free radical generators and are well established as DNA-damaging agents. Several experimental studies suggest that DNA is the most likely target of x-ray damage in the lens. First, x-rays induce profound effects on the lens epithelium (1,59,115,129), including pyknosis and cell death. Epithelial mitotic rates typically are first inhibited and then stimulated following x-ray exposure (129). The rapidly dividing epithelium must be inactivated in order for a cataract to occur (112). Dividing cells have a limited time to repair damaged DNA prior to replication, and thus are likely to pass genetic alterations on to progeny. That the lens demonstrates some form of repair following x-ray exposure is evidenced by dose fractionation studies. For example, the cataractogenic potential of a single x-ray dose is reduced or nullified if the same total dose is given in sufficiently small exposures separated in time (47). Because the damage is not cummulative following repeated exposure, it is apparent that some form of cellular repair occurs between the exposures. Since there is no turnover of cortical fibers, the site of this repair certainly is the epithelium, and by analogy with other cell systems DNA is the most likely target of x-ray damage. This suggestion is supported by findings that fractionation of exposure to high liner energy transfer radiations, which tend to produce irreparable double strand breaks in DNA (33,121), does not reduce cataract development.

A situation related to protecting the germinative zone or to dose fractionation may also be produced by inhibition of lens mitosis. This can be accomplished in frogs by inducing hibernation (160). If frogs are held in the cold for a week following x-ray exposure, they do not later develop cataracts when mitosis is reinduced by warmth (161). Frog lenses can repair DNA damage at $5°C$ (84); thus, it is likely that the week of hibernation after x-ray exposure allows time for repair of DNA damage before cell division occurs in the germinative zone.

UV light, which is also an effective free radical generator, has frequently been associated with cataract formation. Epidemologic studies suggest a relationship between exposure and cataracts in humans (70,77,139). UV radiation has also been found to induce cataracts in experimental animals (171). At 300 nm, UV (in the absence of added photosensitizers) has been shown to induce DNA repair synthesis in lens following irradiation in situ through the cornea (62,81).

In combination with photosensitizers such as chlorpromazine or psoralen (85), longer wavelength UV (UV-A) also induces DNA repair synthesis in the lens epithelium (85), and, also in combination with UV-A, both of these molecules induce cataract formation in experimental animals (29,78,83). The first histologic evidence of damage in the psoralen-UV-A treated animals is marked pyknosis in lens epithelial cell nuclei (83). Autoradiography has demonstrated marked binding of psoralen in the lens epithelium and bow region (99). Thus, it is likely that DNA is a target of psoralen damage in the lens.

EVIDENCE FOR FREE RADICAL DAMAGE IN THE RETINA

Retinal degenerations often begin as disruption of the orderly stacking of the discs in photoreceptor cell outer segments. Several lines of evidence, some direct and others circumstantial, suggest that lipid peroxides damage retinal photoreceptor membranes. The environment and composition of the retina clearly make it an ideal "substrate" for lipid peroxidation reactions. Photoreceptor membrane phospholipids contain the highest concentration of long-chain polyunsaturated fatty acids of any tissue in the body (2,51). Docosahexaenoic acid ($22:6\omega3$), the major polyunsaturated fatty acid of the photoreceptors, is especially susceptible to peroxidation (158). Photoreceptor membranes are constantly exposed to light which may act directly or indirectly through photosensitizers to produce damaging oxygen radicals. The mitochondria of the photoreceptor inner segment has a high demand for oxygen. This results in a constant flux of oxygen from the choriocapillaris circulation across the photoreceptor membranes (see Figs. 2 & 3). These components, polyunsaturated fatty acids, light, and oxygen, favor lipid peroxidation.

The retina has several defense mechanisms that protect against oxidative damage. Very high levels of SOD are found in photoreceptor outer segments (7,64). The product of the dismutase reaction, hydrogen peroxide, can be further degraded by catalase, which is present in the retina. Should peroxidation be initiated, the lipid hydroperoxides are detoxified by glutathione peroxidase which converts the reactive hydroperoxide to the relatively unreactive lipid alcohol. While both catalase (7,8) and glutathione peroxidase (137) are in the whole retina, they have not yet been identified in photoreceptor outer segments.

Lipid peroxidation can proceed as a chain reaction in photoreceptor membranes where a single event is propagated to adjacent molecules. Antioxidants are the main line of defense against auto-oxidation in the retina. Among the naturally occuring antioxidants in the retina is vitamin E, which appears to be located strategically in the photoreceptor outer segment membranes (43). The presence of ascorbate (159) and reduced glutathione (157) in retina suggests they too may serve as retinal antioxidants.

Morphological, Functional, and Chemical Changes in Retinas Exposed to Oxidative Stress

The susceptibility of the eye, especially the retina, to the deliterious effects of oxygen, light, peroxides, drugs, and anti-oxidant deficient diets has been established by a number of investigators. Dietary manipulations promoting antioxidant deficiencies in a wide variety of animals lead to profound changes in retinal morphology. Hayes (68) showed that monkeys maintained for two years on diets deficient in vitamin E developed lesions in the macula that were characterized by disruption of the orderly disc packing in the rod outer segments. The retinal pigment epithelial cells were distended and had accumulated large amounts of fluorescent materials. Hayes (68) reasoned that the high levels of polyunsaturated fatty acids in these membranes rendered these structures more susceptible to oxidizing reactions. Vitamin E deficiency in dogs was found to produce retinopathy as early as three months (120). Morphological analysis revealed prominent photoreceptor outer segment damage and thinning of the outer nuclear layer, especially in the central retina.

Katz et al. (88) found massive accumulation of lipofuscin in the pigment epithelium of pigmented rats fed diets deficient in both vitamin E and selenium. Other animals on this diet demonstrated losses of polyunsaturated fatty acids in both rod outer segments and pigment epithelium (48), the greatest loss being in the retinal pigment epithelium. In rats fed diets deficient in vitamin A and/or vitamin E, the degree of photoreceptor damage was influenced by both vitamins, whereas accumulation of lipofuscin in the pigment epithelium was most affected by vitamin A (122).

The retina of the premature human infant is especially vulnerable to the high levels of oxygen that are necessary for infant survival. Following prolonged oxygen exposure, the part of the retina that is still developing produces abnormal blood vessels that eventually become fibrotic and non-functional. That region of the retina then dies due to lack of an adequate blood supply. Subsequently, the fibrotic tissue can contract, detaching the entire retina. That this effect may be mediated by oxidative mechanisms was suggested from studies showing vitamin E to be efficacious in preventing permanent damage in premature infants (76,89).

Continuous exposure of albino rats to light of intensity far below that necessary to cause thermal burn results in progressive photoreceptor-specific degeneration, with concomitant loss of visual function (58,108).

Studies have shown that the spectrum of light which induces maximal photoreceptor damage approximates that of the absorption spectrum of the visual pigment rhodopsin (58,108,156). The superior retina also demonstrates greater susceptibility to light damage compared to other retinal regions (118). Light damage has been correlated with duration and intensity of light exposure, body temperature, nocturnality, albinism, and age (95).

Human and primate retinas are also at risk for damage by light. For example, prolonged exposure to sunlight or high-intensity incandescent light prior to scheduled enucleation for intraorbital tumor resulted in morphological damage to photoreceptors (116,143). In more controlled experiments, Sperling et al. (135) have shown that monkeys can be permanently blue-blinded if exposed intermittently to blue light of intensity below the threshold for thermal burns. Likewise, Ham et al. (65) have demonstrated that monkeys exposed to short wave-length irradiation undergo irreversible ophthalmoscopic retinal changes.

Noell and associates (108,109) have postulated that photosensitized oxidations, forming toxic photoproducts, could play a role in the pathogenesis of light damage. The proposal that light damage to the retina is due in part to singlet oxygen (38,39) is supported by the finding that singlet oxygen is produced when the retina is irradiated (100). That the effect of light on retina occurs via free radical oxidation of photoreceptor membranes is evidenced by a study demonstrating increases in the concentration of hydroperoxides in whole retinas and in rod outer segments of frogs after only 30 minutes exposure in vivo (86). In addition, lipid hydroperoxide formation in ROS in vitro showed an action spectrum having a maximum near 500 nm, the absorption maximum of rhodopsin. In another study, Kagan et al. (87) found increased lipid hydroperoxides in the retinas of rats following 24 hours of intense constant illumination. Wiegand et al. (153), using less intense illumination, found an increase in lipid conjugated dienes in rat retinas by 3 days. Recent studies by Shvedova et al. (132) suggest that both singlet oxygen and the accumulation of lipid peroxidation products are involved in light-induced damage to retina.

Based on the rationale that vitamin E is a scavenger of free radicals and free radicals may be formed in retinal cells during light exposure, several groups have tested the hypothesis that dietary antioxidant supplementation affords protection against light damage. Neither supplemented nor deficient diets altered the amount of lipofuscin pigment in the RPE of rats under acute light stress (136). Similarly, albino rats raised on vitamin E-deficient diets were not more susceptible to light damage (138). Thus, it appears that dietary vitamin E is not a major factor in protecting the retina and RPE from damage following acute light stress. Alternatively, the decreased levels of vitamin E in photoreceptors of animals raised on E-deficient diets may still be sufficient to give some protection against light stress.

Retinas of rabbits exposed either to x-irradiation or intravitreal injections of linoleic acid hydroperoxides exhibit reduced amplitudes of the electroretinogram (ERG), marked pyknosis of the visual cell nuclei, and eventual photoreceptor cell death (72). Similar studies by Armstrong et al. (6) confirmed that intravitreal injection of hydroperoxides of several polyunsaturated fatty acids causes retinal degeneration concomitant with loss of the ERG amplitudes. Lipid peroxides have been demonstrated in retinas of rabbits (71) and chick embryos (162) exposed in vivo to high concentrations of oxygen.

Clinical reports have shown that iron-containing foreign bodies retained in the eye may result in retinal degeneration resembling retinitis pigmentosa (30). Several studies have documented the degenerative retinal changes associated with experimental ocular siderosis (36,103). Following surgical implantation of an iron nail in rabbit vitreous, retinal lipid peroxides increased concomitant with ERG amplitude reduction (73). Kagan et al. (86) showed that lipid hydroperoxides were formed in isolated frog retina incubated with ferrous ion and ascorbate. The appearance of peroxides coincided with decreases in retinal electrical activity.

Rhodopsin regenerability is reduced under conditions that produce peroxidation of rod outer segment lipids in vitro (34,49). Crouch (34) observed the greatest loss in regenerability in the presence of ferrous ion, which also resulted in the highest production of MDA. Farnsworth and Dratz (49) demonstrated that isolated rod outer segments exposed to pro-oxidant conditions displayed substantial disruption in disc membrane morphology which coincided with the appearance of MDA.

Kretzer and Mehta (90) have observed that adriamycin given as a single intraperitoneal injection to sexually mature rats produced retinal degeneration specific to the photoreceptors. The drug is also cardiotoxic, an action thought to occur via formation of lipid peroxides (16).

Thus, there is ample evidence that the lipid phase of the photoreceptor membranes is the major site for retinal lipid peroxidation and reasonable support for the postulate that peroxidative mechanisms may be involved in certain types of retinal degeneration.

EXPERIMENTAL RETINAL DEGENERATION: STUDIES FROM THE AUTHORS' LABORATORY

Over the past several years, our laboratory has investigated the role of lipid peroxides in retinal degenerations. We have utilized two experimental conditions: constant illumination and intravitreal injection of ferrous ion.

Methodology

Constant Illumination - Experimental Design

Female albino rats (Sprague Dawley) weighing between 225-275 grams were maintained in our vivarium under cyclic light (12L:12D) of about 10-15 foot-candles (ft-c) at the front of the cage nearest the light. Under these conditions, there was no retinal degeneration as long as the light remained cyclic. One week prior to any experiment, the rats were placed in a metabolic chamber (Freas Model 818, Precision Scientific, Chicago, Ill.) at 25° under the same lighting regime at 10-15 ft-c provided with a Vita-Lite fluorescent lamp of 40 watts (Duro-Test Corp., North Bergen, N.J.). The experiment was begun by exposing the rats at the usual time of light onset to constant illumination of either 10-15 ft-c or 100-125 ft-c. At various times thereafter, the animals were sacrificed and their retinas processed for morphological examination and chemical analysis. The data reported herein are from animals sacrificed after 1 hr of illumination at 10-15 ft-c (L-1 Hr), and 1 day (H-1 Dy) and 3 days (H-3 Dy) after exposure to 100-125 ft-c of constant illumination.

Intravitreal Injection of Ferrous Ions

Frogs (Rana pipiens) were immobilized with an intraperitoneal injection of d-tubocurarine chloride (9 µg/g body weight). Using a 32 gauge needle, frogs were then injected intravitreally at the pars plana with 10 µl of an aqueous solution of 20 mM ferrous sulfate or 10 µl solution of 20 mM sodium sulfate.

Morphological Examination

Retinal tissue was processed for light and electron microscopy as follows. Eyecup pieces from rats were fixed by immersion in a mixture of 2.5% glutaraldehyde and 1% formaldehyde buffered to pH 7.35 with 0.09 M sodium cacodylate. Eyecup pieces from frogs were fixed in 0.09 M sodium phosphate buffer (pH 7.4) containing similar concentrations of glutaraldehyde and formaldehyde. After an initial fixation period of 2 hr, the tissue was rinsed several times in cacodylate buffer containing 5% sucrose and postfixed in 1% O_sO_4 for one hour. This was followed by dehydration of the tissue in a graded ethanol series, clearing with propylene oxide and embedding in Epon-Araldite. Orientation of retinal areas along known meridians was maintained during dissection and throughout processing. Thick sections (1 µm) were mounted on glass slides and stained with toluidine blue for light microscopy. Thin sections (500-600 Å) were collected on copper mesh grids and stained with uranyl acetate and lead citrate for examination on a JEOL 100-CX electron microscope.

Electroretinography

The ERG is a non-invasive means of measuring retinal function. The a-wave (see Fig. 7) amplitude denotes photoreceptor cell activity, while the b-wave is a measure of neural retinal function. The ERG results were obtained from anesthetized frogs before intravitreal injections and at various times thereafter. When ERG monitoring was carried out for longer than 4 hr post-injection, frogs were returned to their normal 12L:12D cycle. Any subsequent ERG analyses were preceded by 2 hr dark-adaption. An Ag-AgCl cotton wick electrode was placed on the eye to be stimulated and a reference on the animal's tongue. Signals from the electrodes were coupled to a PAR model 113 pre-amplifier with a time constant to 1.6 seconds. The output of the pre-amplifier was displayed on a storage oscilloscope. Intensity response curves were generated with 20 millisecond flashes of white light from a Xenon lamp (maximum intensity = 10^6 µW/cm^2). Calibrated neutral density filters were placed in the light path to control intensity and a Melles Griot KG-1 water filter was permanently installed in the optical system to eliminate heat.

Isolation and Purity of Photoreceptor Membranes

Isolation of frog (150) and rat (153) ROS was accomplished by discontinuous sucrose gradient centrifugation as described previously. All solutions were saturated with argon to minimize lipid oxidation. The final ROS pellet was transferred in an aliquot of water to a homogenizing tube and resuspended by homogenization. Aliquots of the membrane suspension were taken for Lowry protein quantitation (101), using purified bovine serum albumin as the standard, for membrane purity evaluation by polyacrylamide gel electrophoresis (PAGE) (75), and for lipid extraction (149). The isolation procedure for rats yielded 40-60 μgm ROS protein per retina (cyclic light, 10-15 ft-c), an 85-90% purity of opsin determined by PAGE, and an uncorrected ratio of absorbance at 278 nm to 498 nm of 2.5 to 3.0 with dark-adapted retinas. For frogs, recovery of 150-200 μg ROS protein per retina was routine. Spectral ratios from dark-adapted retinas were on the order of 2.5, uncorrected for light scattering, and PAGE analysis showed that at least 90-95% of the protein band was the visual pigment aproprotein opsin. Membranes prepared from both species showed no recognizable organelles as judged by electron microscopy. We have not attempted to estimate the extent of the contribution of RPE lipids to the apparent lipid composition of the purified ROS membranes. However, we assume the contribution to be negligible.

Lipid Analysis

The lipid extract was evaporated under nitrogen at room temperature and brought to a known volume with chloroform:methanol (19:1, v/v, saturated with H_2O). Aliquots were removed for lipid phosphorus assay (125) as well as for total phospholipid fatty acid quantitation. A portion of the total lipid extract (2-3 μg phosphorus) was applied to a silica gel-HR plate (5 x 20 cm) and developed in hexane:ethyl ether:glacial acetic acid (60:40:1, by vol.). The region on the plate containing the total phospholipids (the origin) was scraped into a screw-capped tube. Methyl esters were prepared with 14% BF_3-methanol (106) and their masses quantitated by gas-liquid chromatography (GLC) using heneicosanoic acid (21:0) as an internal standard (149).

Phospholipid classes were resolved by two-dimensional thin-layer chromatography (TLC) of the lipid extract (4) and recovered from the chromatoplates (149). After the solvent extract containing the purified phospholipids was evaporated to dryness under nitrogen, a known amount of chloroform was added, and aliquots were removed for lipid phosphorus assay or for fatty acid quantitation. A portion of the phosphatidylcholine, phosphatidylethanolamine, and phosphatidylserine extracts was hydrolyzed with phospholipase C to produce 1,2-diglycerides, which were acetylated to 1,2-diglyceride acetates (DGAc), hydrogenated, purified by TLC, and analyzed by high-temperature GLC (151). Each DGAc is defined with a carbon number which is the sum of the combined carbon numbers of the two long-chain fatty acids present in the unhydrolyzed phospholipid. For example, a C-40 DGAc may be composed of a combination of fatty acids containing 16 and 24, 18 and 22, or 20 and 20 carbons.

Conjugated Diene Assay

Conjugated dienes in ROS lipids were estimated according to the procedure described by Buege and Aust (26). A portion of the lipid extract was dried under argon and dissolved immediately in an appropriate volume of argon-purged absolute ethanol. The absorbance at 233 nm was determined against a solvent blank on a recording spectrophotometer; the sample was scanned from 330 nm to 205 nm. Subsequently, phosphorus was determined on the lipid extract and the phospholipid concentration determined. The data are expressed as absorbance units/ml solvent/mg phospholipid.

Results

Retinal Degeneration Following Constant Illumination

The primary effect of constant light on the morphology of the albino rat retina was degeneration of the photoreceptor cells (Fig. 4). This is similar to that observed in numerous other laboratories (see review by Lanum (95). After a 24 hr exposure to 100-125 ft-c, structural changes were evident in the photoreceptor inner and outer segments which showed swollen mitochondria and vesiculated disk membranes, respectively. Photoreceptor losses were not apparent after a 24 hr exposure, although some nuclei were pyknotic. Figure 4 shows the further progression of photoreceptor degeneration with a 3 day exposure to constant illumination of 100-125 ft-c. Light microscopy revealed the disorientation and fragmentation of the photoreceptor inner and outer segments (Fig. 4b) compared to controls maintained for 1 hr at 10-15 ft-c (Fig. 1a). The ONL had been reduced in thickness by approximately 50%, indicating the destruction and removal of photoreceptor cells. Many of the remaining photoreceptor nuclei were pyknotic. Macrophages had invaded the ONL and contained ingested photoreceptor debris. The cells of the inner retina and RPE remained intact although the latter were somewhat edematous.

The outer segment fragments which remained after a 3 day exposure were examined by electron microscopy. In comparison to control outer segments (Fig. 4c), orderly stacking of the disk membranes was disrupted (Fig. 4d). Also, formation of vesicles within the disk membranes was apparent, although considerable variation was evident among outer segments in this regard. The disk membranes seen in these micrographs are representative of those utilized for chemical analysis.

The severity of light damage in the superior and inferior portions of the verticle meridian can be seen in Figure 5. The inferior retina shows moderate damage with slightly more than half of the photoreceptor nuclei missing. In contrast, only an incomplete row of photoreceptor nuclei remain in the mid-peripheral zone of the superior retina, indicating a severe degree of damage in this region.

The PAGE tracings given in Figure 6 show that the ROS membranes isolated after 3 days of constant illumination (H-3 Dy) were similar in protein composition to those isolated after 1 hr exposure to low intensity light (L-1 Hr). The opsin band of the H-3 Dy scan actually comprises a greater percentage of the total protein than was found in

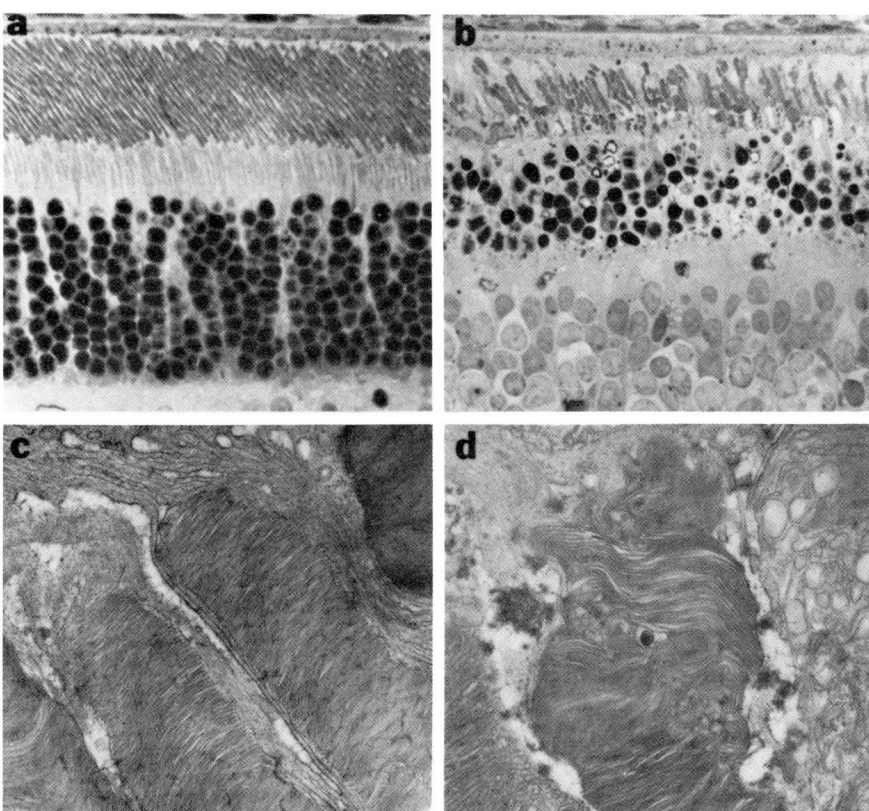

FIG. 4. Morphological changes in the albino rat retina following constant illumination. All micrographs were taken from comparable regions in the midperipheral superior retina along the vertical meridian. a) light micrograph of retina from L-1 Hr control; mag = x800. b) light micrograph of retina from H-3 Dy exposed rat; mag = x800. c) electron micrograph of ROS in a L-1 Hr control; mag = x10,000. d) electron micrograph of ROS in a H-3 Dy exposed rat. Three outer segments can be seen representing various stages of degeneration; mag = x6,700.

the L-1 Hr ROS membranes. The explanation for this is a side issue, and most likely involves light-induced binding of soluble ROS proteins to the photoreceptor discs (93). The protein (Peak 1) binding to ROS membranes prepared from light-adapted retinas occurred very early after light exposure and decreases with time after constant illumination.

In our hands, the opsin band routinely accounted for 85 to 90% of the protein in a typical polyacrylamide gel. The point to be made with regard to ROS purity is that, by their electrophoretic patterns, the membranes prepared after three days of constant illumination and identified as ROS were not significantly contaminated with other membranes and may be considered "minimally damaged." After the purity of the various membrane preparations was established, our attention was concentrated on possible lipid changes in these membranes.

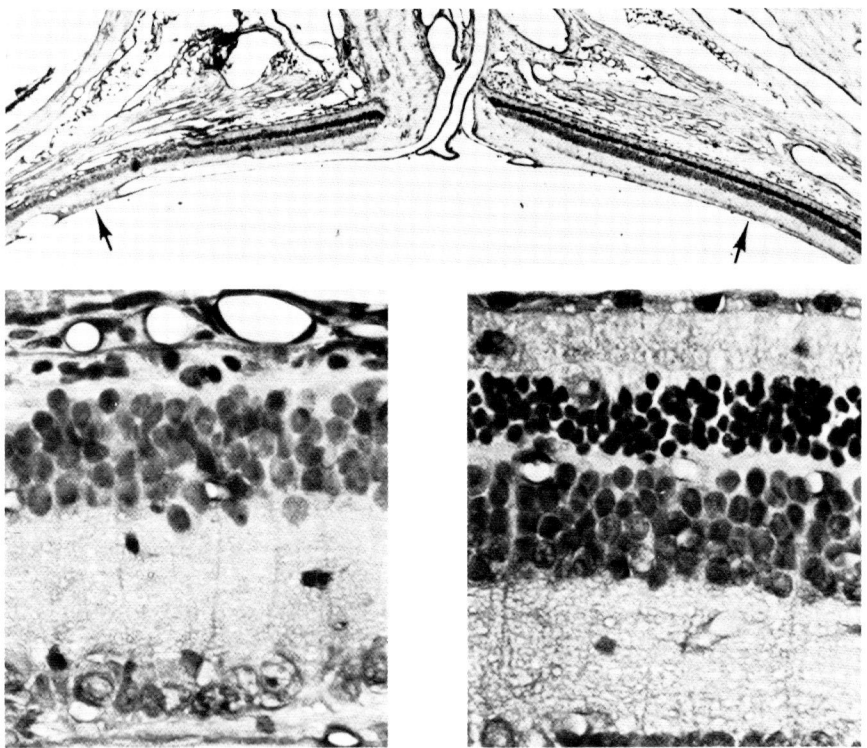

FIG. 5. Micrograph of an albino rat retina exposed to 80 lux for 4 days. Plane of section is through the vertical meridian with the optic nerve head centered, superior retina shown to the left and inferior to the right. Arrows point to regions 1 mm in the peripheral direction from the optic nerve head (x25). Higher magnifications of the regions indicated by the arrows are shown in enlargements below (x400). Left, lesioned area in the superior retina. Right, relatively less affected region in the inferior retina. (From Anderson and Hollyfield, Ref. 3, with permission).

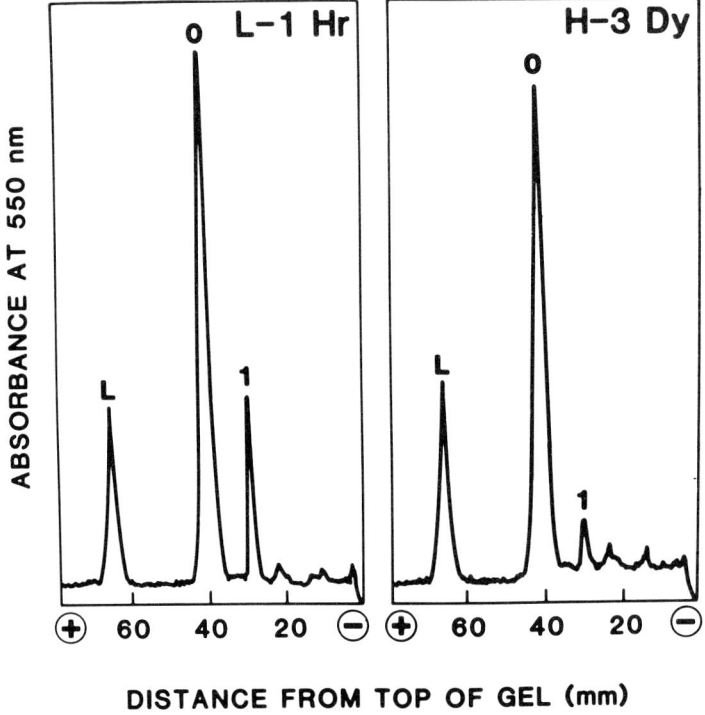

FIG. 6. SDS polyacrylamide gel electrophoresis of ROS of rats exposed for 1 hour to 10-15 foot-candles (L-1 Hr) and for 3 days to 100-125 foot-candles (H-3 Dy). L-lysozyme (internal standard), molecular weight (M.W.) 14.3 K; O-opsin, M.W. 34 K-36 K; Peak 1 - unknown protein with M.W. 53 K-55 K.

Constant illumination for 3 days at 100-125 ft-c reduced by 50% (29.8 versus 60.2 μgm protein per retina) the yield of ROS compared to retinas exposed to the same level of illumination for 1 day or at 10-15 ft-c for 1 hr (Table 1). The reduced yield of ROS membranes per retina may be due in part to degeneration and in part due to loss of lipids from the membranes, resulting in increased membrane density and subsequent alteration of membrane flotation properties on the sucrose gradient.

TABLE I. Protein and lipid content of rat rod outer segments following constant illumination[a]

Light status[b]	ROS yield microgram/retina	Weight ratio protein/phospholipid[c]
L-1 Hr	60.2 ± 13.6 (7)[d]	1.24 ± 0.07 (3)
H-1 Dy	57.3 ± 21.3 (4)	1.31 ± 0.06 (3)
H-3 Dy	29.8 ± 10.4 (8)	1.32 ± 0.11 (5)

[a]Wiegand, Giusto, and Anderson (Ref. 152).
[b]L-1 Hr, 10-15 ft-c for 1 hour; H-1 Dy, 100-125 ft-c for 1 day; H-3 Dy, 100-125 ft-c for 3 days.
[c]Phospholipid calculated by multiplying 25 times the mass of phosphorus.
[d]Mean ± S.D., Value in parenthesis is the number of independent determinations.

The protein-to-phospholipid ratio increased with the time of constant illumination (Table 1), lending support to this notion. Furthermore, PAGE of the 1.140/1.175 interfacial band (ROS usually partition at the 1.115/1.140 interface) showed increased levels of opsin in membranes isolated from retinas exposed to constant illumination for 1 or 3 days. Membranes from this interface were not studied further because they are a heterogenous population derived from many different cellular organelles.

Table 2 shows the levels of the major fatty acids of the total phospholipids expressed as nmoles per mg ROS protein. One of the chemical changes in light-induced retinal degeneration was reduction of the long-chain polyunsaturated fatty acid, $22:6\omega3$, after 3 days of constant illumination. Arachidonic acid ($20:4\omega6$), the other long-chain polyunsaturated fatty acid, as well as the major saturated acids, palmitic and stearic, were not affected. In addition, when the fatty acids of the major phospholipid classes were examined (data not shown), the level of $22:6\omega3$ was reduced at 3 days in each phospholipid class by roughly 10% (absolute) compared to the L-1 Hr or H-1 Dy animals.

Table 3 shows the phospholipid class and carbon number distribution of the diglyceride acetates derived from the major phospholipid classes. Expressed as relative mole %, there appears to be a slight decrease in the amounts of both phosphatidylethanolamine and phosphatidylserine and a concomitant increase in the level of phosphatidylcholine after 3 days of constant illumination.

The levels of lysophospholipids, 1,2-diglycerides, and free fatty acids from the ROS were not affected (data not shown). The possibility exists that photoreceptor degeneration in constant illumination results from activation of phospholipases that hydrolyze membrane phospholipids to free fatty acids and 1,2-diglycerides. These fusogenic lipids could then promote membrane vesiculation and degeneration. However, the constant levels of lysophospholipids, diglycerides, and free fatty acids argue against this possibility.

Analysis of the diglyceride acetates derived from these phospholipid classes showed that those molecular species containing two polyunsaturated fatty acids, namely, C-42 (one 20-carbon and one 22-carbon fatty acid), C-44 (two 22-carbon fatty acids), and C-46 (one 22-carbon and one 24-carbon fatty acid) were decreased after 3 days of constant illumination. Because these carbon number distributions are expressed as relative mole %, there is an apparent increase in the lower molecular weight species. These data suggest that the phospholipids most susceptible to loss of $22:6\omega3$ are those with molecular species containing two of these polyunsaturated fatty acids. Miljanich et al. (105), Wiegand et al. (151), and Aveldano de Calidironi and Bazan (13) have shown that ROS phospholipids contain high levels of dipolyunsaturated fatty acid species. This may be a unique feature of ROS phospholipids, since the usual phospholipid molecular species contain polyunsaturates only at position-2.

TABLE 2. Composition of the major fatty acids of total phospholipids of rat rod outer segments following constant illumination[a]

Fatty acids[b]	Light status[c]		
	L-1 Hr (4)[d]	H-1 Dy (4)	H-3 Dy (5)
	nmol/mg protein[e]		
16:0	302 ± 60	266 ± 31	328 ± 30
18:0	581 ± 23	576 ± 42	560 ± 28
18:1	85 ± 22	66 ± 10	113 ± 24
$20:4\omega6$	108 ± 13	86 ± 20	114 ± 19
$22:6\omega3$	889 ± 86	841 ± 43	654 ± 100

[a]Wiegand, Giusto, Rapp, and Anderson (Ref. 153).

[b]16:0-palmitic acid; 18:0-stearic acid; 18:1-oleic acid; $20:4\omega6$-arachidonic acid; $22:6\omega3$-docosahexaenoic acid.

[c]Refer to legend of Table 1 for explanation of light status.

[d]Values in parenthesis are the number of membrane preparations from at least 10 pooled retinas.

[e]Values reported as mean ± S.D.

TABLE 3. Phospholipid class and carbon number distribution of the diglyceride acetates derived from the major phospholipid classes of rat rod outer segments following constant illumination

Light[a] status	Phospholipid[b] class	Diglyceride acetate carbon number[c]							
		32	34	36	38	40	42	44	46
		Relative mole %							
		Phosphatidylcholine							
L-1 Hr	37.0	15.9	17.7	10.0	14.3	33.6	1.0	6.3	0.6
H-1 Dy	37.8	16.4	18.0	10.0	15.0	33.3	0.9	5.1	0.5
H-3 Dy	40.3	19.1	25.0	14.3	14.4	22.4	0.4	2.9	0.1
		Phosphatidylethanolamine							
L-1 Hr	44.7	0.0	0.3	1.1	14.3	62.7	3.5	17.2	0.6
H-1 Dy	44.5	0.0	0.4	1.1	14.3	64.9	3.3	15.0	0.5
H-3 Dy	41.3	0.1	0.6	2.1	17.4	67.3	1.6	10.3	0.3
		Phosphatidylserine							
L-1 Hr	13.7	0.0	1.4	0.9	4.3	59.5	0.7	24.3	8.5
H-1 Dy	13.8	0.0	1.2	1.1	4.0	58.3	1.0	24.6	9.5
H-3 Dy	12.8	0.0	1.4	1.8	7.9	65.3	0.5	17.8	4.9

[a]Refer to legend of Table 1 for explanation of light status. A single membrane preparation was isolated from 26, 40, or 56 pooled retinas of the L-1 Hr, H-1 Dy, or H-3 Dy animals, respectively.

[b]Values reported as the mean from duplicate thin-layer chromatoplates.

[c]Values reported as the mean of duplicate GLC analyses following hydrogenation. The carbon number is the sum of the carbon atoms of the acyl chains of the parent diglyceride and does not include the five carbons of the acetate and glycerol moieties.

TABLE 4. Lipid conjugated diene content of rat rod outer segments following constant illumination[a]

Group[b]	Conjugated diene[c]
Control	0.49 ± 0.11 (6)[d]
Experimental	1.00 ± 0.41 (8)
	$P < 0.01$

[a]Wiegand, Giusto, Rapp, and Anderson (Ref. 153).
[b]Control values include dark-adapted and L-1 Hr retinas. Experimental values include H-1 Dy and H-3 Dy retinas.
[c]Absorbance at $\lambda = 233$ nm (1 cm optical path) per mg phospholipid dissolved in 1 ml absolute ethanol.
[d]ROS were prepared from pooled retinas of individual rats. Values reported as mean ± S.D. for groups of 6 or 8 animals.

The content of lipid conjugated dienes in ROS lipids is presented in Table 4. Control values include dark-adapted and L-1 Hr retinas; experimental values are from H-1 Dy and H-3 Dy retinas. Animals exposed to constant illumination had twice the content of conjugated dienes as control animals.

Retinal Degeneration Following Intravitreal Injection of Ferrous Ions

Intravitreal injection of frogs with ferrous ions caused dramatic deterioration of the ERG. Figure 7 shows the normal appearance of the ERG pattern of the pre-injection control. However, within 1 hr of injection of ferrous sulfate, amplitude of both the a- and b waves markedly decreased. By 4 hr both a- and b-waves were nearly extinguished and were not detectable 24 hr after injection. In contrast, injection with sodium sulfate produced no change in the ERG compared to pre-injection controls. Figures 8 and 9 show the time course of loss of amplitude of the a- and b-waves, respectively, of the ERG for the 24 hr period following the injection of ferrous sulfate or sodium sulfate. Although it would be difficult from our data to speculate whether the loss of amplitude is exponential or bimodal, it is apparent that the amplitude of both a- and b-waves is reduced by 50% or more within 1 hr after injection of the ferrous ion.

Morphological changes also were observed soon after intravitreal injection of frogs with ferrous ion. Figure 10a is a light micrograph of a retina 24 hr following injection of sodium sulfate. No differences are apparent between this retina and non-injected controls. However, 4 hr

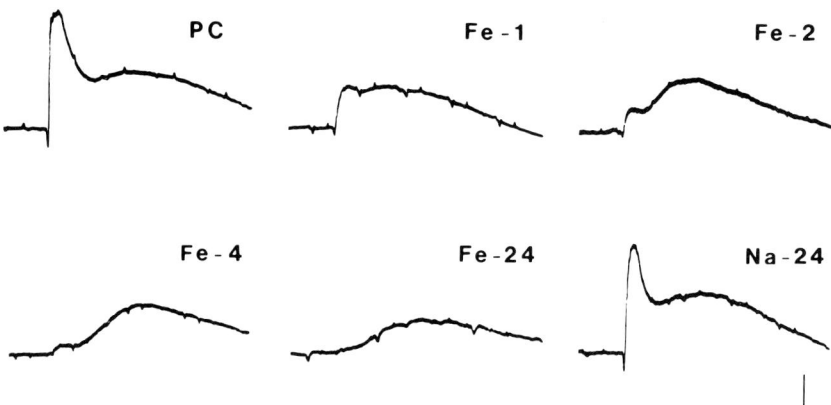

FIG. 7. ERG waveforms recorded at various intervals following intravitreal injections of ferrous sulfate or sodium sulfate. All responses were elicited with saturating flashes. Stimulus duration was 20 milliseconds. The small spikes superimposed on the ERG are due to the frog's heartbeat. Calibration bars denote an amplitude of 200 v and a duration of 0.5 seconds. PC: pre-injection control; Fe-1, -2, -4, -24: 1, 2, 4, 24 hr following $FeSO_4$ injection; Na-24: 24 hr following Na_2SO_4 injection. (From Rapp et al., Ref 119, with permission).

after injection of ferrous sulfate, the orderly arrangement of the ROS disks is clearly disrupted, and the nuclei of most (if not all) of the photoreceptor cells are pyknotic (Fig. 10b). The RPE is somewhat swollen, but appears otherwise intact. After 8 days (Fig. 10c), no rod nuclei are detectable. Between the outer limiting membrane and the pigment epithelium, a layer of disks develops that appears to be made up of photoreceptor remnants and invading macrophages, perhaps containing melanin. Also to be noted at this advanced stage of degeneration is the presence of oil droplets in the subretinal space. These are increased in size and number in the pigment epithelium. With the exception of an occasional darkening of inner nuclear layer cells, the degenerative changes described above were predominantly restricted to photoreceptor cells and pigment epithelium.

Chemical changes were observed in the ROS following ferrous ion injection. The data in Table 5 compare pre-injection controls with animals injected with either sodium or ferrous sulfate. The ROS yield from ferrous ion-injected retinas was always less than from pre-injection or sodium injection controls. PAGE of experimental and control ROS did

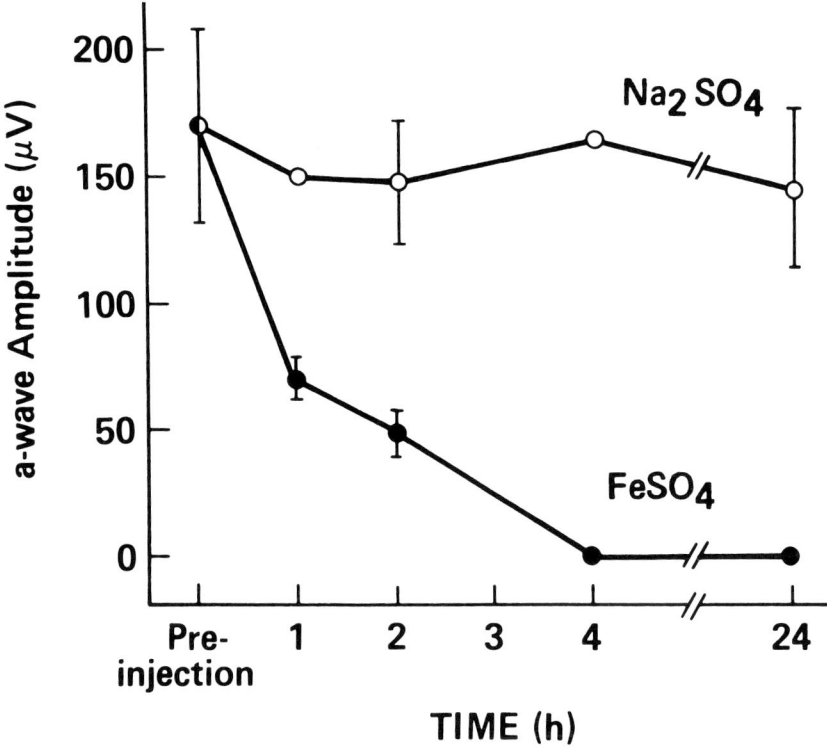

FIG. 8. Time course for changes in a-wave amplitude following intravitreal injections of ferrous sulfate or sodium sulfate into the frog eye. Amplitudes represent the saturation level of the a-wave. Standard deviations are drawn upon mean values for groups of 4 hr or more frogs. (From Rapp et al., Ref. 119, with permission).

not show any significant differences, indicating that our ROS preparations were of equal purity (data not shown). The presence of lipid peroxides, determined by increased absorbance at 233 nm in the experimental group, confirmed that injection of ferrous ion resulted in increased lipid peroxidation. The peroxide levels were 3 to 4 times higher in the ferrous-injected eyes than in the control. Concomitant with the increase in peroxide was a loss of long-chain poly-unsaturated fatty acids from total ROS phospholipids. This was observed as a selective decrease in 22:6ω3. Palmitic acid (16:0), a saturated fatty acid which is not subject to peroxidation reactions, was essentially unchanged over the 24 hr time course.

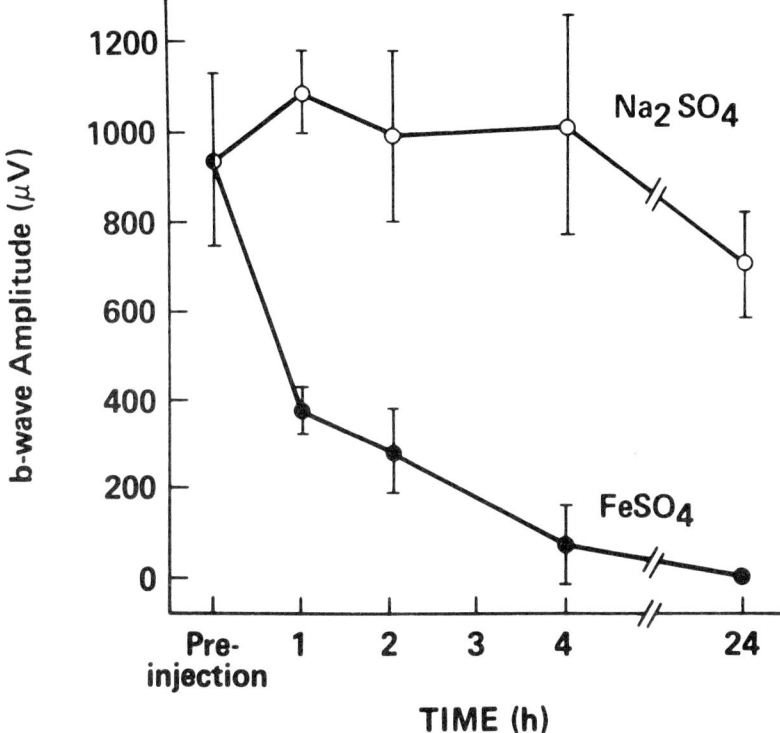

FIG. 9. Time course for changes in b-wave amplitude following intravitreal injections of ferrous sulfate or sodium sulfate into the frog eye. Amplitudes and standard deviations were determined in the same manner as in Fig. 8. (From Rapp et al., Ref. 119, with permission).

Discussion

One result of exposure of rats to constant illumination is progressive loss of polyunsaturated fatty acids and accumulation of lipid hydroperoxides. In the frogs injected intravitreally with ferrous ion, the same biochemical pattern was observed. Our results show that the selective decrease in 22:6ω3 is accompanied by an increase in the levels of lipid conjugated dienes. Of the two determinations (fatty acid versus conjugated diene), the latter is less reliable because the methodology for determining lipid peroxides in these extracts is less sensitive than that for methyl ester analysis. However, both the loss of polyunsaturates and

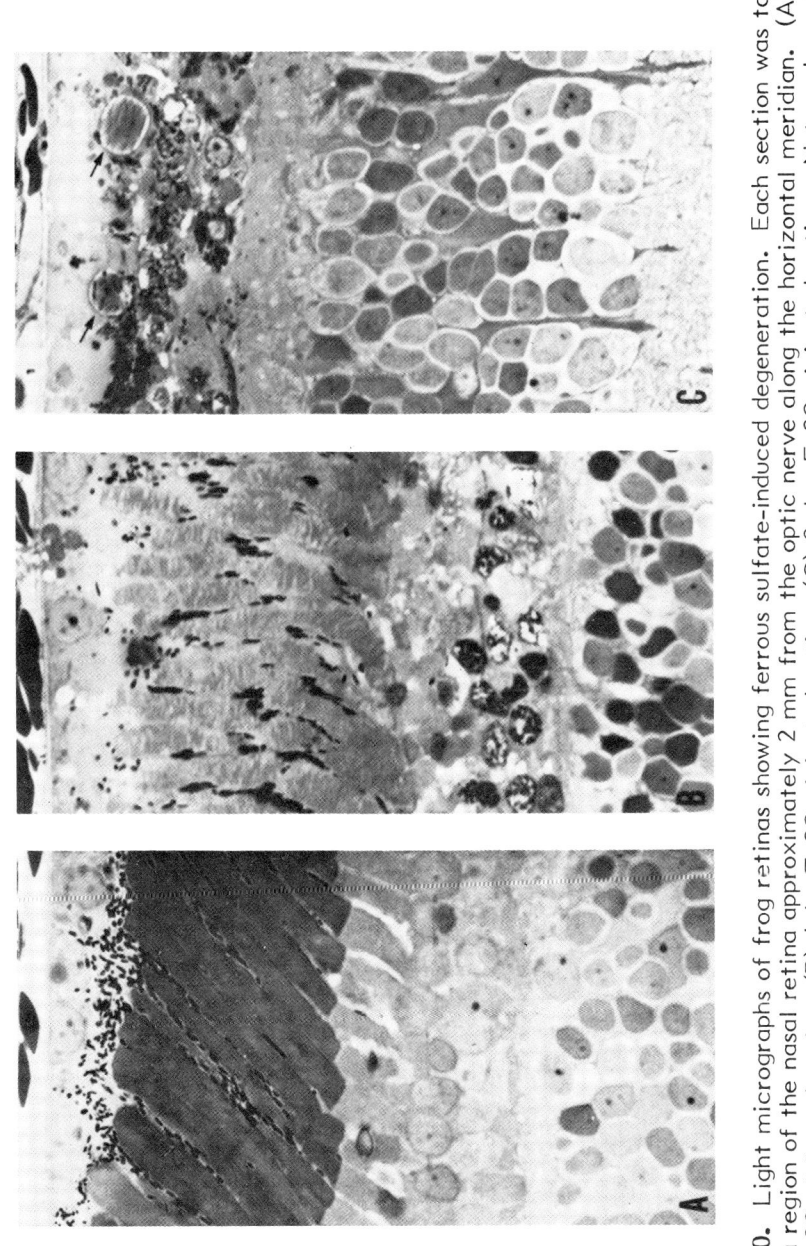

FIG. 10. Light micrographs of frog retinas showing ferrous sulfate-induced degeneration. Each section was taken from a region of the nasal retina approximately 2 mm from the optic nerve along the horizontal meridian. (A) 24 hr Na$_2$SO$_4$-injected retina, (B) 4 hr FeSO$_4$-injected retina, (C) 8 day FeSO$_4$-injected retina. Note enlarged oil droplets (arrows) and compaction of the retina at 8 days (micrographs x 430). (From Rapp et al., Ref. 119, with permission).

TABLE 5. Protein, lipid, and peroxide content of frog rod outer segments[a]

Group[b]	Injection	ROS yield (μg protein/ retina)	Peroxide[c] Levels	Fatty Acids[d] (nmol/mg protein) 16:0	22:6
Preinjection	none	215	0.56	364	950
Preinjection	none	228	0.33	363	958
4 hours	Na$_2$SO$_4$	301	0.55	344	902
4 hours	FeSO$_4$	154	1.98	351	798
24 hours	Na$_2$SO$_4$	214	0.59	327	1012
24 hours	FeSO$_4$	72	1.53	354	684

[a]Rapp, Wiegand, and Anderson (Ref. 119).
[b]In each group, analyses were performed on ROS membranes isolated from 3 pooled retinas. Injected frogs received 10 μl of 20 mM Na$_2$SO$_4$ in the left eye and 10 μl of 20 mM FeSO$_4$ in the right eye.
[c]Absorbance at λ = 233 nm (1 cm optical path) per mg phospholipid dissolved in 1 ml cyclohexane-ethanol (1:1, v/v).
[d]16:0-palmitic acid; 22:6-docosahexaenoic acid.

the increase in conjugated dienes provide evidence for lipid peroxidation. This conclusion was reached by Kagan et al. (86), who showed an increase in lipid conjugated dienes in frog retinas exposed to light for 30 min, relative to controls maintained in darkness. Further studies by Kagan et al. (87) demonstrated increased lipid peroxides in the retinas of rats after 24 hr of constant high intensity illumination. Interestingly, in this latter study, animals deficient in vitamin E (an endogenous antioxidant) had greater amounts of conjugated dienes than controls exposed to the same levels of illumination. Shvedova, et al. (133) examined frog retinas incubated with ferrous ion and ascorbate for up to 20 min and observed a progressive decrease in ERG amplitude that correlated with the appearance of lipid peroxides. No ultrastructural changes were apparent after this short time.

Although the insults in light- and ferrous ion-induced retinal degenerations are obviously different, the mechanism of retinal degeneration may be quite similar. In the case of ferrous ion, lipid peroxides could be generated from hydroxyl radicals which are produced by the oxidation of ferrous ion with hydrogen peroxide (67). Hydrogen peroxide is formed in the ROS by the dismutation of superoxide radicals

(64). Hydroxyl radicals rapidly initiate the chain reactions of lipid peroxidation, ultimately leading to formation of MDA, a bi-functional compound that reacts with primary amines to form a Schiff's base. This bi-functional compound may cross-link amino groups of lipids and proteins in biological membranes (40,107). In ROS, the production of MDA may result in cross-linking between opsin and phospholipids. Crouch (34) reported that illumination of suspensions of bovine ROS led to increased MDA production which correlated with the loss of regenerability of rhodopsin. Undoubtedly, extensive cross-linking would have an effect on the activity of other ROS proteins, directly altering the structure and ion permeability of the photoreceptor membranes. These pertubations would certainly contribute to the deterioration and eventual death of photoreceptor cells.

ACKNOWLEDGEMENTS

This research was supported in part by grants from the Retina Research Foundation (Houston, Texas), National Retinitis Pigmentosa Foundation, Research to Prevent Blindness, Inc., and The National Eye Institute. Dr. Anderson is the recipient of a Dolly Green Special Scholar's Award from Research to Prevent Blindness, Inc. We thank Janice Cason for her patience in typing the manuscript.

REFERENCES

1. Alter, A.J. and Leinfelder, P.J. (1953): Arch. Ophthalmol., 49:257-260.
2. Anderson, R.E. and Andrews, L.M. (1982): In: Visual Cells in Evolution, edited by J.A. Westfall, pp. 1-22. Raven Press, New York.
3. Anderson, R.E. and Hollyfield, J.G. (1983): In: Biochemistry of the Eye, edited by R.E. Anderson, pp. 243-255. American Academy of Ophthalmology, San Francisco.
4. Anderson, R.E., Maude, M.B. and Feldman, G.L. (1970): Exp. Eye Res., 9:281-284.
5. Andrews, J.S. and Leonard-Martin, T. (1981): Invest. Ophthalmol. Vis. Sci., 21:39-45.
6. Armstrong, D., Hiramitsu, T., Gutteridge, J. and Nilsoon, S.E. (1982): Exp. Eye Res., 35:157-171.
7. Armstrong, D., Santangelo, G. and Connole, E. (1981): Curr. Eye Res., 1:225-242.
8. Armstrong, D., Siakotos, A. and Koppang, N. (1979): In: Sensory Systems and Communication in the Elderly, edited by J.M. Ordy and K. Brizzee, pp.115-151. Raven Press, New York.
9. Armstrong, D. and Koppang, N. (1982): In: Ceroid-lipofuscinosis (Batten's disease), edited by D. Armstrong, N. Koppang and J.A. Rider, pp. 159-165. Elsevier, Amsterdam.
10. Augusteyn, R.C. (1975): Ophthalmic Res., 7:217-224.
11. Augusteyn, R.C. (1981): In: Mechanisms of Cataract Formation in the Human Lens, edited by G. Duncan, pp. 71-115. Academic Press, London.

12. Auricchio, G. and Testa, M. (1972): Ophthalmologica, 164:228-235.
13. Aveldano, M.I. and Bazan, N.G. (1983): J. Lipid Res., 24:620-627.
14. Bando, M., Ishii, Y. and Nakajima, A. (1976): Ophthalmic Res., 8:456-463.
15. Benedek, B.G. (1971): Applied Optics, 10:459-473.
16. Berlin, V. and Haseltine, W.A. (1981): J. Biol. Chem., 256:4747-4756.
17. Bhuyan, K.C. and Bhuyan, D.K. (1979): In: Biochemical and Clinical Aspects of Oxygen, edited by W.S. Caughey, pp. 785-796. Academic Press, London.
18. Bhuyan, K.C., Bhuyan, D.K. and Podos, S.M. (1981): IRCS Med. Sci., 9:195-196.
19. Bhuyan, K.C. and Bhuyan, D.K. (1984): Curr. Eye Res., 3:67-81.
20. Bhuyan, K.C., Bhuyan, D.K. and Podos, S.M. (1981): IRCS Med. Sci., 9:126-127.
21. Bhuyan, K.C., Bhuyan, D.K., Kuck, J.F.R., Kuck, K.D. and Kern, H.L. (1983): Curr. Eye Res., 2:597-606.
22. Bhuyan, D.K. and Bhuyan, K.C. (1983): In: Oxygen Radicals in Chemistry and Biology, edited by W. Bors, M. Saran and D. Tart, pp. 349-356. Walter De Orugten and Co., Berlin.
23. Borkman, R.F. and Lerman, S. (1977): Exp. Eye Res., 25:303-309.
24. Broekhuyse, R.M. (1981): In: Mechanisms of Cataract Formation in the Human Lens, edited by G. Duncan, pp. 151-191. Academic Press, London.
25. Buckingham, R.H. and Pirie, A. (1972): Exp. Eye Res., 14:297-299.
26. Buege, J.A. and Aust, S.D. (1978): In: Methods of Enzymology, edited by S. Fleischer and L. Packer, pp. 302-310. Academic Press, New York.
27. Burk, J.M. (1981): Invest. Ophthalmol. Vis. Sci., 20:435-441.
28. Chand, D. and Varma, S.D. (1983): Invest. Ophthalmol. Vis. Sci., Suppl., 24:31.
29. Cloud, T.M., Hakim, R. and Griffin, A.C. (1961): Arch. Ophthalmol., 66:689-694.
30. Cogan, D.G. (1969): Arch. Ophthalmol., 81:45-53.
31. Cohen, A.I. (1972): In: Handbook of Sensory Physiology, Vol. VII, Part 2, edited by M.G.F. Fuortes, pp. 63-110. Springer-Verlag, New York.
32. Cotlier, E., Obara, Y. and Toftness, B. (1978): Biochim. Biophys. Acta, 530:267-278.
33. Cox, A.B., Ainsworth, A.J., Jose, J.G., Lee, A.C., and Lett, J.T. (1983): Adv. Space Res., 3:211-219.
34. Crouch, R. (1980): In: The Effects of Constant Light on Visual Process, edited by T.P. Williams and B.N. Baker, pp. 309-318. Plenum Press, New York.
35. Cunha-Vaz, J.G., Shakib, M. and Ashton, N. (1966): Br. J. Ophthalmol., 50:441-453.
36. Declercq, S., Meredith, P. and Rosenthal, R. (1977): Arch. Ophthalmol., 95:1051-1058.
37. Delamere, N.A., Paterson, C.A. and Cotton, T.R. (1983): Exp. Eye Res., 37:45-53.

38. Delmelle, M. (1978): Photochem. Photobiol., 27:731-734.
39. Delmelle, M. (1977): Biophys. Struct. Mech., 3:195-198.
40. Dillard, C.J. and Tappel, A.L. (1973): Lipids, 8:183-189.
41. Dillon, J. and Spector, A. (1976): Nature, 259:422-423.
42. Dillon, J., Garner, M., Roy, D. and Spector, A. (1982): Exp. Eye Res., 34:651-658.
43. Dilly, R. and McConnell, D. (1970): J. Membrane Biol., 2:317-323.
44. Dische, Z., Elliot, J., Pearson, E. and Merriam, G.R. (1959): Am. J. Ophthalmol., 47:368-379.
45. Dische, Z. and Zil, H. (1951): Am. J. Ophthalmol., 34:104-113.
46. Duncan, G. and Bushell, A.R. (1979): Ophthalmic Res., 11:397-404.
47. Evans, T.C., Richards, R.D. and Riley, E.F. (1960): Radiat. Res., 13:737-750.
48. Farnsworth, C., Stone, W. and Dratz, E. (1979): Biochim. Biophys. Acta, 552:281-293.
49. Farnsworth, C. and Dratz, E. (1976): Biochim. Biophys. Acta, 443:556-570.
50. Feldman, G.L. (1968): In: Biochemistry of the Eye, edited by M.U. Dardenne and J. Nordmann, pp. 348-357. S. Karger, Basel.
51. Fliesler, S.J. and Anderson, R.E. (1983): Prog. Lipid Res., 22:79-131.
52. Francois, J. and Neetens, A. (1962): In: The Eye, Vol. I, edited by H. Davson, pp. 369-416. Academic Press, New York.
53. Fukui, H.N. (1976): Exp. Eye Res., 23:595-599.
54. Fulhorst, H.W. and Young, R.W. (1966): Invest. Ophthalmol., 5:298-303.
55. Garner, M.H. and Spector, A. (1980): Proc. Natl. Acad. Sci. USA, 77:1274-1277.
56. Garner, W.H., Garner, M.H. and Spector, A. (1983): Proc. Natl. Acad. Sci. USA, 80:2044-2048.
57. Giblin, F.J., Chakrapani, B. and Reddy, V.N. (1978): Exp. Eye Res., 26:507-519.
58. Gorn, R.A. and Kuwabara, T. (1967): Arch. Ophthalmol., 77:115-118.
59. Goldman, H. and Liechti, A. (1938): Albrecht von Graefe Arch. Ophthalmol., 138:722-736.
60. Goosey, J.D., Allison, M.E. and Garcia, C.A. (1983): Invest. Ophthalmol. Vis. Sci., Suppl., 24:75.
61. Goosey, J.D., Zigler, J.S. Jr. and Kinoshita, J.H. (1980): Science, 208:1278-1280.
62. Grabner, G. and Brenner, W. (1982): Ophthalmic Res., 14:160-166.
63. Grover, D. and Zigman, S. (1972): Exp. Eye Res., 13:70-76.
64. Hall, M. and Hall, D. (1975): Biochem. Biophys. Res. Comm., 67:1199-1204.
65. Ham, W.T. Jr., Mueller, H.A., Ruffolo, J.J. Jr. and Clarke, A.M. (1979): Photochem. Photobiol., 29:735-743.
66. Harding, J. J. and Dilley, K. J. (1976): Exp. Eye Res., 22:1-3.
67. Hochstein, P., Nordenbrand, K. and Ernster, L. (1964): Biochem. Biophys. Res. Commun., 14:323-328.
68. Hayes, K.C. (1974): Invest. Ophthalmol., 13:499-510.

69. van Heyningen, R. (1973): Exp. Eye Res., 17:137-147.
70. Hiller, R., Giacometti, L. and Yueu, K. (1977): Am. J. Epidemiol., 105:450-459.
71. Hiramitsu, T.Y., Hasegawa, K., Hirata, K., Nishigaki, I. and Yagi, K. (1976): Experientia., 32:662-623.
72. Hiramitsu, T., Majima, Y., Hasegawa, Y. and Hirata, K. (1974): Acta Soc. Opthalmol. Jap., 78:819-825.
73. Hiramitsu, T., Majima, Y., Hasegawa, Y., Hirata, K. and Yagi, K. (1976): Experientia, 32:1324-1325.
74. Hogan, M.J., Alvarado, J.A. and Weddell, J.E. (1971): Histology of the Human Eye, pp. 393-522. W.B. Saunders Co., Philadelphia.
75. Hsieh, W.-C. and Anderson, R.E. (1975): Anal. Biochem., 69:331-338.
76. Hittner, H.M. Godio, L.B., Rudolph, A.J., Adams, J.M., Garcia-Prats, J.A., Friedman, Z., Kautz, J.A. and Monaco, W.A. (1981): N. Eng. J. Med., 305:1365-1371.
77. Hollows, F. and Moran, D. (1981): Lancet, 2:1249-1251.
78. Howard, R.O., McDonald, C.J., Dunn, B. and Creasey, W. (1969): Invest. Ophthalmol., 8:413-421.
79. Jernigan, H.M. Jr., Fukui, H.N., Goosey, J.D. and Kinoshita, J.H. (1981): Exp. Eye Res., 32:461-466.
80. Jose, J. (1978): Ophthalmic Res., 10:52-62.
81. Jose, J. (1982): Invest. Ophthal. Vis. Sci., Suppl., 22:198.
82. Jose, J.G. (1983): In: Biochemistry of the Eye, edited by R.E. Anderson, pp. 111-144. American Academy of Ophthalmology, San Francisco.
83. Jose, J.G., Koch, H.R. and Respondak, A. (1982): Graefes Arch. Clin. Exp. Ophthalmol., 219:44-53.
84. Jose, J.G. and Yielding, K.L. (1977): Photochem. Photobiol., 26:549-551.
85. Jose, J.G. and Yielding, K.L. (1978): Invest. Ophthalmol. Vis. Sci., 17:687-691.
86. Kagan, V., Shvedova, A., Novikov, K. and Kozlov, Yu. (1973): Biochim. Biophys. Acta, 330:76-79.
87. Kagan, V.E., Kuliev, I.Ya., Spirichev, V.B., Shvedova, A.A. and Kozlov, Yu. P. (1981): Bull. Exp. Biol. Med., 91:144-147.
88. Katz, M.L., Stone, W.L. and Dratz, E.A. (1978): Invest. Ophthalmol. Vis. Sci., 17:1049-1058.
89. Kretzer, F.L., Hittner, H.M., Johnson, A.T., Mehta, R.S. and Godio, L.B. (1982): Ann. N.Y. Acad. Sci., 393:145-166.
90. Kretzer, F. and Mehta, R. (1981): Invest. Opthalmol. Vis. Sci., Suppl., 20:40.
91. Kuck, J.F.R. Jr. (1970): In: Biochemistry of the Eye, edited by O.N. Graymore, pp. 319-371. Academic Press, London.
92. Kuck, J.F.R. Jr. (1976): Invest. Ophthalmol., 15:405-407.
93. Kuhn, H. (1980): Neurochem. Internat., 1:269-285.
94. Kurzel, R.B., Wolbarsht, M., Yamanashi, B.S., Staton, G.W. and Borkman, R.F. (1973): Nature, 241:132-133.
95. Lanum, J. (1978): Surv. Ophthalmol., 22:221-249.
96. Latarjet, R. and Caldas, L.R. (1955): J. Gen. Physiol., 35:455-470.

97. Lerman, S. (1980): Radiant Energy and the Eye, pp. 115-186. Macmillan, New York.
98. Lerman, S. and Borkman, R. (1976): Ophthalmic. Res., 8:335-353.
99. Lerman, S., Megaw, J.M., Gardner, K., Takei, Y. and Willis, I. (1981): Ophthalmic. Res., 13:106-116.
100. Lion, Y. Delmelle, M. and van de Vorst, A. (1976): Nature, 263:442-443.
101. Lowry, O.H., Rosebrough, N.J., Farr, A.L. and Randall, R.T. (1951): J. Biol. Chem., 193:265-275.
102. Maraini, G. and Fasella, P. (1970): Exp. Eye Res., 10:133-139.
103. Masciulli, L., Anderson, D. and Charles, S. (1972): Am. J. Opthalmol., 74:638-661.
104. Michaelson, I.C. (1954): Retinal Circulation in Man and Animals. Charles C. Thomas Co., Springfield, Ill.
105. Miljanich, G., Sklar, L., White, D. and Dratz, E. (1979): Biochim. Biophys. Acta, 552:294-306.
106. Morrison, W.R. and Smith, L.M. (1964): J. Lipid Res., 5:600-608.
107. Nielsen, H. (1981): Lipids, 16:215-222.
108. Noell, W.K., Walker, V., Kang, B. and Berman, S. (1966): Invest. Ophthalmol., 5:450-473.
109. Noell, W.K. (1980): Vision Res., 20:1163-1171
110. Pirie, A. (1965): Nature, 205:500-501.
111. Pirie, A. (1968): Invest. Ophthalmol., 7:634-650.
112. Pirie, A. and Howard Flanders, P. (1967): Arch. Ophthalmol., 57:849-854.
113. Pirie, A. and Rees, J.R. (1970): Exp. Eye Res., 9:198-203.
114. Pirie, A., Rees, J.R. and Holmberg, N.J. (1970): Exp. Eye Res., 9:204-218.
115. Puntenney, I. and Shoch, D. (1954): Am. J. Ophthalmol., 38:673-682.
116. Radnot, M., Jobbagyi, P., Heszberger, I. and Lovas, B. (1969): Ophthalmologica, 159:460-471.
117. Radnot, M. (1969): At. Energy Rev., 7:129-166.
118. Rapp, L.M. and Williams, T.P. (1980) In: The Effects of Constant Light On Visual Processes, edited by T.P. Williams and B.N. Baker, pp.135-159. Plenum Press, New York.
119. Rapp, L.M. Wiegand, R.D. and Anderson, R.E. (1982) In: Problems of Normal and Genetically Abnormal Retinas, edited by R.M. Clayton, J. Haywood, H.W. Reading and A. Wright, pp.109-119. Academic Press, London.
120. Riis, R.C., Sheffy, B.E., Loew, E., Kern, T.J. and Smith, J.S. (1981): Am. J. Vet. Res., 42:74-86.
121. Ritter, M.A., Cleaver, J.E. and Tobias, C.A. (1977): Nature, 266:653-655.
122. Robison, W.G. Jr., Kuwabara, T. and Bieri, J.G. (1980): Invest. Opthalmol. Vis. Sci., 19:1030-1037.
123. Rosenfeld, L. and Spector, A. (1982): Exp. Eye Res., 35:69-75.
124. Rosenkranz, J. (1977): Int. Rev. Cytol., 50:25-185.
125. Rouser, G., Fleischer, S. and Yamamoto, A. (1970): Lipids, 5:494-496.

126. Ruppel, H. and Hagins, W.A. (1973): In: Biochemistry and Physiology of the Visual Pigments, edited by H. Langer, pp.257-261. Springer-Verlag, Berlin.
127. von Sallmann, L. (1951): Arch. Ophthalmol., 45:149-164.
128. von Sallmann, L., Grimes, P. and McElvain, N. (1962): Exp. Eye Res., 1:449-456.
129. von Sallmann, L., Tobias, C.A., Anger, H.O., Welch, C., Kimura, S.F., Munoz, C.M. and Drungis, A. (1955): Arch. Ophthalmol., 54:489-514.
130. Schocket, S.S., Esterson, J., Bradford, B., Michaelis, M. and Richards, R.D. (1972): Isr. J. Med. Sci., 8:1596-1601.
131. Sery, T.W. and Petrillo, R. (1984): Curr. Eye Res., 3:243-252
132. Shvedova, A.A., Alekseeva, O.M., Kuliev, I.Ya., Muranov, K.O., Kozlov, Yu. P. and Kagan, V.E. (1982): Curr. Eye Res., 2:683-689.
133. Shvedova, A., Sidorov, A., Novikov, K., Galushchenko, I. and Kagan, V. (1979): Vision Res., 19:49-55.
134. Spector, A. and Garner, W.H. (1981): Exp. Eye Res., 33:673-681.
135. Sperling, H., Johnson, C. and Harwerth, R. (1980): Vision Res., 20:1117-1125.
136. Stone, W.L., Katz, M.L., Lurie, M., Marmor, M.F. and Dratz, E.D. (1979): Photochem. Photobiol., 29:725-730.
137. Stone, W.L. and Dratz, E.A. (1980): Biochim. Biophys. Acta, 631:503-506.
138. Sykes, S.M., Robison, W.G., Jr., and Bieri, J.G. (1981): DHHS Pub. FDA, 81:8156.
139. Taylor, H.R. (1980): Br. J. Ophthalmol., 64:303-310.
140. Testa, M. Fiore, C., Bocci, N. and Calabro, S. (1968): Exp. Eye Res., 7:276-290.
141. Tomita, T. (1972): In: Handbook of Sensory Physiology, Vol. VII, Part 2, edited by M.G.F. Fuortes, pp. 635-655. Springer-Verlag, Berlin.
142. Truscott, R.J.W. and Augusteyn, R.C. (1977): Exp. Eye Res., 24:159-170.
143. Tso, M.O.M., LaPiana, F.G. and Appleton B. (1974): Trans Am. Acad. Ophthalmol. Otolaryngol., 78:677-678.
144. Varma, S.D., Kumar, S. and Richards, R.D. (1979): Proc. Nat. Acad. Sci. USA, 76:3504-3506.
145. Varma, S.D. (1981): Intl. J. Quant. Chem., 20:479-484.
146. Varma, S.D., Beachy, N.A. and Richards, R.D. (1982): Photochem. Photobiol., 36:623-626.
147. Varma, S.D., Chand, D., Sharma, Y.R. and Kuck, J.F.R. (1983): Invest. Ophthalmol. Vis. Sci., Suppl., 24:31.
148. Varma, S.D., Srivastava, V.K. and Richards, R.D. (1982): Ophthalmic. Res., 14:167-175.
149. Wiegand, R.D. and Anderson, R.E. (1982): In: Methods in Enzymology, Vol. 81, edited by L. Packer, pp. 297-304. Academic Press, New York.
150. Wiegand, R.D. and Anderson, R.E. (1983): Exp. Eye Res., 36:389-396.

151. Wiegand, R.D and Anderson, R.E. (1983): Exp. Eye Res., 37:159-173.
152. Wiegand, R.D., Giusto, N.M. and Anderson, R.E. (1982): In: Problems of Normal and Genetically Abnormal Retinas, edited by R.M. Clayton, J. Haywood, H.W. Reading and A. Wright, pp.121-128. Academic Press, London.
153. Wiegand, R.D., Giusto, N.M., Rapp, L.M. and Anderson, R.E. (1983): Invest. Ophthalmol. Vis. Sci., 24:1433-1435.
154. Weiter, J.J. and Subramanian (1978): Invest. Ophthalmol. Vis. Sci., 17:869-873.
155. Weiter, J.J. and Finch, E.D. (1975): Nature, 254:536-537.
156. Williams, T.P. and Howell, W.L. (1983): Invest. Ophthalmol. Vis. Sci., 24:285-287.
157. Winkler, B.S. and Giblin, F.J. (1983): Exp. Eye Res., 36:287-297.
158. Witting, L. (1965): J. Am. Oil Chemist's Soc., 42:908-913.
159. Woodford, B.J., Tso, M.O.M. and Lam, K.-W. (1983): Invest. Ophthalmol. Vis. Sci., 24:862-867.
160. Worgul, B.V., Merriam, G.R. Jr., Szechter, A. and Srinivasan, B.D. (1976): Arch. Ophthalmol., 94:996-999.
161. Worgul, B.V. and Rothstein, H. (1975): Ophthalmic Res., 7:21-32.
162. Yagi, K., Matsuoka, S., Ohkawa, H., Ohishi, N., Takeuchi, Y. and Sakai, H. (1977): Clin. Chim. Acta, 80:355-360.
163. Young, R.W. (1976): Invest. Ophthalmol., 15:700-725.
164. Young, R.W. (1974): Exp. Eye Res., 18:215-223.
165. Yu, N-T., Bando, M. and Kuck, J.F.R. Jr. (1983): Invest. Ophthalmol., 24:1157-1161.
166. Zigler, J.S. and Goosey, J.D. (1981): Photochem. Photobiol., 33:869-874.
167. Zigler, J.S. and Hess H.H. (1983): Invest. Ophthalmol. Vis. Sci., Suppl., 24:75.
168. Zigman, S., Datiles, M. and Torczynski, E. (1979): Invest. Ophthalmol. Vis. Sci., 18:462-467.
169. Zigman, S., Griess, G., Yulo, T. and Schultz, J. (1973): Exp. Eye. Res., 15:255-264.
170. Zigman, S., Groff, J., Yulo, T. (1977): Photochem. Photobiol., 26:505-509.
171. Zigman, S. and Vaughn, T. (1974): Invest. Ophthalmol., 13:462-465.
172. Zigman, S., Yulo, T. and Schultz, J. (1974): Ophthalmic Res., 6:259-270.
173. Zinn, K.M. and Benjamin-Henkind, J.V. (1979): In: The Retinal Pigment Epithelium, edited by K.M. Zinn and M.F. Marmor, pp. 3-31. Harvard University Press, Cambridge, MA.

Free Radicals in Inflammatory Disease

M. V. Torrielli and M. U. Dianzani

Institute of General Pathology, Faculty of Medicine, University of Turin, 10125 Torino, Italy

In the course of the inflammatory reactions the respiratory burst in the phagocytic cells is coincident with the generation of very reactive oxygen species (3,98,133,135) that are thought to play an important role in conditioning the defense capability of the host against microbial aggression throughout oxidative killing mechanisms.

The primary reactive oxygen metabolites produced by phagocytes are represented by monovalent and divalent products, namely superoxide anion (O_2^-) and hydrogen peroxide (H_2O_2). These toxic oxygen products, then, cooperate in the formation of additional and more reactive oxidants, including hydroxyl radical ($OH^{\bullet}$) and halide derived oxidant products, which in turn, amplify oxygen-dependent microbicidal system, as well as are effective as tumoricidal agents. Furthermore, since the extreme reactivity of the oxygen species it is not surprising, they also can be implicated in the modulation of cell function, cellular injury and tissue destruction.

It appears clear now that one of the most important mechanisms in driving the sequence of reactions underlying these alternative processes can be the peroxidative decomposition of cell membrane lipids following attack by oxygen metabolites. Involvement of oxygen radical species as well as non radical species ($^1O_2, O_3$), acting as potent molecular initiators of lipid peroxidation, has been firmly established (4,31,34,51,52,65,69,79).

LIPID PEROXIDATION AND OXYGEN FREE RADICAL SPECIES

Mechanisms of Lipid Peroxidation

Lipid peroxidation is the oxidative decomposition of polyunsaturated lipids (22,78,91,114,122,138) primarily involving free radical mechanisms. It is a complex process characterized by three distinct phases: an induction period, a rapid propagation phase (autocatalytic), and a termination phase. Free radicals capable of abstracting hydrogen atoms from unsaturated fatty acids initiate lipid peroxidation, whereas transitional metal ions catalyze the propagation reactions. The termination phase is paralleled by formation of low molecular weight substances. Among them, aldehydes are very abundant. They, in turn, act as modulators or damaging factors for cell biochemistry and function. The peroxidative processes "in vivo" have been said to be of primary importance in the mechanism of aging, of injury to cells by air pollution, of various forms of hepatic damage, of some stages of atherosclerosis and of oxygen toxicity (123). However, much evidence of lipid peroxidation is indirect and mostly based on "in vitro" findings by determination of the intensity of a colour reaction with thiobarbituric acid (TBA test). There is a general agreement that this test is a reliable index of peroxidative decomposition in tissue extracts, although failure to detect TBA-reacting material is not an indication of the absence of lipid peroxidation. In fact, it has been recently described that malonaldehyde (MA) content accounts for about 95% of TBA-reactivity, but this substance is neither the major nor the most toxic aldehyde so far produced during lipid peroxidation (27). In addition, it is now clear that the final pattern of aldehydes is strongly influenced by their metabolism, that results mostly from oxidative reactions as catalyzed by aldehyde dehydrogenase, as well as by very active reductases (12,89). Moreover, increased serum levels of MA have been observed in patients with recent brain infarction (106), as well as in diabetics (107). The interrelationship between lipid peroxidation and degenerative diseases appears also shown by the finding that during chronic active hepatitis, as well as in liver cirrhosis, serum level of TBA-reacting material are sometimes elevated (121).

The most important substrate for lipid peroxidation is arachidonic acid. It is to be noted that during inflammatory reactions, this substance undergoes three types of decomposition involving free radical mechanisms. The first one in lipid peroxidation that begins on arachidonic acid (as well as on other unsaturated fatty acids) is a radical mechanism involving the rupture of the covalent binding between a carbon atom and a hydrogen. As this usually occurs when the fatty acid is still bound to lipid macromolecules, a lipid free radical is produced. In the presence of oxygen, this takes up a molecule of O_2, so forming lipoperoxide free radicals (FIG. 1). The decomposition continues by internal rearrangements of the molecule and ends with the formation of low molecular weight substances, and especially of aldehydes. These provided different toxic potentials, the most active among them belonging to the series of 4-hydroxyalkenals (FIG. 2), especially 4-hydroxynonenal and 4-hydroxyundecenal.

The second and third pathway of arachidonic acid decomposition are respectively the prostaglandin cascade and the leukotriene biosynthetic pathway. The first one starts by the activity of a microsomal cycloxygenase acting on free arachidonic acid. Leukotriene biosynthesis is ef-

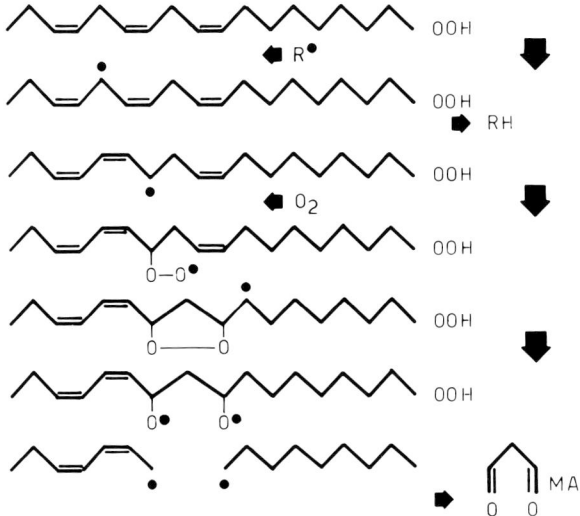

FIG. 1. Scheme for the oxidative breakdown of an unsaturated fatty acid.

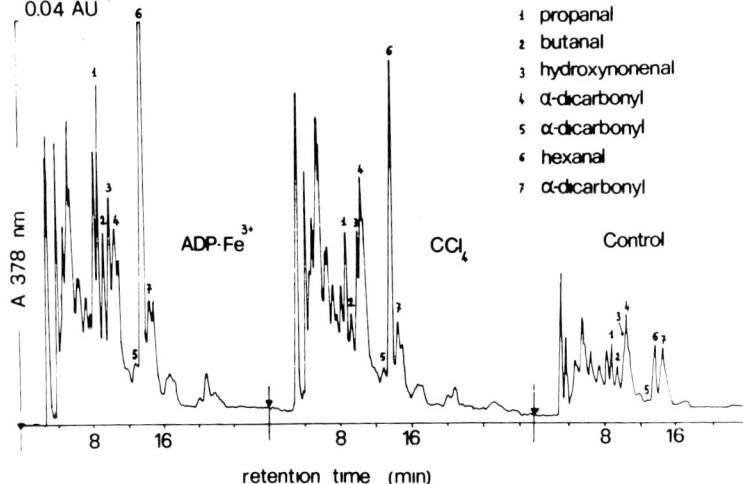

FIG. 2 Non-polar (1,2,6,7) and medium polar (3,4,5) aldehyde production by isolated rat hepatocytes after peroxidative stimulation with an ADP-FE^{3+} complex, or carbon tetrachloride (CCl_4).

fected by another enzymatic system referred to as lipoxygenase. The three metabolic lines probably have interconnections. For instance, it has been demonstrated that during autoxidation of polyunsaturated fatty acids, prostaglandin endoperoxide-like substances as well as hydroperoxides are formed (93). It has also been shown that a precursor of SRS.A (5-HPETE) originates when arachidonic acid is exposed to atmospheric oxygen for 48 hours (90). Another interaction between cycloxygenase dependent and independent lipid peroxidation is the finding that biosynthesis of prostaglandins is triggered by lipid peroxides (44) which could be generated by prostaglandin synthetase itself (13). Evidence for this is the inhibition of cycloxygenase by glutathione peroxidase (117), an enzyme which catalyzes the conversion of hydrogen peroxide to water and requires reduced glutathione as an electron donor. Recently, the endoperoxide PGG_2 has been considered as the secondary cycloxygenase-produced activating agent (43). However, in the presence of cosubstrates, any other peroxidase, including prostaglandin peroxidase could reduce the levels of peroxides necessary to trigger prostaglandin biosynthesis. In addition, besides the enzymic removal of lipid peroxides, agents which prevent the occurrence of lipid peroxidation (antioxidants), or non enzymically reduced peroxides, may contribute to decrease the cellular peroxide steady state and consequently the biosynthesis of prostaglandins. On the basis of these considerations, since PGG_2 can be an activator, once the reaction is triggered by small amounts of peroxides, an accelerative process of peroxide production results in continued activation. In this context, other processes could lead to the formation of other stimulatory agents. For instance, the perturbation of the plasmalemma of phagocytic cells leads to the endogenous generation of H_2O_2 and O_2^- that can initiate membrane lipid peroxidation (51,79).

Oxygen Species and Their Biological Reactivity.

Superoxide Anion. After appropriate stimulation of the plasmamembrane of phagocytes by means of different surface-active agents such as deoxycholate, fatty acids, digitonin (35,50), phorbolmyristate acetate (PMA) (96), during phagocytosis or immune complex contact (133), a membane bound oxidase dependent on NAD(P)H is activated and phagocytic cells increase the uptake of oxygen to produce a substantial amount of superoxide anion (O_2^-) with the concomitant production of hydrogen peroxide according to the following reaction:

$$2O_2 + \begin{array}{c} NADH \\ or \\ NADPH \end{array} \xrightarrow{\text{oxidase}} 2O_2^- + \begin{array}{c} NAD^+ \\ or \\ NADP^+ \end{array} + H^+$$

As a consequence, an oxygen molecule is formed that contains an extra impaired electron (29). In this situation the reactive molecule may function as an oxidant or a reducing agent; in the first case it gains an electron producing H_2O_2; in the second one, O_2^- loses its impaired electron and then it is oxidized to oxygen. Two superoxide radicals, in an aqueous medium, can interact in such a way that one molecule is oxidized and the other reduced (dismutation reaction) with the formation of H_2O_2 and O_2:

$$\begin{array}{c} O_2^- \\ + \\ O_2^- \end{array} + \begin{array}{c} H \\ \\ H \end{array} \longrightarrow H_2O_2 + O_2$$

Since the superoxide anion is a weak base the dismutation reaction is influenced by the pH. In an acid environment, therefore, as is found in the phagolysosomes of phagocytizing cells, the spontaneous formation of H_2O_2 from O_2^- may occur at a relevant rate (48). However, at a neutral or alkaline pH in oxygen using cells, the dismutation reaction is catalyzed by the metallo-enzyme superoxide dismutase (SOD) with the consequence of decreasing the steady state concentration of O_2^- at higher pH as in the cytosol (32). SOD is present in the cytosol of all phagocytes and exerts a protective effect against oxidative damage. As far as it regards the mechanism of formation of superoxide anion, it has been recently shown that the NADPH-dependent enzymatic line which generates this substance, involves at least four identified catalytic substances. The reaction is started in the presence of a quinol dehydrogenase oxidizing NADPH; the hydrogen is then transferred to cytochrome c, and from this to a chromophore absorbing at 450-455 nm. Cytochrome c is reduced by a cytochrome c reductase, whereas the 450-455 chromophore is reduced by a chromophore reductase. Although quinol dehydrogenase is detectable both in resting and in activated neutrophils, the other components become evident only in stimulated cells (36). Segal and Meshulam (109) have, however, rediscussed the whole problem of superoxide formation. They have cast some doubt on the physiology of the sequence, and suggest that it might be an experimental artifact occurring on addition of external cytochrome c to the measurement mixture. It is interesting to note that chemotactic factors, especially N-formyl-methionyl-leucyl-phenylalanine, stimulate superoxide anion formation. This stimulation is inhibited by prostaglandins E_1 and I_2 (28).

Perhydroxyl Radical. On the basis of the high reactivity of free radicals, superoxide has been considered as a mediator of phagocyte-dependent microbicidal action and cytotoxicity. However, it now appears clear that its reactivity in aqueous systems is limited (2,56) although in some circumstances it can act as an oxidant (29,131). In this connection, it is noteworthy that the protonated form of O_2^-, perhydroxyl radical (HO_2), is a much stronger peroxidative agent than superoxide (7,108) and may be found in phagocytic vacuoles. Thus, the pathobiological importance of superoxide anion not only resides in its ability to produce H_2O_2, but to interact with hydrogen peroxide to generate singlet oxygen (see below) and hydroxyl free radical and to the possibility to be protonated to form the potent oxygen agent perhydroxyl radical.

Hydrogen Peroxide. The most studied oxygen metabolite for its cytotoxic oxidative potential is, perhaps, hydrogen peroxide. This oxidant in phagocytes may be generated by dismutation of O_2^-, or may be released directly by cell-oxidase activity. According to Root and Metcalf (100) the major part (80%) of H_2O_2 produced by neutrophils arises from O_2^- dismutation. Hydrogen peroxide is a very strong oxidant, but its ability to interact with organic molecules is restricted (99). By contrast, H_2O_2 can rapidly react with transitional metal ions and their complexes to generate highly reactive oxygen species. A large number of investigations supports the view that activated phagocytes produce sufficient amounts of H_2O_2 to be able to oxidize intracellular constituents of target cells: erytrocytes, endothelial cells, and tumor cells. In addition, hydrogen peroxide can damage or modulate the functions of leucocytes directly. The potential damaging effect of hydrogen peroxide is modulated by the antioxidant content in the target cell populations, represented primarily by the glutathione GSH redox system (FIG. 3), a critical component in determining the susceptibility of cells to H_2O_2-induced damage. The GSH redox "unit",

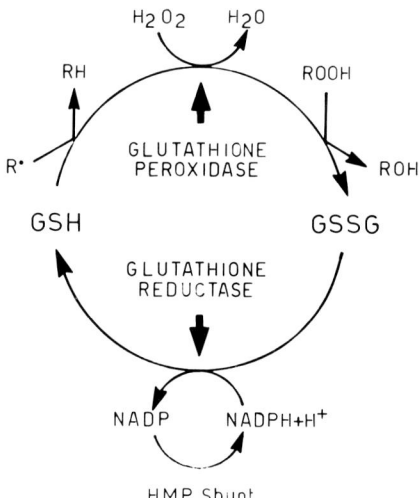

FIG. 3   Scheme of the GSH redox "unit".
(Adapted after Roos et al., 1979; ref. 99)

present in the cytosol of neutrophils (82,85,95,120) contains reduced glutathione, glutathione peroxidase and glutathione reductase, three components that operate "in Harmony" to detoxify the intracellular environment and to repair peroxidatic injury of biomolecules. In fact, GSH not only reacts with H_2O_2, but also with organic peroxides and free radicals. In addition, the GSH cycle is able to restore the functional activity of important protein molecules when sulphydryl groups have been oxidized and form disulfide bridges. In this connection, it should be noted that the microtubule system has been identified as one of the targets for the peroxidative damaging effect of H_2O_2 and the level of GSSG may modulate the induction of microtubule disassembly or the inhibition of their assembly (10,84,85,86).

Phagocytic cells contain oxygen-independent (72) as well as oxygen-dependent antimicrobial systems. Among the oxygen-derived active microbicidal species, H_2O_2 appears to be one of the most important. Its oxidative microbicidal activity as well as its potentially cytotoxic effect is substantially increased in systems in which it interacts with other agents by an enzymatic pathway. The most important enzymatic system is represented by myeloperoxidase (MPO), H_2O_2 and a halide (55,59). Myeloperoxidase, a heme enzyme, is present in the primary lysosomal granules (granules A) of neutrophils and mononuclear phagocytes from a variety of species (57,59,98,101). After appropriate stimuli, phagocytes discharge myeloperoxidase into the phagosome, or into extracellular milieu where the heme-enzyme interacts with H_2O_2 formed by the respiratory burst. In this situation, myeloperoxidase catalyzes the oxidation of halides: chloride (Cl^-), iodide (I^-), or bromide (Br^-) by transferring two electrons from the halide to H_2O_2 (electron acceptor) to generate an oxidized halide:

$$H_2O_2 + Cl^- \xrightarrow{\text{MPO}} HOCl + H_2O$$
$$(I^-, Br^-) \qquad\qquad (HOI, HOBr)$$

The halide utilized "in vivo" is not well known, but Cl^- is present at much higher concentrations than other halides (I^-, Br^-). This system has been demonstrated to be effective in killing bacteria (53,77), mycoplasmas (47), viruses (6), yeasts (54,67) and tumor cells (59).

It has been recently described that myeloperoxidase can be replaced by Fe^{2+} only when the halide present is iodide (58). It is possible that this additional system may work "in vivo". The precise mechanism responsible for the antimicrobial effect of MPO-H_2O_2-halide system is, at present, unknown. One of the proposed mechanisms is the final iodination (or halidation) of bacterial proteins by a iodine (or halogen) free radical:

$$H_2O_2 + I \longrightarrow OH^- + OH^\cdot + I^\cdot$$

Klebanoff (58) points however, to the killer properties of the formed hydroxyl free radical.

Some evidence indicates that the system is able to cleave peptide bonds (110) and to oxidatively decarboxylate amino acids with the consequent formation of aldehydes, CO_2, and ammonia (140). The latter reaction seems to occur through the production of chloroamine intermediates (118). The unstable chloramines, in aqueous systems, spontaneously decompose to generate aldehydes, CO_2 and NH_4Cl. Furthermore, some reports (88,103) suggest the possibility that singlet oxygen can be formed by MPO-H_2O_2-halide system. Being a potent oxidant, $^1O_2$ may participate in oxidative cleavage of peptide bond and in membrane lipid peroxidation. In this connection, it should be noted that another potent oxidant generated by the MPO-H_2O_2-halide system, namely HOCl, reacts strongly and rapidly with amines, amino acids, sulphydryl compounds, thioethers, and aromatic substances (41,42) and appears to be able to react with other oxygen metabolites such as O_2^- to form an extremely powerful oxidant, the hydroxyl radical (70):

$$O_2^- + HOCl \longrightarrow OH^{\bullet} + Cl^- + O_2$$

Extensive investigations have been carried out using model systems consisting of H_2O_2, purified peroxidases and halide (57,59). Findings from these studies indicate that the oxidants originated by the activity of the MPO-H_2O_2-halide system, can modify cell function and mediate cell injury. Experimental evidence supports the view that the myeloperoxidase system plays an important role in producing a fall in the respiratory burst, as well as the inactivation of lysosomal enzymes and autotoxicity in activated phagocytes. Studies with myeloperoxidase deficient neutrophils show that after appropriate stimulation, these phagocytes present an increased rate of oxygen consumption and O_2^-, H_2O_2 production, such as is observed in normal neutrophils (102). However, whereas in control cells, the O_2^- generation was deeply depressed after about 60 min., in MPO deficient neutrophils, it continued for much more time. This finding is indicative of the inhibitory effect on the respiratory burst by myeloperoxidase. As mentioned above, the myeloperoxidase system is an important component also in the inactivation of phagocytic lysosomal enzymes, as confirmed by a recent report (132). Moreover, the MPO system, through its own pro-oxidant products can be inactivated. In the absence of oxidable substrate in fact, HOCl can irreversibly or reversibly deactivate myeloperoxidase (40). In addition, phagocytes are very sensitive to myeloperoxidase-derived oxidants which induce irreversible injury in these cells leading to autotoxicity. From experimental data (128) it appears that compounds interacting with HOCl can block the autotoxicity process. However, damaging effects of the myeloperoxidase system is not limited to the cells producing sufficient H_2O_2 and myeloperoxidase, since the MPO-H_2O_2-halide can be released into the extracellular milieu and consequently attack adjacent cell targets. Also in this situation, the mediators of cytotoxicity appear to be hypo-halous acids. It has been demonstrated that HOCl is tumoricidal (116) and its cytotoxicity can be depressed by HOCl scavengers (135).

The mechanism of action of this oxidant is not well known, but it is thought that it could interact with virtually any components of plasma-membrane or intracellular constituents (1).

Hydroxyl Free radical. A number of evidences indicate that during phagocytosis, H_2O_2 and O_2^- may combine to form the much more cytotoxic $OH^{\bullet}$ radicals. This free radical is an extremely reactive oxidant substance capable of interacting with many organic and inorganic molecules (139). It has been proposed that such a radical could be originated from the reduction of H_2O_2 by superoxide anion (38):

$$O_2^- + H_2O_2 \longrightarrow OH^{\bullet} + OH^- + O_2 \text{ (Haber-Weiss reaction)}$$

Subsequently (75,97), however, it became clear that the Haber-Weiss reaction occurs physiologically at very slow rates since the rate constant for the interaction of O_2^- with H_2O_2 is below that of the competing reaction concerning the spontaneous dismutation of O_2^-. To resolve this difficulty, it was suggested and demonstrated that iron salts and their complexes could effectively catalyze an interaction between O_2^- and H_2O_2 (39) with the formation of $OH^{\bullet}$. This would occur according to the following reaction:

$$H_2O_2 + Fe^{2+} \longrightarrow OH^{\bullet} + OH^- + Fe^{3+}$$

Fe^{3+} would then be reduced to Fe^{2+} on the presence of superoxide anion (136):

$$O_2^- + Fe^{3+} \longrightarrow O_2 + Fe^{2+}$$

In this reaction, superoxide anion is acting only as a reducing agent and therefore it may be replaced by other compounds able to reduce the metal. From these findings, it is clear that in the presence of O_2^- and H_2O_2 and of a properly coordinate transitional metal, a potent oxidant agent identical to, or with characteristics very similar to $OH^{\bullet}$ is generated.

A further possibility in the generation of hydroxyl free radical, other than that related to the interaction of HOCl with O_2^- after activation of $MPO-H_2O_2$-halide system, is connected with the highly reactive oxidizing species derived from the prostaglandin biosynthetic pathway. During the arachidonate cascade, as initiated by cycloxygenase it has been observed that reducing agents such as epinephrine and luminol, undergo oxidation (73). Investigations on the enzymic dynamics of prostaglandin peroxidase, which catalyzes the biotransformation of PGG_2 to PGH_2, showed that the species responsible for the co-oxygenation of adrenalin and luminol was an oxidant product released during the peroxidatic reaction (24). In addition, this oxidizing reactant appeared to be able to deactivate enzymes of the prostaglandin biosynthetic pathway (25). Furthermore, it shows a wide spectrum of biological damaging effects since it can attack a variety of organic compounds including proteins. On this basis, it was postulated that the catalytic action of PG hydroperoxidase is coupled with potential generalized effects on cells and may be related, in part, to the cellular damage found during inflammatory reactions (62). This is in contrast to the physiological functions of other peroxidases (glutathione peroxidase, catalase), that reduce hydroperoxides to less potentially toxic compounds.

The identity of the oxidizing agent resulting from the enzymatic reduction of hydroperoxy acids is unclear at present. It may be $OH^{\bullet}$, $^1O_2$ or O_2^-. However, some experimental findings may be interpreted to favour the evidence of this oxidizing species as the hydroxyl free radical (63). If so, the injurious effects of hydroperoxy acids that are considered potent oxidizing agents would be amplified by a more lethal oxidant than hydroperoxy acid substrates.

The OH• free radicals have been considered a more important factor in phagocyte-mediated antimicrobial activity, than a potential cytotoxic agent capable to injury cell targets. Findings suggesting a role for OH in cytotoxic events are limited to phagocyte auto-injury. However, in considering its very high reactivity, it is reasonable to expect that hydroxyl free radical might cause severe damage to biological molecules. One of the main mechanisms involved in OH•-induced cytotoxicity appears to be related to the peroxidative deterioration of cell lipids. In fact, the hydroxyl free radical is a potent initiator of lipid peroxidation by its very ability to abstract hydrogen atoms (92) from the allylic position of polyunsaturated fatty acids to yield hydroperoxides. Moreover, the hydroxyl radical is capable to rapidly reacting with nucleic acids (11,69) through the abstraction of hydrogen atoms from saturated carbon atoms. In addition, the biological reactivity of OH• appears to be indiscriminate in reacting with purines and pyrimidines without substantial selectivity.

Singlet Oxygen. The respiratory burst of phagocytizing leucocytes is associated with the emission of photons over a broad range of wavelengths (chemiluminescence). Several evidences indicate that such a process is accompanied by the generation of singlet oxygen ($^1O_2$); another toxic oxygen product with a distorted electron configuration giving to it a high reactivity towards molecules that contain regions of high electron density, such as carbon-carbon double bonds. Several reactions, recently reviewed by Klebanoff (56) seem to be capable of producing singlet oxygen under conditions that may be important in phagocytic leucocytes. Among them, of particular interest is the formation of hypochlorous acid in the myeloperoxidase reaction:

$$H_2O_2 + Cl^- \xrightarrow{MPO} OCl^- + H_2O$$
$$OCl^- + H_2O_2 \longrightarrow {^1O_2} + Cl^- + H_2O$$

and the reaction between superoxide anion and hydrogen peroxide (modified Haber-Weiss reaction) which is believed to be responsible for the production of both $^1O_2$ and hydroxyl radical:

$$O_2^- + H_2O_2 \longrightarrow {^1O_2} + OH^• + OH^-$$

Singlet oxygen and hydroxyl free radicals have been implicated as causative agents in lipid peroxidation (31,69). However in contrast with OH•, $^1O_2$ can react directly with polyunsaturated fatty acids to produce hydroperoxides. Since singlet oxygen acts as a strong electrophil in solution, guanine could be the preferential target of singlet oxygen being the best electron donor among the bases constituent nucleic acids.

LIPID PEROXIDATION, FREE RADICALS AND INFLAMMATION

As previously noted, lipid peroxidation is thought to play one of the key roles in promoting cellular damage in many conditions and it appears as a process that can be involved in a variety of pathobiological situations. Several lines of evidence point out to the possibility that cell and molecular injury in tissues after a phlogogen stimulus, are related to lipid peroxidative events and to the end-products and/or intermediates of the oxidative decomposition of polyunsaturated fatty acids (9,81, 112,123). In addition, a number of evidences have emphasized the significance of free radicals for inflammatory and arthritic diseases (74,105) as well as the anti-inflammatory activity of free radical scavengers and antioxidants in both clinical and experimentally-induced inflammatory processes (23,45, 87,94,124). Additional indications for an involvement of lipid peroxidation during the development of inflammatory reactions come from the findings showing the presence of final products of peroxidative deterioration of lipids at the phlogotic site. Extensively used for this purpose is the determination of TBA-reacting material, mostly represented by malonaldehyde (MA) in inflammatory exudates and tissues. The presence of MA in inflamed tissues, however, is not solely indicative for the presence of lipid peroxidation. In fact, it may be produced even during the arachidonate cascade leading to prostaglandin biosynthesis and catabolism. Moreover, aldehydes may be produced by powerful oxidizing reactants generated in the course of the transformation of PGG_2 to PGH_2 (FIG. 4).

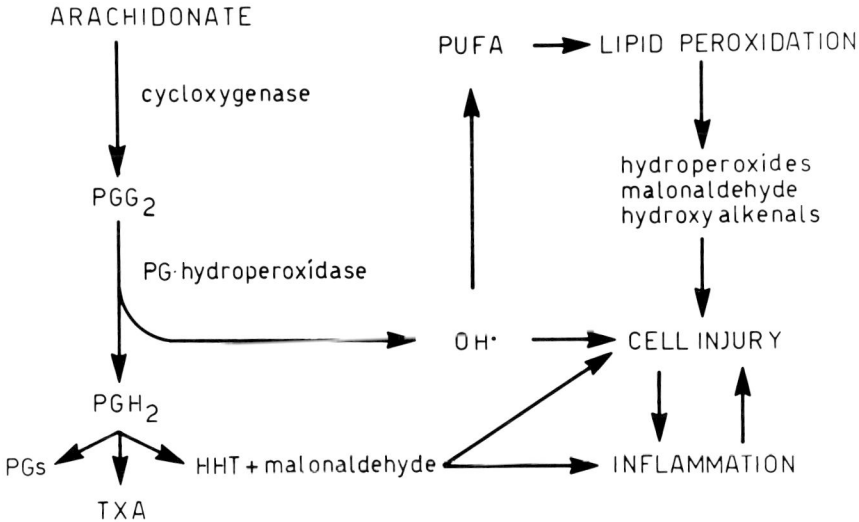

FIG. 4 Relationship between prostaglandin biosynthetic pathway, generation of reactive oxidants (OH•?), production of MA and lipid peroxidation.
    PGs: stable prostaglandins
    TXA: thromboxane A_2
    HHT: 12L-hydroxy-5,8,10-heptadecatrienoic acid

The most substantial activation of lipid peroxidation is, however, probably represented by the initiating attack of polyunsaturated fatty acids by free radical oxygen species produced during the phagocytic burst. Very recently, for instance, it has been proved by "in vitro" experiments that a lactoperoxidase/hydrogen peroxide/halide system can initiate lipid peroxidation of linoleate (51). Several authors (31,60,64,65,66) have suggested that the real effector of linoleate attack is the hydroxyl free radical coming from the H_2O_2 Haber-Weiss decomposition. Other authors (79) have, however, cast some doubt on this possibility. In any case the only real proofs of an involvement of lipid peroxidation in inflammation are, at present, the demonstration of increased TBA-reacting material, the demonstration of the protective effect by antioxidants—that, however, may act even by preventing prostaglandin biosynthesis (125)—and the demonstration that aldehydes display some of the features of inflammatory agents. They increase vascular permeability (131) and induce foot-paw oedema in the rat (23,124).

As we have noted above, the TBA-test is only indicative of the production of malonaldehyde that can also be generated in the prostaglandin biosynthetic pathway (FIG. 4). No profound attempt to study the aldehyde pattern of produced aldehydes by more sophisticated methods has been performed as yet. Moreover, one cannot disregard that in stimulated phagocytes, substantial amounts of aldehydes may be formed through the hypochlorite pathway following the activation of myeloperoxidase-H_2O_2-halide system:

$$H_2O_2 + Cl^- + H^+ \xrightarrow{\text{MPO}} HOCl + H_2O$$

$$CH_3\text{-}CH\text{-}NH_3^+\text{-}COOH \longrightarrow$$

$$CH_3\text{-}CH\text{-}NH\text{-}COOH + H_2O + H^+$$
$$\underset{|}{Cl} \quad + H_2O$$

$$CH_3\text{-}CHO + NH_4 + CO_2 + Cl^-$$

At any rate, among the experimental findings supporting the production of aldehydes in the inflammatory exudate, one has to quote experiments by Bragt et al. on kaolin-induced granuloma pouch inflammation (9). In inflammatory exudates of these animals, the level of TBA-reacting substances, but not that of PGs-like compounds, although increased, appeared to be related to the severity of the induced phlogotic process. In this connection, it is not surprising that the major contribution to malonaldehyde formation must be ascribed to non-enzymatic peroxidative decomposition of lipids.

An increase of TBA-reacting material has also been demonstrated in plasma of patients suffering of rheumatoid arthritis (81). In this pathological condition the plasma level of aldehydes appeared much more elevated in those patients most severely diseased and higher level of malonaldehyde were related to the more active form of this chronic inflammatory disorder. It is noteworthy, that rheumatoid arthritis patients present an increased level of prostaglandins in plasma and in synovial fluid of affected joints (128).

The production of aldehyde compounds has also been observed "in vitro" in phagocytes obtained from human blood (119). Under these experimental conditions, the TBA-reacting material formed was particularly elevated in monocytes, whereas neutrophils were capable of producing aldehydes only during the endocytosis of particles containing linoleate. These findings suggest that phagocytizing mononuclear cells are capable of inducing the peroxidation of a higher amount of endogenous lipid substrate than polymorphonuclear leucocytes do. On the other hand, neutrophils have the possibility to peroxidize engulfed lipids. Can the different behaviour of monocytes and neutrophils be related to a minor quantity of peroxidable substrate present in the latter cells? It should be noted that the neutrophil, a phagocytic cell essential for the initiation of acute inflammatory reactions, seems to be a relatively poor producer of prostaglandins (19). On the other hand, the monocyte is somehow more productive, and the macrophage, another phagocytic cell similarly derived, generates and releases large amounts of prostaglandins in response to phlogotic stimuli (46). It appears clear that lipid peroxidative derangement occurs during the development of inflammatory reactions and as a consequence, the formation of end-products and intermediates of the oxidative degradation of lipids takes place. However, the intimate role of these later products in inflammation remains in part, an open problem. Peroxidative degradation of polyunsaturated fatty acids results in severe alteration of the highly organized structure of lipid-rich biomembranes through a number of separate mechanisms which are dependent on the presence of the various molecules (intermediates) formed during the different stages of peroxidative chain reactions. Initially, lipid peroxidation causes damage to the structures and functions of cell membranes by affecting metabolic integrations which are dependent on intracellular compartmentation. Further, the hydroperoxides and the peroxy radicals produced during lipid degradation attack susceptible enzymes and proteins by oxidizing thiol groups and by degrading haem groups (114,115).

Moreover, chain termination of lipid peroxidation results in the production of a variety of compounds, some of which display strong biological reactivity. Notwithstanding the fact that malonaldehyde has been found able to induce cross-linking reactions in proteins (26), it is probably one of the less reactive aldehydes formed during lipid peroxidation. Much more powerful biological reactivity is displayed by other aldehydes being produced at rather high amounts during the oxidative decomposition of unsaturated lipids (27), both saturated and unsaturated. The most interesting pathobiological properties are found among the 4-hydroxy-2,3-unsaturated aldehydes, the most abundant of which seems to be 4-hydroxynonenal. These aldehydes rapidly interact with thiol groups, thereby strongly inhibiting several enzyme activities (20,21). Moreover, they provoke vascular damage, increase the capillary permeability (124, 131) and are also able to induce platelet aggregation (131). Other investigations (119) suggest that aldehydes and hydroperoxides, during inflammatory reactions, cooperate in the defense mechanisms with the phagocytes in potentiating their oxidative microbicidal action. The mechanism by which these compounds alter the biomembrane integrity of cellular structures may represent a component of microbial toxicity, tumoricidal action and indiscriminate cell damage as exerted by phagocytic cells at the inflammatory site. Moreover, it should be noted that experimental findings (113) have demonstrated that after ingestion of pneumococci by polymorphonuclear cells, extensive peroxidative derangement of arachidonic acid in the bacterial membranes takes place. This type of reaction can be initiated by hydroxyl free radicals (31). In substance, lipid peroxidation associated with other mechanisms adopted by activated phagocytes in defense actions may result in beneficial effects. At the same time, all the reactive agents generated during the respiratory burst can also cause detrimental effects towards cell and tissue. In the course of the inflammatory reactions, massive invasion of granulocytes and/or macrophages at the site of injury or infection is observed. Both cell types are capable of producing very toxic oxygen products which either alone or in combination with the myeloperoxidase system can escape into the extracellular space, thereby being responsible, in collaboration with lysosomal enzymes, for the subsequent damage to surrounding tissue and adjacent cells. The injurious effect of these products has been shown by McCord (105) by the denaturation of bovine synovial fluid and by the depolymerization of hyaluronate underlying a decrease in the viscosity of this joint lubricant.

Oxygen-mediated injury to endothelial cells in situations where granulocytes accumulate and initiate the respiratory burst has been reported (104). This type of cytotoxic reaction appeared to be independent of the release of lysosomal enzymes. It was also suggested that in myocardial infarctions, cholesterol particles released from atheromata may be engulfed by metabolically activated cells at the site of the necrotic area and then stimulate granulocytes to generate substantial amount of toxic oxygen products which might initiate a pathobiological chain reaction that support further cellular injury (37).

Moreover, phagocytes have been found able to release into the extracellular milieu very reactive products derived from molecular oxygen that attack neighbouring cells as well as artificial membranes (33,76), indiscriminately injuring protein and lipid substrates. They also generate biologically active products from membrane derived arachidonic acid such as endoperoxides in parallel with reactive oxidants, prostaglandin and thromboxanes. Concomitantly, they form products in the lipoxygenase pathway, including hydroperoxy fatty acids and leukotrienes, that are all powerful mediators of the inflammatory response (8,30,68,71,111). Finally, phagocytes release a number of enzymes which provided a strong toxic potential towards cells and mesenchymal structures. In the latter instance, collagenase, elastase, and cathepsin G, in cooperation with the mentioned reactive products, can modulate the defense mechanisms as well as amplify tissue and cell injury (FIG. 5).

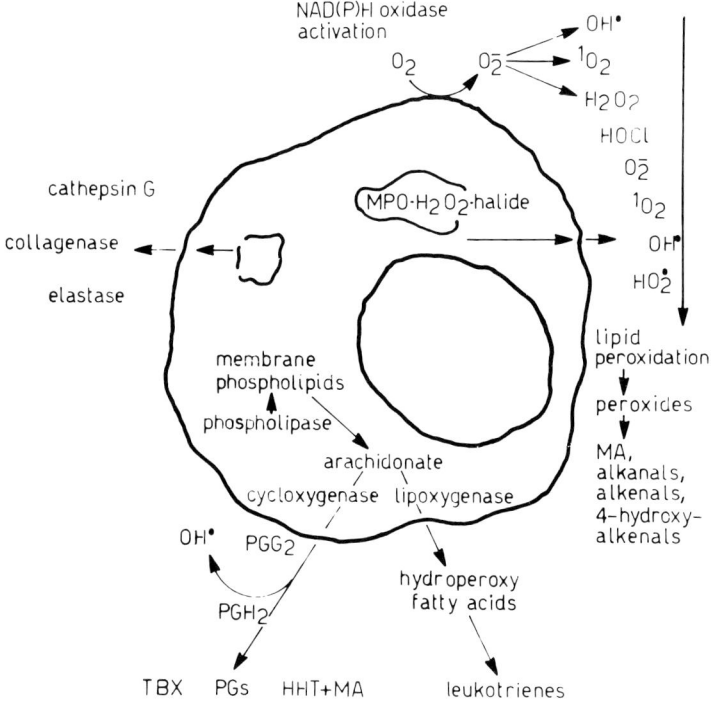

FIG. 5 The "Arsenal" of phagocyte.
     TBX: thromboxanes
     PGs: stable prostaglandins
     HHT: 12L-hydroxy-5,8,10-heptadecatrienoic acid
     MA: malonaldehyde

FINAL PRODUCTS OF LIPID PEROXIDATION AND CELL BIOLOGY OF INFLAMMATION

Another interesting aspect of the interrelationships between end-products of peroxidative degradation of lipid substrates and the pathobiology of the inflammation cells is the capability of these compounds, independently to their toxic power, in the modulation of some important functions of phagocytes.

It has been demonstrated in our laboratory (83) that some 4-hydroxy-alkenals are able to depress, at different degrees, the phagocytic activity of rat polymorphonuclear leucocytes without affecting their viability (FIG. 6).

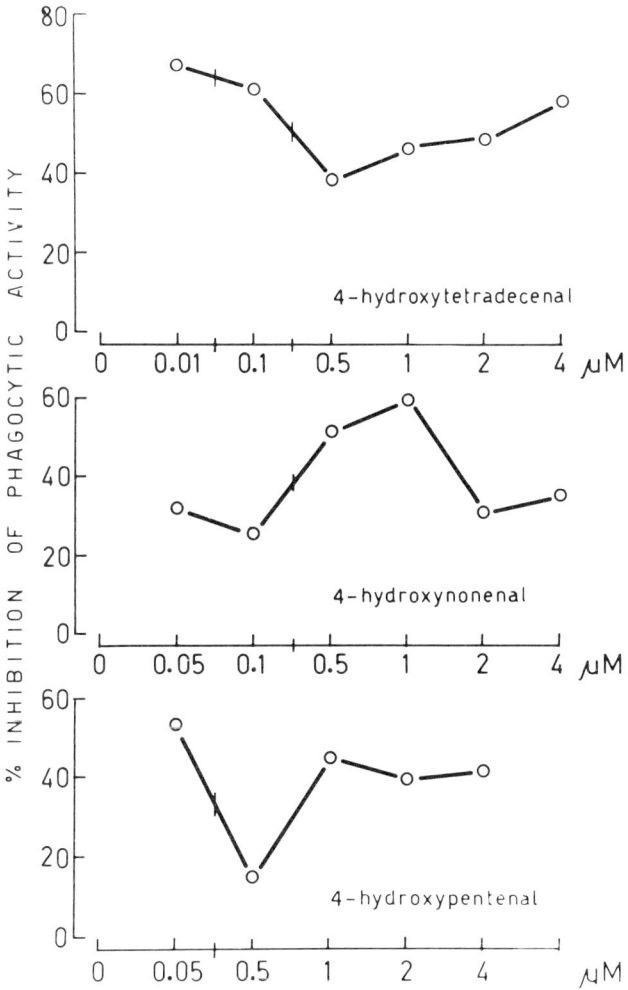

FIG. 6   Effect of 4-hydroxyalkenals on phagocytic activity of rat polymorphonuclear leucocytes.
(Adapted after Negro et al., 1979; ref. 83)

With regard to the capability of oriented locomotion of these cells, studies of Turner et al. (130) carried out with human granulocytes indicated that compounds derived by the oxidation of unsaturated fatty acids are variously chemotactic for phagocytes. These findings suggest that the high amount of arachidonic acid present in biomembranes can represent a wide reserve of precursors of chemotactic messengers.

In the Institute of General Pathology of Turin, a series of investigations on the chemotactic activity of aldehydes started a few years ago (14-18) on the background that red cells killed by a microlaser beam attract polymorphonuclear leucocytes. Light is a well known producer of free radicals, so the idea that leucocyte attraction in this system might be mediated through the release of substances coming from the peroxidative deterioration of plasma membrane lipids was put forward. We decided, therefore, to test the effect of aldehydes on chemotaxis by the Boyden chamber method. Among the number of aldehydes so far assayed for chemotactic power, we obtained positive results only with the 4-hydroxy-2,3-enals. While the synthetic unnatural compound 4-hydroxypentenal was found to be practically inactive, very good stimulation of chemotaxis was obtained with 4-hydroxy-2,3-transnonenal (15,16), a compound that is produced in rather high amounts during lipid peroxidation of liver homogeates. This substance was found to be active up to 10^{-9}M, the 50% effective dose being at 1×10^{-8}M. It is interesting to note that the same concentration is able, during the very first minutes of incubation, to inactivate the adenylate cyclase of isolated plasmamembrane from liver cells (5,21).

Adenylate cyclase activation in plasmamembranes is considered to be important for Ca^{2+} intake. N-formyl-methionyl-leucyl-phenylalanine, as well as leukotriene B_4 and arachidonate, are known to increase Ca^{2+} and Na^+ influx into rabbit neutrophils (80,138). Calmodulin, a protein needed for Ca^{2+} transport and known to be connected with cellular movements related to chemotaxis, is not influenced at all by 4-hydroxynonenal.[1] The same aldehyde does not affect $Ca^{2+} - Mg^{2+}$-dependent adenosine triphosphatase of rat liver isolated plasmamembranes, but does inhibit Na^+-K^+-stimulated adenosine triphosphatase at higher concentrations (10^{-5}M) than those shown to be active in the case of adenylate cyclase. Such concentrations are not very far from the active concentrations of other well studied physiologic chemoattractants, such as tripeptide N-formyl-methionyl-leucyl-phenylalanine (10^{-9}M) and leukotriene B_4 (10^{-7}M) (49). 4-hydroxynonenal is, however, not the most active aldehyde of the series under study. More active in stimulating the oriented locomotion of granulocytes appear to be other alkenals, such as 4-hydroxyundecenal (present in aldehyde patterns from incubated liver homogenates), 4-hydroxytetradecenal and 4-hydroxypentadecenal (FIG. 7). The aldehyde being by far the more active, is the last one, whose 50% effective dose is 4×10^{-12}M (Table 1), but with significant stimulatory activity demonstrable until 10^{-13}M. 4-hydroxytetradecenal and 4-hydroxypentadecenal do not seem to be produced by liver homogenates. As far as it regards 4-hydroxyoctenal, it seems possible that it is formed at a concentration 5 to 10 times less than 4-hydroxynonenal, but the relative peak on the aldehyde pattern has not been surely identified as 4-hydroxyoctenal by mass spectrophotometry as yet.

---

[1] G. Barrera, personal communication.

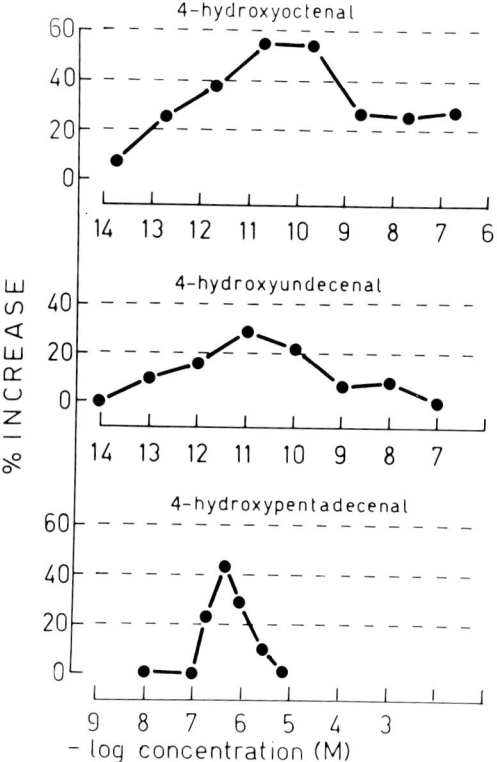

FIG. 7 Chemotactic effects of 4-hydroxyalkenals on rat polymorphonuclear leucocytes.
(Adapted after Curzio et al., 1983; ref. 18)

TABLE 1. Chemotactic activity (ED_{50} of 4-hydroxyalkenals towards rat neutrophils

Aldehydes	Chemotactic power ED_{50} ($\pm$ SD)
4-hydroxyoctenal	$4.6 \pm 2.8 \times 10^{-13}$
4-hydroxynonenal	$1.0 \pm 1.5 \times 10^{-8}$
4-hydroxyundecenal	$4.9 \pm 1.7 \times 10^{-7}$
4-hydroxytetradecenal	$2.4 \pm 3.5 \times 10^{-9}$
4-hydroxypentadecenal	$4.0 \pm 0.7 \times 10^{-12}$

(Adapted after Curzio et al., 1983; ref. 18)

The continuous increase in activity from 4-hydroxypentenal to 4-hydroxypentadecenal has led us to the preliminary conclusion that the chemo-attractive power of the 4-hydroxyenal series was increasing in parallel with their lipophilic properties, but the finding that 4-hydroxyoctenal is more active than any other tested aldehydes suggests that either this conclusion is not true, or that the activity of the different aldehydes takes place through at least two different receptor types on the surface of leucocytes (18). The same aldehydes are devoid of any effect on random locomotion (chemokinesis). At concentrations of $10^{-4} - 10^{-5}$, chemokinesis is inhibited, suggesting a possible influence of the aldehydes on the at random movement system. From this point of view, one may remember that 4-hydroxynonenal, as well as other aldehydes, blocks $^H2$-colchicine binding to hepatocyte tubulin (20,21).

For the moment it is unclear whether the inhibition of phagocytosis by 4-hydroxyenals, as reported above, is only as aspect of the block of at random movements. In any case, it takes place at the same concentrations. No effect either on chemokinesis or on phagocytosis was found at concentrations as low as those used to stimulate the oriented migration of leucocytes.

At the low concentrations stimulating chemotaxis, 4-hydroxynonenal is able to produce oriented changes in leucocyte shape, as well as myeloperoxidase release from polymorphonuclear cells. These phenomena are known to be strictly connected with chemotaxis. In addition, 4-hydroxyenal-induced chemotaxis displays the deactivation phenomenon. Neutrophils preincubated with stimulatory amounts of 4-hydroxyenal do not respond after washing, to a second stimulation by the same aldehyde, or by another 4-hydroxyenal, but are still able to respond to N-formyl-methionyl-leucyl-phenylalanine. In the cross experiments, leucocytes preincubated with N-formyl-methionyl-leucyl-phenylalanine do not respond after washing, to the same tripeptide, but retain their ability to respond to an aldehyde of the 4-hydroxy-2,3-enal series. So, if the chemotactic response to aldehydes is mediated through the presence on leucocyte surface receptors, these are different from those known to be needed for the tripeptide response. The final effector pathway, however, seems to be the same for 4-hydroxyenals and tripeptide. In fact, ineffective concentrations of the tripeptide used in combination with ineffective doses of 4-hydroxyenals, result in a positive response. It is not known if the deactivation phenomenon is displayed by 4-hydroxyenals in cross experiments with leukotrienes. So, we cannot state, for the moment, if receptors for aldehydes, if they exist, are the same recently demonstrated to work in the case of leukotrienes (49). At any event, it appears very interesting that substances coming from arachidonic acid by a pathway different from that leading to the production of leukotrienes, display similar biological properties at doses identical or lower than those being active in the case of leukotrienes.

The biological importance of the observed phenomena requires deeper investigation. In fact, the existing knowledge on the formation of 4-hydroxyenals is derived from the study of the products of peroxidizing liver homogenates. So, we have not the certainty that the same aldehydes are produced during the inflammatory response, either by leucocytes or by the tissue cells. For the moment, we only know that pleural fluid evoked by acute inflammation and containing high amount of leucocytes are rich in aldehyde material, but we cannot state where such substances come from. If such compounds might be produced during the development of the inflammatory reactions, their importance for recruitment of new phagocytic cells in the phlogotic area might be relevant, especially if we consider that no other known substances acts at such low concentrations on neutrophil chemotaxis. Of course, the experiments must be extended to macrophages, that seem to display a much bigger activity than polymorphonuclear leucocytes in arachidonic acid metabolic pathways.

REFERENCES

1. Albrich, J.M., McCarthy, C.A. and Hurst, J.K. (1981): Proc. Natl. Acad. Sci., 78:210-214.
2. Babior, B.M. (1982): In: Pathology of Oxygen, edited by A.P. Autor, pp. 45-58. Academic Press, New York-London.
3. Badwey, J.A. and Karnovsky, M.L. (1980): Ann. Rev. Biochem., 49:687-726.
4. Baird, M.B., Massie, H.R. and Ppiekielniak, M.J. (1977): Chem. Biol. Interact., 16:145-153.
5. Barrera, G., Parola, M., Amoroso, L., Paradisi, L. and Dianzani, M.U. (1982): Boll. Soc. It. Biol. Sper., 58:1582-1588.
6. Belding, M.E., Klebanoff, S.J. and Ray, G.C. (1970): Science, 167:195-196.
7. Bielsky, B.H.J. and Shiue, G.G. (1979): CIBA Found. Symp., 65:43-
8. Borgeat, P. and Samuelsson, B. (1979): J. Biol. Chem., 254:7865-7869.
9. Bragt, P.C., Schenkelaars, E.P.M. and Bonta, I.L. (1979): Prostaglandins Med., 2:51-61.
10. Burchill, B.R., Oliver, J.M., Pearson, C.B., Leinbach, E.D. and Berlin, R.D. (1978): J. Cell Biol., 76:439-447.
11. Cadet, J. and Teoule, R. (1978): Photochem. Photobiol., 28:661-667.
12. Canuto, R.A. and Garcea, R., to be published.
13. Cook, H.W. and Lands, W.E.M. (1975): Biochem. Biophys. Res. Comm., 65:464-471.
14. Curzio, M., Roch-Arveiller, M., Negro, F., Giroud, J.P., Esterbauer, H., Torrielli, M.V. and Dianzani, M.U. (1981): Boll. Soc. It. Bol. Sper., 57:2479-2485.
15. Curzio, M., Torrielli, M.V. and Dianzani, M.U. (1982): Boll. Soc. It. Biol. Sper., 58:1672-1678.
16. Curzio, M., Torrielli, M.V., Giroud, J.P., Esterbauer, H. and Dianzani, M.U. (1982): Res. Comm. Chem. Path. Pharmacol., 36:463-476.
17. Curzio, M., Esterbauer, H. and Dianzani, M.U. (1983): IRCS, Med. Sci., 11:521.
18. Curzio, M., Esterbauer, H. and Dianzani, M.U. (1983): Hoppe-Seyler's Z. Physiol. Chem., in press.

19. Davies, P., Bonney, R.J., Humes, J.L. and Kuehl, F.A. (1977): Inflammation, 4:335-
20. Dianzani, M.U. (1979): In: Submolecular Biology of Cancer, edited by G.E.W. Wolstenholme, D.W. Fitzsimons and J. Whelan, pp. 245-279. Excerpta Medica, Amsterdam.
21. Dianzani, M.U. (1982): In: Free Radicals, Lipid Peroxidation and Cancer, edited by D.C.H. McBrien and T.F. Slater, pp. 129-158. Academic Press, New York, London.
22. Dianzani, M.U. and Ugazio, G. (1978): In: Biochemical Mechanisms of Liver Injury, edited by T.F. Slater, pp. 669-708. Academic Press, New York-London.
23. Dianzani, M.U., Torrielli, M.V., Paradisi, L. and Franzone, J.S. (1978): Eur. J. Rheum. Inflamm., 1:187-196.
24. Egan, R.W., Paxton, J. and Kuehl, F.A. (1976): J. Biol. Chem., 251:7329-7335.
25. Egan, R.W., Gale, P.H. and Kuehl, F.A. (1979): J. Biol. Chem., 254:3295-3302.
26. Esterbauer, H. (1982): In: Free Radicals, Lipid Peroxidation and Cancer, edited by D.C.H. McBrien and T.F. Slater, pp. 102-128. Academic Press, New York-London.
27. Esterbauer, H., Cheseman, K., Dianzani, M.U., Poli, G. and Slater, T.F. (1982): Biochem. J., 208:129-140.
28. Fantone, J.C. and Kinnes, D.A. (1983): Biochem. Biophys. Res. Comm., 113:506-512.
29. Fee, J.A. and Valentine, J.S. (1977): In: Superoxide and Superoxide Dismutases, edited by A.M. Michelson, J.M. McCord and Fridovich, I., pp. 19-60. Academic Press, New York-London.
30. Feinmark, S.J., Lindgren, J.A., Claesson, H.E., Malmsten, C. and Samuelsson, B. (1981): FEBS Letters, 136:141-144.
31. Fong, K.L., McCay, P.B., Poyer, G.L., Keele, B.B. and Misra, H.P. (1973): J. Biol. Chem., 248:7792-7797.
32. Fridovich, I. (1978): Science, 201:875-880.
33. Goldstein, I.M. and Weissmann, G. (1977): Biochem. Biophys. Res. Comm., 75:604-609.
34. Goldstein, B.D., Balchum, O.J., Demopoulos, H.B. and Dukes, P.S. (1968): Arch. Environ. Hlth., 17:46-49.
35. Graham, R.C., Karnovsky, M.I., Shafer, A.W., Glass, E.A. and Karnovsky, M.L. (1967): J. Cell Biol., 32:629-647.
36. Green, T.R., Wirtz, M.K. and Wu, D.E. (1983): Biochem. Biophys. Res. Comm., 110:873-879.
37. Greenberg, C.S., Hammerschmidt, D.E., Ceaddock, P.R. and Jacob, H.S. (1979): Trans. Assoc. Am. Phys., 92:130-135.
38. Haber, F. and Weiss, J. (1934): Proc. R. Soc. London, Ser. A, 147:332-351.
39. Halliwell, B. (1981): Bull. Eur. Physiopathol. Resp., 17 (Suppl.):21-29.
40. Harrison, J.E. and Schultz, T. (1976): J. Biol. Chem., 251:1371-1374.
41. Harrison, J.E., Watson, B.D. and Schultz, T. (1978): FEBS Letters, 92:327-332.

42. Held, A.M. and Hurst, J.K. (1978): Biochem. Biophys. Res. Comm., 81:878-885.
43. Hemler, M.E., Lands, W.E.M. and Graf, G. (1978): Biochem. Biophys. Res. Comm., 85:1325-1331.
44. Hemler, M.E., Cook, H.W. and Lands, W.E.M. (1979): Arch. Biochem. Biophys., 193:340-345.
45. Huber, W. and Menander-Huber, K.B. (1980): Clinics Rheum. Dis., 6:465-498.
46. Humes, J.L., Bonney, R.J., Pelus, L., Dahlgren, M.E., Sadowsky, S.J., Kuehl, F.A. and Davies, P. (1977): Nature (London), 269:149-151.
47. Jacobs, A.A., Low, I.E., Paul, B.B., Strauss, R.R. and Sbarra, A.J. (1972): Infect. Immun., 5:127-131.
48. Jensen, M.S. and Bainton, S.F. (1973): J. Cell Biol., 56:379-388.
49. Jubitz, W. (1983): Biochem. Biophys. Res. Comm., 110:842-850.
50. Kakinuma, K. (1974): Biochim. Biophys. Acta, 348:76-85.
51. Kanner, J. and Kinsella, J.E. (1983): Lipids, 18:204-210.
52. Kellog, E.W. and Fridovich, I. (1975): J. Biol. Chem., 250:8812-8817.
53. Klebanoff, S.J. (1982): In: Advances in Host Defense Mechanisms, edited by J.I. Gallin and A.S. Fauci, pp. 111-162. Raven Press, New York.
54. Klebanoff, S.J. (1970): Science, 169:1095-1097.
55. Klebanoff, S.J. (1980): In: The Reticuloendothelial System, edited by A.J. Sbarra and R. Strauss, pp. 279-308. Plenum Publishing Corp., New York.
56. Klebanoff, S.J. (1980): In: Mononuclear Phagocytes. Functional Aspects, edited by R. Van Furth, part II, pp. 1105-1141. Martinus Nijhoff Publ., The Hague.
57. Klebanoff, S.J. (1980): Ann. Intern. Med., 93:480-489.
58. Klebanoff, S.J. (1982): J. Exp. Med., 156:1262-1267.
59. Klebanoff, S.J. and Clark, R.A. (1978): In: The Neutrophil: Function and Clinical Disorders, pp. 410-434. Elsevier North Holland, Amsterdam.
60. Koster, J.F. and Slee, R.G. (1980): Biochim. Biophys. Acta, 620:489-497.
61. Kreisle, R.A. and Parker, C.W. (1983): J. Exp. Med., 157:628-641.
62. Kuehl, F.A., Humes, G.L., Egan, R.W., Ham, E.A., Beveridge, G.C. and Van Arman, C.G. (1977): Nature (London), 265:170-173.
63. Kuehl, F.A., Ham, E.A., Egan, R.W., Dougherty, H.W., Bonney, R.J. and Humes, J.L. (1982): In: Pathology of Oxygen, edited by A.P. Autor, pp. 175-190. Academic Press, New York-London.
64. Lai, C.S. and Piette, L.H. (1977): Biochem. Biophys. Res. Comm., 78:51-59.
65. Lai, C.S. and Piette, L.H. (1978): Arch. Biochem. Biophys., 190:27-38.
66. Lai, C.S., Grover, T.A. and Piette, L.H. (1979): Arch. Biochem. Biophys., 193:373-382.
67. Lehrer, R.I. (1969): J. Bacteriol., 99:361-365.

68. Lewis, R.A., Goetzl, E.J., Draen, J.M., Soter, N.A., Austen, K.F. and Corey, E.J. (1981): J. Exp. Med., 154:1243-1248.
69. Lynch, R.E. and Fridovich, I. (1978): J. Biol. Chem., 253:1838-1845.
70. Long, C.A. and Bielsky, B.H.J. (1980): J. Phys. Chem., 84:555-557.
71. Lundberg, U., Radmark, O., Malmsten, C. and Samuelsson, B. (1981): FEBS Letters, 126:127-132.
72. Mandell, G.L. (1974): Infect. Immun., 9:337-341.
73. Marnett, L.J., Wlodawer, B. and Samuelsson, B. (1975): J. Biol. Chem., 250:8510-8517.
74. McCord, J.M. (1974): Science, 185:529-531.
75. McCord, J.M. and Day, E.D. (1978): FEBS Letters, 86:139-142.
76. McCord, J.M., Stokes, S.H. and Wong, K. (1979): In: Advances in Inflammation Research, edited by G. Wiessmann, R. Paoletti and B. Samuelsson, pp. 273-280. Raven Press, New York.
77. McRipley, R.J. and Sbarra, A.J. (1967): J. Bacteriol., 94:1425-1430.
78. Mead, J.F. (1976): In: Free Radicals in Biology, edited by W.A. Pryor, pp. 51-68. Academic Press, New York-London.
79. Morehouse, L.A., Tien, M., Bucher, J.R. and Aust, S.D. (1983): Biochem. Pharmacol., 32:123-127.
80. Molski, T.F.P., Naccache, P.H., Borgeat, P. and Sha'Afi, R.I. (1981): Biochem. Biophys. Res. Comm., 103:227-232.
81. Muus, P., Bonta, I.L. and Den Oudsten, S.A. (1979): Prostaglandins Med., 2:63-65.
82. Noseworthy, J. and Karnovksy, M.L. (1972): Enzymes, 13:110-131.
83. Negro, F., Curzio, M., Torrielli, M.V., Esterbauer, H. and Dianzani, M.U. (1981): Boll. Soc. It. Biol. Sper., 57:2472-2478.
84. Oliver, J.M. (1978): Am. J. Pathol., 93:221-259.
85. Oliver, J.M., Albertini, D.F. and Berlin, R.D. (1976): J. Cell Biol., 71:921-932.
86. Oliver, J.M., Spielberg, S.P., Pearson, C.B. and Shulman, J.D. (1978): J. Immun., 120:1181-1186.
87. Oyanagui, Y. (1976): Biochem. Pharmacol., 25:1473-1480.
88. Piatt, J.F., Cheema, A.S. and O'Brien, P.J. (1977): FEBS Letters, 74:251-254.
89. Poli, G., Esterbauer, H., Dianzani, M.U. and Slater, T.F., in preparation.
90. Porter, N.A., Wolff, R.A., Yarbro, E.M. and Weenen, H. (1979): Biochem. Biophys. Res. Comm., 80:1058-1064.
91. Pryor, W.A. (1973): Fed. Proc., 32:1862-1869.
92. Pryor, W.A. and Tang, R.H. (1978): Biochem. Biophys. Res. Comm., 81:498-503.
93. Pryor, W.A., Stanley, J.P. and Blair, E. (1976): Lipids, 11:370-372.
94. Puig-Parellada, P. and Planas, J.M. (1978): Biochem. Pharmacol., 27:535-537.
95. Reed, P.W. (1969): J. Biol. Chem., 244:2459-2464.

96. Repine, J.E., White, J.G., Clawson, C.C. and Holmes, B.M. (1974): J. Lab Clin. Med., 83:911-920.
97. Rigo, A., Stevanato, R., Finazzi-Agro, A. and Rotilio, G. (1977): FEBS Letters, 80:130-132.
98. Roos, D. (1980): In: The Cell Biology of Inflammation, edited by G. Weissmann, pp. 337-386. Elsevier/North Holland, Amsterdam.
99. Roos, D., Weening, R.S. and Loos, J.A. (1979): In: Inborn Errors of Immunity and Phagocytosis, edited by F. Guttler, J.W.T. Seakins and R.A. Harkness, pp. 261-286. MTP Press Ltd., Lancaster.
100. Root, R.K. and Metcalf, J.A. (1977): J. Clin. Invest., 60:1266-1279.
101. Root, R.K. and Cohen, M.S. (1981): Rev. Infect. Dis., 3:565-598.
102. Rosen, H. and Klebanoff, S.J. (1976): J. Clin. Invest., 58:50-60.
103. Rosen, H. and Klebanoff, S.J. (1977): J. Biol. Chem., 252:4803-4810.
104. Sacks, T., Moldow, C.F., Craddock, P.R. and Jacobs, H.S. (1978): J. Clin. Invest., 61:1161-1167.
105. Salin, M.L. and McCord, J.M. (1975): J. Clin. Invest., 56:1319-1323.
106. Satoh, K. (1978): Clin. Chim. Acta, 90:37-43.
107. Satoh, K., Sakamoto, N., Matsuoka, S., Ohishi, N. and Yagik, F. (1979): Biochem. Med., 21:104-107.
108. Sawyer, D.T. and Gibian, M.J. (1980): Tetrahedron, 35:1471-1475.
109. Segal, A.W. and Meshulam, T. (1979): FEBS Letters, 100:27-32.
110. Selvaraj, R.S., Paul, B.B., Strauss, R.R., Jacobs, A.A. and Sbarra, A.J. (1974): Infect. Immun., 9:255-260.
111. Serhan, C.N., Radin, A., Smolen, J.E., Korchak, H., Samuelsson, B. and Wissmann, G. (1982): Biochem. Biophys. Res. Comm., 107:1006-1012.
112. Sharma, O.P. (1976): Biochem. Pharmacol., 25:1811-1812.
113. Shohet, S.B., Pitt, J., Baehner, R.L. and Poplack, D.G. (1974): Infect. Immun., 19:1321-1328.
114. Slater, T.F. (1972): Free Radical Mechanisms in Tissue Injury. Pion Press, Ltd., London.
115. Slater, T.F. (1978): In: Biochemical Mechanisms of Liver Injury, edited by T.F. Slater, pp. 1-44. Academic Press, New York-London.
116. Slivka, A., Lo Buglio, A.F. and Weiss, S.J. (1980): Blood, 55:347-350.
117. Smith, W. and Lands, W.E.M. (1972): Biochemistry, 11:3276-3285.
118. Stelmaszynka, T. and Zgliczynski, J.M. (1974): Eur. J. Biochem., 45:305-312.
119. Stossel, T.P., Mason, R.J. and Smith, A.L. (1974): J. Clin. Invest., 54:638-645.
120. Strauss, R.R., Paul, B.B., Jacobs, A.A. and Sbarra, A.J. (1969): Arch. Biochem. Biophys., 135:265-271.

121. Suematsu, T., Kamada, T., Abe, H., Kikuchi, S. and Yagy, K. (1977): Clin. Chim. Acta, 79:267-270.
122. Swern, D. (1961): In: Autoxidation and Antioxidants, edited by W.O. Lundberg, pp. 2-49, Wiley and Sons, New York.
123. Tappel, A.L. (1973): Fed. Proc., 32:1870-1874.
124. Torrielli, M.V. (1981): In: Recent Advances in Lipid Peroxidation and Tissue Injury, edited by T.F. Slater and A. Garner, pp. 267-296. Brunel Printing Service, Uxbridge.
125. Torrielli, M.V., Franzone, J., Natale, T. and Sena, L. (1980): In: Inflammation: Mechanisms and Treatment, edited by D.A. Willoughby and J.P. Giroud, pp. 323-325. MTP Press, Lancaster.
126. Torrielli, M.V., Giroud, J.P., Curzio, M. and Negro, F. (1981): In: L'Immunita nella patogenesi delle malattie. Atti XVI Congr. Soc. Ital. Patol., pp. 337-354. Copisteria Scientifica Universitaria, Torino.
127. Trang, L.E., Granstrom, E. and Lovgren, O. (1977): Scand. J. Rheum., 6:151-154.
128. Tsan, M.F. and Denison, R.C. (1980): Inflammation, 4:371-382.
129. Turner, S.R., Campbell, J.A. and Lynn, W.S. (1975): J. Exp. Med., 141:1437-1441.
130. Ugazio, G., Torrielli, M.V., Burdino, E., Sawyer, B.C. and Slater, T.F. (1976): Biochem. Soc. Transact., 4:353-356.
131. Valentine, S. (1979): In: Biochemical and Clinical Aspects of Oxygen, edited by W.S. Caughey, pp. 659-681. Academic Press, New York-London.
132. Voetman, A.A., Weening, R.S., Hamers, M.N., Meerhof, L.J., Bot, A.A.A.M. and Roos, D. (1981): J. Clin. Invest., 67:1541-1549.
133. Ward, P.A., Duque, R.E., Sulavik, M.C. and Johnson, K.J. (1983): Am. J. Pathol., 110:297-309.
134. Weiss, S.J. and Lo Buglio, A.F. (1982): Lab. Invest., 47:5-18.
135. Weiss, S.J. and Slivka, A. (1982): J. Clin. Invest., 69:255-262.
136. Weiss, S.J., Rustagi, P.K. and Lo Buglio, A.F. (1978): J. Exp. Med., 147:316-323.
137. White, J.R., Naccache, P.H., Molski, T.F.P., Borgeat, P. and Sha'Afi, R.I. (1983): Biochem. Biophys. Res. Comm., 113:44-50.
138. Willson, R.L. (1978): In: Biochemical Mechanisms of Liver Injury, edited by T.F. Slater, pp. 123-224. Academic Press, New York-London.
139. Willson, R.L. (1979): CIBA Found. Symp., 65:19-42.
140. Zgliczynski, J.M., Stelmaszynska, J., Domanski, J. and Ostrowski, W. (1971): Biochim. Biophys. Acta, 235:419-424.

Free Radicals in Molecular Biology, Aging, and Disease, edited by D. Armstrong et al.
Raven Press, New York © 1984.

Free Radicals and Lung Injury

Clark T. Bishop, Bruce A. Freeman, and James D. Crapo

Division of Allergy, Critical Care and Respiratory Medicine, Duke University Medical Center, Durham, North Carolina 27710

The lung's unique physiologic role allows interaction with a wide variety of substances. The human lung alveolar surface area is estimated to be 120 x m², as large as a tennis court (1,2). This provides a vast membrane available for release of carbon dioxide (CO_2) and uptake of oxygen. The large pulmonary capillary bed allows the entire cardiac output of blood returning from the systemic circulation to exchange carbon dioxide for oxygen. This ensures that blood cells returning to the body have the appropriate CO_2 and oxygen content. The large alveolar surface area and capillary volume of the lung which permits normal pulmonary physiology also makes the lung susceptible to injury from toxic substances when present in either circulating blood or inspired air. Many of these toxins including drugs, xenobiotics and oxygen metabolites, exert their deleterious effect via free radicals. This chapter will address free radicals and lung injury. Oxygen toxicity will be addressed as a prototype of free radical injury. Oxygen metabolism, protective mechanisms, and morphometric effects of free radical damage will be discussed. Biochemical correlates to the morphometric changes seen in oxygen toxicity will also be addressed.

Oxygen Centered Free Radicals

In 1775 Priestly postulated that purified oxygen, which he termed "dephlogistated air" might be useful for treating some disease processes; but he also suggested that there was a potential risk to oxygen therapy (3). Indeed, subsequent experience proved that most mammals die within a few days when breathing 100% oxygen (4). In the last 15 years a large body of evidence has accumulated implicating free radicals in oxygen toxicity.

Evidence to support this model was provided from studies of oxygen metabolism. Molecular oxygen is fully oxidized, and is capable of accepting a total of four electrons to become fully reduced as H_2O (5). In porcine and rat lung under normoxic conditions, 95% of oxygen metabolism occurs via cytochrome c oxidase of the mitochondrial respiratory chain, where oxygen accepts 4 electrons to become fully reduced (6) (Figure 1).

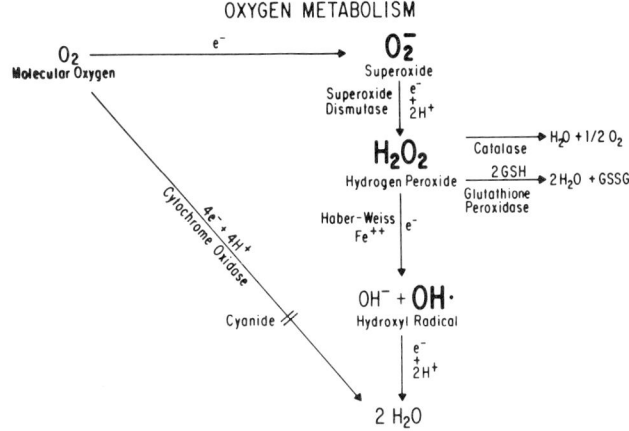

Figure 1. Oxygen Reduction. Molecular oxygen may become fully reduced to H_2O by a 4 electron transfer at the cytochrome c oxidase site of the respiratory chain, or it may form partially reduced species, free radicals, via single electron transfers.

This transfer of electrons to oxygen at cytochrome c oxidase can be blocked by cyanide. The remaining 5% of oxygen metabolism is cyanide resistant and has been found to correlate with partially reduced oxygen species (free radicals) formation from partial reduction of oxygen (7,8,9,10). In lung slices maintained in 85% oxygen, the ratio changes from 95% cyanide sensitive:5% cyanide resistant to 80% cyanide sensitive:20% cyanide resistant (6). It is therefore implied that during hyperoxia more partially reduced oxygen species are generated in lung tissue.

In this discussion the partially reduced oxygen species superoxide (O_2^-), hydrogen peroxide (H_2O_2) and hydroxyl radical (OH·) are all referred to as free radicals. Hydrogen peroxide is not strictly speaking a free radical since all of its electrons are paired. But since it can readily form OH· in the presence of Fe(II) and can initiate chain reactions with polyunsaturated fatty acids (PUFA) to form lipid peroxides (11,12), it is not unreasonable to generalize that H_2O_2 belongs to the group of oxygen-centered free radicals, since it is functionally a free radical. In addition to the cyanide resistant measurement of hyperoxia augmented free radical formation, which is indirect, direct measurement of increased O_2^- and H_2O_2 generation in hypoxic lung tissue been performed. Hyperoxia augments O_2^- and H_2O_2 formation from porcine lung isolated mitochondria (9,10); hyperoxia increases H_2O_2 generation from porcine lung microsomes (9,10). Moreover, recently hyperoxia has been shown to increase H_2O_2 release from intact cultured porcine endothelial cells (Figure 2, reference 13). This suggests that during hyperoxia H_2O_2 is able to escape antioxidant defenses and enter the extracellular milieu.

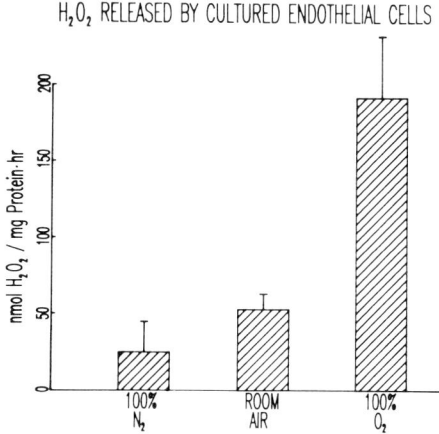

Figure 2. Medium containing intact endothelial cells in 5% oxygen, room air and 95% oxygen were assayed for H_2O_2. The cells in 100% oxygen released significantly greater amounts of H_2O_2 ($p<0.05$).

Oxygen centered free radicals, once formed, can enter into a wide variety of toxic reactions (Table I). They react with protein, lipid, membranes, DNA base pairs and the ribose phosphate backbone of DNA. Lipid peroxidation is particularly devastating since chain reactions can be initiated whereby one free radical can initiate oxidation of many polyunsaturated fatty acid chains. This oxidation of cellular constituents disrupts membrane integrity, inhibits cellular enzymatic processes and results in cell damage or death (11).

TABLE I

Cellular Consequences of Free Radicals

Target	Result
Macromolecules	
Protein	Denaturation, chain scission
DNA	Strand scission, base modification, cell cycle disruption
"Small" Molecules	
Carbohydrates	Cell surface receptor changes
Nucleic acid	Cell cycle changes, mutations
Unsaturated lipids	Fatty acid oxidation organelle and cell membrane permeability changes
Amino Acids	Enzyme inhibition, cross linking
Antioxidants	Decreased availability of GSH, α-tocopherol and β-carotene
Cofactors	Decreased availability and activity of nicotinamide and flavin-containing cofactors

Antioxidant Defense Mechanisms

The above mentioned toxic effects to the cell are possible even under normoxic conditions since approximately 1-5% of normal cellular oxygen metabolism yields these potentially reactive free radicals. It is therefore important for the cell to be able to detoxify the O_2^- and H_2O_2 and $OH\cdot$. Enzymatic scavengers are present in all aerobic cells and remove these partially reduced oxygen species. Superoxide dismutase (SOD), a ubiquitous enzyme among aerobic organisms, combines hydrogen ions with two superoxide molecules:

$$2O_2^- + 2H^+ \xrightarrow{SOD} H_2O_2 + O_2 \qquad \text{Reaction \#1}$$

Hydrogen peroxide is detoxified in eukaryotic cells, via catalase or glutathione peroxidase. Catalase promotes dismutation of H_2O_2 in the reaction below:

$$2H_2O_2 \xrightarrow{Catalase} 2H_2O + 2O_2 \qquad \text{Reaction \#2}$$

Glutathione peroxidase (GPx) oxidizes reduced glutathione (GSH) during the reduction of H_2O_2 via the following reaction:

$$H_2O_2 + 2GSH \xrightarrow{GPx} 2H_2O + GSSG \qquad \text{Reaction \#3}$$

Oxidized glutathione is reduced in the presence of glutathione reductase (GR) and NADPH + H, which is provided by glucose-6-phosphate dehydrogenase-mediated reduction of $NADP^+$ in the pentose phosphate shunt.

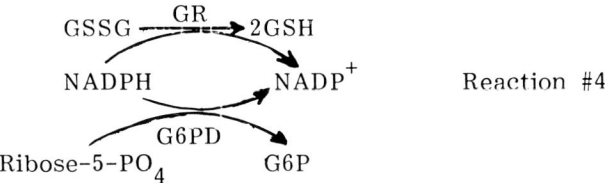

Reaction #4

The glutathione peroxidase system is thought to be important in reducing low levels of cellular H_2O_2, while at greater rates of cellular H_2O_2 generation, catalase becomes more important (14). Hydroxyl radical can be scavenged by tocopherol, ascorbate and β-carotene, but this radical specie is so reactive it has a short diffusion distance and will react with virtually any adjacent molecule (11). Thus, under normoxic conditions there are a variety of defense mechanisms to detoxify the oxygen free radicals.

As noted, during hyperoxia free radicals are produced at increased rates, ultimately causing cell injury by overwhelming the antioxidant defenses (11). Support to this free radical theory of oxygen toxicity is given by experiments which specifically augment intracellular defense enzymes. Increased specific activity of antioxidant enzymes was first noted in rats exposed to 85% oxygen for seven days, then placed in 100% oxygen. These rats were able to survive the 100% oxygen exposure while room air treated controls died after 60-70 hrs of 100% oxygen exposure. The oxygen adapted rats were found to have increased specific activity of SOD (15,16). However, subsequent attempts to ameliorate oxygen toxicity with aerosolized and intravenously administered antioxidant enzymes did not protect experimental animals from oxygen toxicity (17,18). Recent studies employing liposome entrapped antioxidant enzymes have resulted in increased resistance to oxygen toxicity. Liposome entrapment mediates organ uptake and intracellular delivery of normally membrane-impermeable macromolecules such as enzymes (19,20). Liposomes containing SOD and catalase were prepared and injected intravenously into rats. Treated and untreated rats were then exposed to 100% oxygen. The rats injected with SOD and catalase liposomes had increased survival time and decreased pleural effusion volume compared with the untreated rats (21). In vitro experiments were also performed: cultured endothelial cells were treated with liposomes containing SOD, empty liposomes, or free enzymes then exposed to 95% oxygen. The cells treated with liposomes containing SOD suffered less cell damage as measured by release of LDH or prelabeled ^{51}Cr compared to control cells (22). These data suggest that augmenting antioxidant defenses can attenuate oxygen toxicity; but the enzymes must be delivered at appropriate sites in the tissue.

Morphometric Changes from Oxygen Exposure to Lung Tissue

It has been established that oxygen centered free radicals are produced in the lung in increased amounts during hyperoxia and scavenging the radicals ameliorates the toxicity. Correlating pathologic morphometric observations with biochemical data has provided further insight into the mechanisms of oxygen radical toxicity.

Pathologic and morphometric examination of rat lungs exposed to hyperoxia reveals that the pulmonary capillary endothelial cell is the main target of injury, although other cell types also appear damaged (16). A morphometric analysis of cell number and mean cell volume after 60 hr of exposure to 100% oxygen revealed over 30% of the total capillary lung endothelium is destroyed (16). Alveolar type I cells are not affected by 60 hr of 100% oxygen, and decrease in average volume only slightly after 7 and 14 days of exposure to 85% oxygen (16). Alveolar type II cells do not change after a 60 hr exposure to 100% oxygen but proliferate, increasing in mean cell volume (51%) and in cell number (102%) after 7 days of exposure to 85% oxygen. The interstitial cells, including fibroblasts, alveolar septal cells, monocytes, macrophages, polymorphonuclear leukocytes, pericytes and intermediate cells proliferate after exposure of 100% oxygen for 60 hrs and after a 7 day exposure to 85% oxygen (16,23).

It has been recently noted that primate lungs differ somewhat from rat lungs in response to oxygen. Baboons are more resistent to oxygen toxicity than rats but the baboon lungs do have destruction of the capillary bed and exhibit interstitial and alveolar type II cell proliferation (24).

Biochemical Correlates to Morphometry

Recent studies have addressed the reasons for the relative vulnerability of the endothelium to oxygen toxicity. The total cyanide resistant respiration of cultured endothelial cells under both normoxic and hyperoxic conditions is greater than that of whole lung slices (6). This suggests that endothelial cells may endogenously produce more free radicals than other lung cells. The observation that H_2O_2 is released from intact endothelial cells during hyperoxia suggests that H_2O_2 is produced in sufficient quantities to overwhelm the antioxidant enzymes during oxidant stress (13). H_2O, crossing the endothelial cell membrane, can react with a variety of substances in serum and can begin an inflammatory response. Lipids, in the cell membrane, can become oxidized resulting in cell membrane dysfunction (11,12). The lipid peroxides formed from such reactions are chemotactic to phagocytic cells (25). The role of inflammatory cells in amplifying the initial insult of oxygen toxicity is complex and has been recently reviewed (26,27,28). Arachidonate metabolites (eicosanoids)

(11,30), hydrolases (29), and complement (30,31) may all participate in the inflammatory response which culminates in destruction of the capillary bed (Figure 3).

Recent investigation of cultured endothelial cell antioxidant enzymes has also provided insight into the reason for the vulnerability of the endothelial cell. Comparison of antioxidant enzyme specific activities of endothelial cells and fibroblasts reveal lower specific activities for glutathione peroxidase and glucose-6-phosphate dehydrogenase in cultured endothelial cells compared to cultured fibroblasts [40% and 30% of the fibroblast specific activity respectively (32)]. In other experiments, endothelial cells were incubated in Medium 199 and 10 μM H_2O_2, then harvested and antioxidant enzymes measured. Glutathione peroxidase was induced and increased by 200% over cells not treated with H_2O_2. However, when the same experiment was performed with 10 μM H_2O_2, Medium 199 and with the addition of 1% serum, glutathione peroxidase was not induced, but rather the specific activity fell by 20% (32) (Figure 4). Thus H_2O_2 produced during hyperoxia can enter the vascular lumen, react with constituents of serum and form products which can enter the cell and inhibit essential defense enzymes.

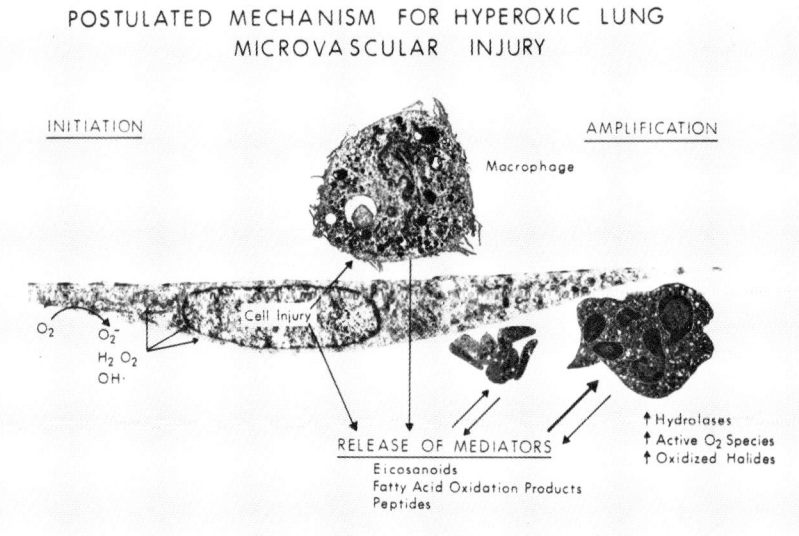

Figure 3. Oxygen toxicity is initiated by the formation of free radicals which cause the influx of inflammatory cells. Amplification is mediated via oxygen metabolites, proteolytic enzymes and eicosanoids.

MODULATION OF GLUTATHIONE PEROXIDASE

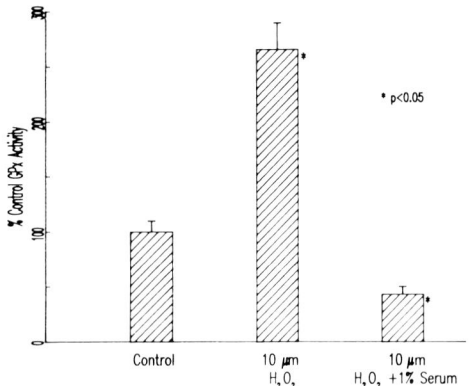

Figure 4. Glutathione peroxidase was assayed in cells treated with 10 μM H_2O_2, or 10 μM H_2O_2 with 1% serum. The combination of serum plus H_2O_2 resulted in significant inhibition of glutathione peroxidase p<0.05.

In summary, the pulmonary capillary endothelium is the major site of acute lung injury due to oxygen. This site is particularly vulnerable to oxygen toxicity for the following reasons: (1) endothelial cells have relatively low specific activities of key antioxidant enzymes, (2) the endothelial cells appear to make more partially reduced species of oxygen than the rest of the lung, (3) the endothelial cell is in an anatomic location which renders its free radical production particularly disastrous since the H_2O_2 can exit the cell and react with blood constituents to mount inflammatory response, and (4) inhibitors of a key antioxidant enzyme, glutathione peroxidase, may be formed in the vascular space and enter the endothelial cell, thus allowing more H_2O_2 to escape.

Xenobiotic Free Radical Reactions

In addition to oxygen producing acute lung injury, there is a whole host of acute and chronic pathologic states in the lung involving free radicals. Superoxide, H_2O_2, OH·, and hypochlorous radical may play a role in any disease process which involves activated leukocytes (29). Air, or fat embolus, and shock lung from trauma or sepsis are examples of acute lung injury in which PMNs marginate in the lung and begin an inflammatory response which includes release of free radicals (29). Chronic exposure to inhaled free radicals is present in persons who breath smoke fumes and polluted air. Cigarette smoke, for example, is estimated to contain 2×10^{14} free radicals per puff including the substances in both the gas and tar phase (33). A wide variety of xenobiotics can interrupt the normal metabolic flow of electrons and catalyze one electron transfers to oxygen resulting in free radical formation (33). Paraquat, bleomycin, and nitrofurantoin are examples of the drugs which can result in free radical damage to the lung (33,35). In some cases such as that of bleomycin, oxygen acts synergistically to augment damage to the lung (36). Several recent review articles have been published which discuss mechanisms of xenobiotic free radical formation (33,34,37).

In summary, the alveolar and capillary surface areas of the lung are exposed to a number of substances which result in free radical formation and damage to the lung. It is hoped that understanding the mechanisms of free radical lung injury will eventually prove useful in preventing and treating this diverse group of disease processes.

References

1. Weibel, E.R. (1983): Am. Rev. Resp. Dis., 128:752-760.
2. Weibel, E.R. (1983): Chest 83:657-665.
3. Priestly, J.: Alembic Club Reprints No. 7 (1894), William F. Clay Publisher, Edinburgh, pp. 5-55.
4. Clark, J.M. and Lamberston, C.J. (1971): Pharmacol. Rev., 23:37-133.
5. McCord, J.M. (1983): Chest 83:35(S)-36(S).
6. Crapo, J.D., Freeman, B.A., Barry, B.E., Turrens, J.F. and Young, S.L. (1983): The Physiol. 26:170-176.
7. Freeman, B.A. and Crapo, J.D. (1981). J. Biol. Chem. 256:10986-10992.
8. Freeman, B.A., Topolosky, M.K., and Crapo, J.D. (1982) Arch. Biochem. Biophys. 216:477-484.
9. Turrens, J.F., Freeman, B.A., and Crapo, J.D. (1982): Arch. Biochem. Biophys. 217:411-421.
10. Turrens, J.D., Freeman, B.A., Levitt, J.G., and Crapo, J.D. (1982): Arch. Biochem. Biophys. 217:401-410.
11. Freeman, B.A. and Crapo, J.D. (1982): Lab. Invest. 47:412-426.
12. Morehouse, L.A., Tien, M., Bucher, J.R., and Aust, S.D. (1983): Biochem. Pharmacol. 32:123-127.
13. Bishop, C.T., Crapo, J.D., and Freeman, B.A. (1984): Clin. Res., In Press.
14. Jones, D.P., Eklow, L., Hjordis, T., and Orrenius, S. (1981): Arch. Biochem. Biophys., 210:505-516.
15. Crapo, J.D. and Tierney, D.F. (1974): Am. J. Physiol. 226:1401-1407.

16. Crapo, J.D., Barry, B.E., Foscue, H.A., and Shelburne, J. (1980): Am. Rev. Resp. Dis. 122:123-142.
17. Yam, J. and Roberts, R.J. (1979): Toxicol. Appl. Pharmcol. 47:367-375.
18. Crapo, J.D., Delony, D.M., Sjostrom, K., Hasler, G., and Drew, R.T. (1977): Am. Rev. Resp. Dis. 115:1027-1033.
19. Roozemond, R.C. and Urli, D.C. (1982): Biochim, Biophys. Acta 689:499-412.
20. LaVelle, D., Paxton, W.B., Blaustein, D.I., Ostro, M.J., and Giacomoni, D. (1982): Arch. Biochem. Biophys. 215:486-497.
21. Turrens, J.F., Crapo, J.D., and Freeman, B.A. (1984): J. Clin. Invest. 73:87-95.
22. Freeman, B.A., Young, S.L., and Crapo, J.D. (1983): J. Biol. Chem. 258:12534-12542.
23. Crapo, J.D., Peters-Golden, M., Marsh-Salin, J., and Shelburne, J.S. (1978): Lab. Invest. 39:640-653.
24. Knapp, M.J., Cole, P.H., Wolfe, W.G., and Crapo, J.D. (1983): Am. Rev. Resp. Dis., In Press.
25. Perez, H.D., Weksler, B.B., and Goldstein, I.M. (1980): Inflammation 4:313-328.
26. Shasby, D.M., Fox, R.B., Harada, R.N., and Repine, J.E. (1982): J. Appl. Physiol. 52:1237-1244.
27. Fox, R.B., Hoidal, J.R., Brown, D.M., and Repine, J.E. (1981): Am. Rev. Resp. Dis. 123:521-523.
28. Fox, R.B., Shasby, M., Harada, R., and Repine, J.E. (1981): Chest 80:3(S)-4(S).
29. Tate, R.M. and Repine, J.E. (1983): Am. Rev. Resp. Dis. 128:552-559.
30. Hammerschmidt, D.E., Hudson, L.D., Weaver, L.J., Craddock, P.R., and Jacob, H.S. (1980): Lancet 1:947-949.
31. Zimmerman, G.A., Renzetti, A.D., and Hill, H.R. (1983): Chest 83:87(S)-89(S).
32. Bishop, C.T., Crapo, J.D., and Freeman, B.A. (1984): In Vitro, In Press.
33. Pryor, W.A. (1982): Ann. N.Y. Acad. Sci., 393:1-22.
34. Mason, R.P. and Chignell, C.F. (1982): Pharmacol. Rev. 533:189-211.
35. Martin, W.J. (1983): Chest 83:515-525.
36. Toledo, C.H., Ross, W.E., Hood, C.I., and Block, E.R. (1982): Cancer Treat. Rep. 66:359-362.
37. Mason, R.P. (1982): Free Radicals in Biology, Vol. V., edited by W. Pryor, Academic Press, New York.

Cytotoxicity and Somatic Mutation Induced in Mammalian Cells in Culture by Hyperoxia and Ionizing Radiation: Effect of Superoxide Dismutase and Catalase Inhibitors

S. Lesko, L. Trpis, and S. Yang

Division of Biophysics, Johns Hopkins University, Baltimore, Maryland 21205

It is now generally accepted that many carcinogens must undergo metabolic activation in order to initiate neoplastic transformation. When given at sufficient dosage most genotoxic chemicals have tumor-promoting effects as well as tumor-initiating activity (18). Phorbol ester tumor promoters have been reported to stimulate the formation of various activated oxygen species by polymorphonuclear leukocytes (7,21). This enhanced respiratory burst has been associated with extensive DNA strand scission (1) and chromosomal aberrations (3). This has led to the hypothesis that DNA damage may be related to the action of phorbol-12-myristate-13-acetate as a skin tumor promoter in mice (1). The environmental carcinogen, benzo(a)pyrene (BP), undergoes metabolic conversion to reactive electrophiles (6) and to intermediates which generate free radicals and reactive oxygen species (12,14). These metabolites, BP-diones and 6-hydroxy-BP, induce DNA strand scission in vitro (14) and are cytotoxic to mammalian cells in culture (13). Conditions which modify the biological and biochemical activity of BP-diones and 6-hydroxy-BP indicate that reduced oxygen species propagate the free-radical reactions responsible for the observed effects. The role of free radical intermediates and reactive oxygen species in the initiation or promotion phases of the carcinogenic process needs to be thoroughly investigated.

Oxygen is also reduced univalently by all aerobic cells during normal metabolic processes (4). Oxygen toxicity probably results from the action of one or more of these species that arise by partial reduction. This is attested to by the ubiquitous distribution of protective enzymes, viz., superoxide dismutase, catalase and glutathione peroxidase in aerobes. An enormous amount of oxygen must be reduced to meet daily energy requirements. A steady state normally exists in respiring cells between the rate of production of reactive oxygen metabolites and their rate of inactivation. In case of an altered inactivation efficiency, there is a potential for higher risk to cells due to the interaction of oxygen radicals with critical macromolecules.

This chapter will compare the response of mammalian cells in culture to hyperoxia and ionizing radiation. The role of free radical damage in radiation-induced carcinogenesis is widely accepted and Gerschman (6) has pointed out that damage caused by excess radiation and that caused by increased pressures of oxygen have much in common. The relationship of the functional state of the cellular oxygen defense system to cellular responses induced by reduced oxygen species will also be considered.

Oxygen Susceptibility of Syrian Hamster Embryo Fibroblasts

Two Syrian hamster embryo cell lines, BP6T (neoplastically transformed by BP) and SHE 13/3/O$_2$ (derived from exposure of early passage cells to 70% O$_2$-48 hours), were exposed to hyperoxia and cytotoxicity measured by reduction in growth rate. The two cell lines showed a differential sensitivity to 40% O$_2$ with E 13/3/O$_2$ being more sensitive than BP6T cells. In air, E 13/3/O$_2$ fibroblasts went through 5.0 population doublings during a 48 hour period compared to 2.6 in 40% O$_2$ while the values for BP6T cells were 4.1 and 2.8 respectively. Neither cell line grew very well when the oxygen tension was increased to 50% or above. The same differential sensitivity was observed when the cell lines were exposed to 95% O$_2$ for 24 hours and assayed for cytotoxicity by reduction in cloning efficiency. The 10,000 g supernatants from sonicates of BP6T and E 13/3/O$_2$ contained 6.4±1.2 and 12.1±1.8 units of catalase activity and 6.7±0.6 and 8.2±1.1 units of superoxide dismutase activity per mg of protein, respectively. Thus, the differential sensitivity to hyperoxia cannot be explained on a decreased level of activity of either of these two enzymes. In fact, the level of catalase and superoxide dismutase activity correlates with the growth rate of these two cell lines.

Somatic Mutation Induced by Hyperoxia and γ-Irradiation

Mutation frequency at the HGPRT locus was measured by resistance to 6-thioguanine (3×10^{-5}M) after exposure of BP6T cells to hyperoxia, γ-irradiation and N-methyl-N'-nitro-N-nitrosoguanidine (MNNG). The number of induced mutants per viable cell observed with 500 rads of γ-irradiation (3×10^{-5}, 48% survival) is 3 to 4 times greater than that found after exposure to 95% O$_2$ for 36 hours (8×10^{-6}, 28% survival) and 3 times greater than that after exposure to 70% O$_2$ for 48 hours (1×10^{-5}, 43% survival). The spontaneous mutation frequency was approximately 1.5×10^{-6}. The alkylating agent, MNNG, served as a positive control; the induced mutation frequency observed after exposure to 1µM MNNG for 2 hours was 3.7×10^{-5} at a survival of 63%. Thus, at nearly equitoxic doses, hyperoxia is only 3 to 4 times less mutagenic than γ-irradiation or MNNG in cells with very active defense system against reduced oxygen species.

Effect of Diethyldithiocarbamate (DDC), an Inhibitor of Superoxide Dismutase

BP6T cells were exposed to γ-irradiation or for various time intervals to 95% O_2 before and after treatment with 3mM DDC, a copper chelator which inhibits cellular superoxide dismutase activity about 85% after 1.5 hours of incubation. Cytotoxicity was measured by reduction in cloning efficiency and survival curves plotted. With control cultures the exposures required to reduce survival to 37% were found to be 550 rads of γ-irradiation or 38-39 hours in 95% O_2. Pretreatment with DDC enhances the sensitivity of BP6T cells to both γ-irradiation and hyperoxia. The doses required to obtain survivals of 37% were reduced to 240 rads of γ-irradiation and about 13 hours exposure to 95% O_2. DDC treatment alone is cytotoxic; survival is reduced to 70% after exposure of BP6T cells to 3mM DDC for 1.5 hours. It is postulated that the enhanced sensitivity of DDC-treated cells to γ-irradiation, hyperoxia and normoxia is due to inhibition of superoxide dismutase. This is substantiated by an experiment which demonstrated that the amount of [^{3}H]thymidine incorporated into cellular DNA at various time intervals in DDC-free medium, subsequent to exposure to 3mM DDC, is correlated with the level of cellular superoxide dismutase activity. After 3 hours in DDC-free medium, the superoxide dismutase activity is inhibited about 80% and incorporation of [^{3}H]thymidine into acid insoluble material is inhibited 70%. Both gradually return to normal and after 22 hours, the level of superoxide dismutase activity and the amount of [^{3}H]thymidine incorporated into acid insoluble material are the same as that found in controls. The data indicate that superoxide dismutase provides an effective defense against oxygen radicals formed in normal metabolic processes or by γ-irradiation and that inhibition of this enzyme results in a toxic response.

Exposure of BP6T cells to 3mM DDC for 1.5 hours results in a 13-fold increase in mutation frequency over background at the HGPRT locus. It is postulated that mutagenicity is due to oxidative damage to DNA resulting from a compromised cellular oxygen defense system. Rannug, et al. (8,15) have reported that several thiurames and dithiocarbamates are mutagenic in S. typhimurium strains TA 1535 and TA 100. They postulate a similar mechanism of action for DDC-induced mutagenicity since increased oxygen tension enhanced the effect of tetramethylthiurame disulfide and no distinct DNA alkylating activity could be found (15).

Effect of Aminotriazole, an Inhibitor of Catalase Activity

Treatment of BP6T cells with 0.1M 3-amino-1,2,4-triazole for 4 hours results in over a 90% inhibition of cellular catalase activity with little or no reduction in cloning efficiency. However, these cells become extremely sensitive to H_2O_2. Exposure to only 10μM H_2O_2 for 1/2 hour results in the cloning efficiency being reduced to 4% compared to 70-80% for cells not treated with aminotriazole. The cells treated with aminotriazole are also more sensitive to hyperoxia; a survival of 37% is obtained after exposure to 95% O_2 for 17-18 hours compared to 38-39 hours of control cells. There is at least a 2-fold increase in the mutation frequency over background at the HGPRT locus as a result of treatment with aminotriazole. Thus, there appears to be a risk to mammalian cells associated with exposure to a compound which inhibits catalase activity.

Effect of Liposome-Encapsulated Catalase and Superoxide Dismutase on Cytotoxicity Induced by Hyperoxia and H_2O_2

Cationic liposomes composed of phosphatidyl choline-stearylamine-cholesterol (7:2:1) were prepared by reverse phase evaporation (20) and contained encapsulated superoxide dismutase, catalase or phosphate-buffered saline. Hamster embryo cells were exposed to liposomes (~150-850 n moles of lipid) in Hank's balanced salt solution containing Ca^{2+} and Mg^{2+} for 2 hours. Liposomes were aspirated off and fresh growth medium was added for 2 or 18 hours. The monolayers were then washed several times before exposure to air, hyperoxia or H_2O_2. Liposomes containing encapsulated catalase provided protection, although not complete, against the cytotoxic effect of H_2O_2 while free enzyme or free enzyme plus liposomes containing only encapsulated buffer were not nearly as effective. The degree of protection was dependent upon the number of liposomes added, but only to a certain point because the lipid itself is cytotoxic. Cationic liposomes containing catalase or superoxide dismutase enhanced the survival of E 13/3/O_2 cells exposed to 95% O_2 for 24 hours and increased the growth rate of these cells in 40% O_2. As in the H_2O_2 study, free enzyme or free enzyme plus liposomes containing encapsulated buffer were not very effective. The major problem with this technology is that the lipid itself is cytotoxic. The data do indicate that liposomes containing encapsulated catalase or superoxide dismutase can be useful in probing the role of the superoxide radical anion and H_2O_2 in oxygen toxicity.

Concluding Remarks

The data presented indicate that cells in culture are at risk as a result of exposure to hyperoxia. Exposure to 95% O_2 for 38-39 hours is equivalent in cytotoxic effect to 550 rads of γ-irradiation. The induced mutation frequency at the HGPRT locus resulting from 500 rads of γ-irradiation is only 3-fold greater than that observed when cells are exposed to 70% O_2 for 48 hours; cytotoxicity is approximately equivalent in both cases. The cytotoxic effect of hyperoxia or γ-irradiation is enhanced when cells are pretreated with diethyldithiocarbamate which inhibits superoxide dismutase activity. Pretreatment with aminotriazole, which inhibits cellular catalase activity, will also enhance the cytotoxic effect of hyperoxia. Increased oxygen tension is not required to obtain a mutagenic response with DDC or aminotriazole. Thus, a malfunction or inhibition of the oxygen defense system by environmental agents may be important factors in the induction of damage to the genetic apparatus that could lead to spontaneous mutation, aging and cancer.

In vitro, reduced oxygen species have been shown to induce DNA strand scission (11), DNA-protein crosslinks (10), DNA interstrand crosslinks (10) and saturate the 5,6-double bond of thymine (2,11,16). These various types of DNA damage observed in vitro are caused by hydroxyl radicals generated in Fenton type reactions which involve chelated transition metal ions and H_2O_2 (10,11). Superoxide plays an important role by supplying the reducing power to recycle the oxidized metal ion. The protection afforded hamster embryo cells in culture by liposomes containing encapsulated catalase or superoxide dismutase against the cytotoxic effects of hyperoxia implicate H_2O_2 and superoxide radical anion in the responsible mechanisms.

Benzoyl peroxide and other free radical generating compounds have been reported to be effective skin tumor promoters (17). Phorbol ester tumor promoters stimulate the production of reduced oxygen species in polymorphonuclear leukocytes (7,21) and decrease both superoxide dismutase and catalase activities in mouse epidermus (19). A low molecular weight compound with superoxide dismutase activity inhibited the biological and biochemical actions of phorbol ester in mouse epidermis implicating reactive oxygen species in the tumor promotion process (9). Chromosomal damage and DNA strand scission are associated with the enhanced production of active oxygen species by phorbol esters and may be related to the action of these compounds as skin tumor promoters in animals (1,3). The metabolic conversion of BP to reactive electrophiles (5) as well as intermediates which generate reactive oxygen species (12,14) may account for its ability to serve as a complete carcinogen, i.e., having both initiating and promoting activities. The role of reactive oxygen species in tumor promotion has recently come under active investigation. A major problem with initiation-promotion studies is that activity in either of these stages can only be demonstrated in the presence of the other activity.

REFERENCES

1. Birnboim, H. (1982): Science 215:1247-1249.
2. Demple, B., and Lin, S. (1982): Nucleic Acid Res. 10:3781-3789.

3. Emerit, I., and Cerutti, P. (1981): Nature 293:144-146.
4. Fridovich, I. (1978): Science 201:875-880.
5. Gelboin, H. (1980): Physiol. Rev. 60:1107-1166.
6. Gerschman, R. (1964): In: Oxygen in the Animal Organism, edited by F. Dickens and E. Neil, pp. 475-492. Macmillan, New York.
7. Goldstein, B., Witz, G., Amoruso, M., Stone, D. and Troll, W. (1981): Cancer Lett. 11:257-262.
8. Hendenstedt-Rannug, A., Rannug, U., Ramel, C., and Wachmeister, C. (1979): Mutat. Res. 68:313-325.
9. Kensler, T., Bush, D., and Kozumbo, W. (1983): Science 221:75-77.
10. Lesko, S., Drocourt, J., and Yang, S. (1982): Biochemistry 21:5010-5015.
11. Lesko, S., Lorentzen, R., and Ts'o, P. (1980): Biochemistry 19:3023-3028.
12. Lorentzen, R., Caspary, W., Lesko, S., and Ts'o, P. (1975). Biochemistry 14:3970-3977.
13. Lorentzen, R., Lesko, S., McDonald, K., and Ts'o, P. (1979): Cancer Res. 39:3194-3198.
14. Lorentzen, R., and Ts'o, P. (1977): Biochemistry 16:1467-1472.
15. Rannug, A., Rannug, U., and Ramel, C. (1983): In: Int. Symp. Occupational Hazards Related to Plastics and Synthetic Elastomer, in press. Alan Liss, New York.
16. Schellenberg, K. (1979): Fed. Amer. Soc. Exp. Biol. 38:501.
17. Slaga, T., Klein-Szanto, A., Triplett, L., and Yotti, L. (1981): Science 213:1023-1025.
18. Slaga, T., Sivak, A., and Boutwell, R., editors (1978): Carcinogenesis, Vol. 2, Mechanisms of Tumor Promotion and Cocarcinogenesis, Raven Press, New York.
19. Solanki, V., Rana, R., and Slaga, T. (1981): Carcinogenesis 2:1141-1146.
20. Szoka, F., and Papahadjopoulos, D. (1981): Proc. Nat. Acad. Sci. 75:4194-4198.
21. Witz, G., Goldstein, B., Amoruso, M., Stone, D., and Troll, W. (1980): Biochem. Biophys. Res. Commun. 97:883-888.

Closing Remarks

T. F. Slater

Department of Biochemistry, School of Biological Sciences, Brunel University, Uxbridge, Middlesex UB8 3PH, United Kingdom

We live in an increasingly industrialized world surrounded by a large number of chemical hazards that can produce tissue malfunctions and cell injuries of diverse kinds. If we are to adopt rational approaches to the recognition and control of such hazards, to develop effective protection against the types of injury involved and where necessary to apply specific therapies, it is essential to understand more of the basic biochemical mechanisms that are involved in injuries to cells. Such studies are important in their own right, not only with regard to the applications noted above, but because they can often give insight into normal metabolic processes as these are perturbed, controlled, or go out of control. This underlying philosophy can be traced back to Claud Bernard's book "La Science Experimentale" (1878), but its extension to free radical mediated injuries is much more recent (see Slater: "Free Radical Mechanisms in Tissue Injury," Pion Ltd., London, 1972).

When the book cited above was written, it was possible to give a number of examples where free radical intermediates were clearly involved in a significant manner in producing cell injury. For example: in CCl_4-induced liver injury, in transitional metal overload, in high energy radiation damage, in photosensitization, and in defects of normal protective mechanisms. Now, however, some 15 years after that book was first started, the list of suitable examples can be expanded greatly. We now know that damaging free radical intermediates can be produced from a wide range of substances including some chemical carcinogens, food additives, certain clinically used drugs under particular conditions, many industrial chemicals and so on. Tissue dysfunctions that are known or purported to involve free radical mediated disturbances cover such disparate conditions as rheumatoid arthritis, pancreatic diseases, fibrosis of the lung, inflammation, chemical carcinogenesis and promotion, ethanol-induced liver disease, cytotoxicities of various quinones, lung damage by gases such as ozone, nitrogen oxides and high oxygen tensions, liver damage by halogenated hydrocarbons, some disturbances of the immune system, the cell killing mechanisms of leucocytes, disturbances to the prostaglandin and lipoxygenase pathways, changes in membrane fluidity and receptor-enzyme coupling, and in ageing. Many of these aspects of free radical biochemistry are recent developments and thus hard to put into clear and unequivocal focus. Symposia of the type that has led to this volume are of considerable significance and importance in helping researchers to achieve such a focus, and in stimulating the interested outsider to make important contributions of their own.

The origin of this Symposium came from the unanimous view of the organizers that a comprehensive review was required of the ideas so cogently developed by Denham Harman in the 1950's that free radical mechanisms play an important role in ageing. After a long period in which the importance of free radical reactions in biology and pathology was given scant attention, or regarded with indifference, we have witnessed in recent years an explosion of interests in free radical intermediates, both in relation to physiological processes as well as to the diverse perturbations associated with cell injuries. New methods and techniques have been developed, with much improved resolution and sensitivity, that permit serious and decisive testing of important hypotheses based on free radical reactions. Moreover, this last decade has witnessed major advances in our knowledge of how endogenous defence mechanisms operate normally to control deleterious free radical species, and how they can on occasion be overwhelmed or supplemented.

Thus is was felt that the time was right for an evaluation in depth of the significance and contributions of free radical reactions to ageing, and the biological consequences and significance of modifications of natural defence mechanisms.

Hand in hand with the developments in free radical biochemistry mentioned above, has been the dramatic advances in molecular biology that promise to revolutionise our viewpoints of the management and treatment of disease. How do all these developments affect our appreciation and acceptance of an important role for free radicals in ageing? In my view, the contributions to this Symposium, based on recent developments in molecular biology, biochemistry and free radical mechanisms, do much to strengthen the view that free radical reactions and disturbances contribute significantly to the ageing process. In addition, this Symposium firmly establishes the view that the study of free radical mechanisms in relation to ageing is of fundamental interest and justifies considerable expansion of effort.

Pathology is the science or study of disease. Research advances made in the last 20 years or so, have fully justified the belief that an important section of Pathology is that which concerns damaging free radical intermediates: careful study of the proceedings of this Symposium will show that Free Radical Pathology has come of age.

Subject Index

A

Acetates, diglyceride, 339–340
Adrenocortical cells, bovine, as cell aging models
  and ACTH, need increase, 206
  aging in, 218–219
  autoradiograph, 207
  bovine, advantages of, 204
  cortisol, 207–208
  culture lifespan
    deficiency, problems with, 214
    experiment design, 215–216
    fibronectin, 214
    passage of cells, 214
    proteases, 214
    selenium in, 216,217
    vitamin E, 216,217
  cumene hydroperoxide, toxicity of
    experiment, 212–213
    glutathione peroxidase activity versus selenium, 215
    and vitamin E, 214
  differentiation, 208
  FGF, effects on culture lifespan, 216, 218,219
    absence of, 216
    doublings, 216
    keratinocytes, 218
  fibroblasts, disadvantages, 204–205
  glutathione peroxidase, 204, 211–212,215
    activity, 212
    effects of culture, 212
    selenite, 212
  Hayflick limit, 203
  11-β-hydroxylase, induction of, 205, 206
  21-hydroxylase, 208
  life history in culture, 205
  long-term growth rate versus deficiencies, 210–211,213
    effects, selenium and vitamin E compared, 213
    medium, 210
    selenium, 210
    selenite, 211
    serums, 211
    vitamin E, 210,211
  mitochondrial function
    aminooxyacetate, 208,210
    butylated hydroxyanisole, 210
    glutamine/pyruvate ratios, versus population doublings, 209
    oxidation ratios, 208
    phenols, binding of, 210
    pyruvate, 208
    selenium against toxicity, 209
    tricarboxylic acid cycle, substrate oxidation in, 208
    vitamin E against toxicity, 209
  oxidative damage, 218
  receptor loss with age, 206
  reinduction of 11-β-hydroxylase, 206
  selenium, action of, 204,218
  selenium, usual deficiency of, in culture, 203–204
  steroid pseudosubstrates, 206
  steroidogenesis, cholesterol-dependent, 206
  superoxide release, 206–207
  tritium uptake, 206
  and vitamin E supply in culture, 203
Aging, and antioxidants; *see also under* Antioxidants entries
  age at death, types, 237
  age-dependent dysfunctions, 236
  death hormones, 239
  decline, 235
  differentiation, loss of, 241
  diseases, effects of destruction, 235, 236
  evolutionary reasons against aging, 239
  exogenous source pathologies, 236
  genetic hypothesis, 238–239
  metabolism versus lifespan, 240
  oxygen metabolism, 241
  optimum capacity, limit of, 235
  passive hypothesis, 239–240
    metabolic byproducts, 240
    similarity, nature, 240
    and tradeoffs, 239

Aging, and antioxidants *(cont'd)*
  piecemeal approach to cure, 237
  recent improvement in health, 235
  recognition of importance, 237
  sudden death syndrome, 239
  toxic byproducts, 241
Aging, and free radicals; *see also*
    Oxygen, species of free radicals of,
    and biological reactivity
  and arachidonic acid cascade, 32
  and atherosclerosis, 32
  chronic diseases, 30
  and emphysema, 30
  human rates, 31
  and medical progress, 29
  prostacyclin synthetase, 32
  survivorship curves, 29
  and thromboxane synthetase, 32
  tissue injury, 32
Aging, phenomena in
  autoimmune diseases, 5
  and decreased ATP production, 5
  free radical diseases
    autopsies, above age 85, lack of
      specific cause in one-third, 6
    autosomal recessive, 6
    dietary lipids, 6
    environmental influences, 6
    generation of, 6
    lipofuscin, 6
    other diseases, 7
    systemic lupus erythematosus, 6
  X chromosome, and G6PD gene, 5
Aldehydes, 357,368
Alkenals, *see* 4-Hydroxyalkenals
Alloxan, 307,312,313
Antioxidants, concentrations of, in
    humans, compensational
    nature of
  compensation, mutal, 261,262
  dietary problems, 261
  E deficiency, increase of other
    antioxidants by, 261
  E supplement, decrease of other
    antioxidants by, 262
  mouse strains, 262
  net amount, 261,262
  overlaps, 261
Antioxidants, lack of, harmful effects in
    eye and retina; *see also* Vitamin E
  absence of photoreceptors versus
    lipofuscin accumulation, 171

cell height in retina, 175
cells with pigment in outer segment, 174
cells in subretinal space, 174
dietary, 173
disk membranes, disruption of, 170
and exposure of RPE cells, 175
lipid droplets, 175
logarithm of vitamin E intake versus
  amount of lipofuscin accumulation, 171
lysosomal enzyme activity, 175
phagosomes in RPE, 172,174
photoreceptor cell death, 171
photoreceptor cell loss, versus type of
  diet, 172
photoreceptors, 170
RCS strain, photoreceptor genetic
  defect in, 172–173
RPE role, problems with, 170–171
and vitamin A, connection to vitamin
  E, 171
vitamin A, loss of, vitamin E
  deficiency as cause, 172
vitamin A versus RPE lipofuscin, 173
and vitamin E, 173
vitamin E deficiency, example, 170
Antioxidants, and lipofuscin accumulation acceleration; *see also* Lipofuscin; Lipopigments, free radicals in formation
  autofluorescence, 168
  ethane in breath, 167
  microsomes, autotoxicity of, 168
  pentane in breath, 167
  vitamin E, 167
Antioxidants, longevity-determining
  formulas, 244
  loss, 243
Arachidonic acid, metabolism of
  cascade, 32
  cyclooxygenase pathway, 63,64
  HPETE formation, possible route, 65
  inflammation, 356
  lipooxygenase pathway, 63,64
  LTA_4, possible formation route, 65
  microvascular envelope, 96
  PGG_2, possible formation route, 64
Ascorbic acid; *see also* Vitamin C
  ascorbate control, 256
  and iron, 256
  loss of, with time, 256

plasma levels, by species, 255
tissues, levels in, 256
Ascorbyl radicals in cancer
  immobilized, 286
  lyophilized blood cells, 286,287
  quinone system, 286
Atherosclerosis, as free radical disease
  chain length, 8–9
  glutathione, 8
  high-density lipoprotein, 8
  hydroperoxides in plaques, 8
  lesions on vessel wall, 8
  low-density lipoprotein, 8
  peroxidation, 8
  plaque components, peroxidizibility, 8
  platelet aggregation, 8
  polymerization, 8
  serum lipid levels, 8
ATP, 5,106,107,108
Autoxidation
  in air, 17
  azo compound as source, 16
  conjugated diene hydroperoxide, 17
  conjugated dienyl radical, 17
  initiation sequence, 17
  linoleic acid in SDS micelles, 16–17
  peroxyl radical, action of, 17
  peroxyl radical, formation of, 16
  potentiators, 169
    double bonds in fatty acids, 169
    oxygen flux in retina, 169
    vitamin A, 169
  propagation reactions, 17
  propagation sequence, 17
  terminations, 17
Autoxidation, similarity of, to aging in
  retina and RPE
  irregularities of size and shape in
    cells, 177
  lipid droplets, 177
  lipofuscin, 177
  and photoreceptor loss, 177
  senescence, 177–178
Autoxidation, techniques for
  inhibition of
  inhibitor radical, 21
  kinetics of
    bilayer system, 23
    formulas, 23
    initiaton versus inhibitor, 24
    lag time, 24
    and linoleic acid, 23

long-chain approximation, failure,
  23
rate formula, 24–25
and vitamin E constant, 25
and vitamin E effects, 23
and peroxyl radicals, 22
stoichiometry, 21
and vitamin E, 21–22,24–25

B

Bilayer, lipid, 57–58
Bleomycin, possible mechanism of
  action of, 278
Bloom syndrome, 6
Brain, aging of, possible involvement of
  iron and oxygen free radicals in
  aging of
    ascorbate, 146
    catalase, 146
    enzymes, lack of, 146
    fatty acids, 146
    glutathione oxidase, 146
    iron, 146
    vitamin, E, 146
  animals, study of
    age groups, 152
    cytochrome c in mitochondria, 149
    heme a basis, 154
    inhibitors, effects, 152
    iron in brain, 147
    iron chelators, 150
    iron content of brain regions, 149
    iron-ascorbate, 151,153
    lipid peroxidation, 149
    malate-glutamate, 153,155,156
    mitochondria of brain, 147,153
    nonsynaptic mitochondria, 154,
      155,156
    versus oxygen tension, 150
    oxygen tension versus iron
      ascorbate, effects, 153
    OH˙ scavengers, 150
    peroxidation versus total iron
      content, 151
    by region of brain, 149
    scheme, 148
    singlet oxygen scavengers, 150
    species, 146–147
    substrates, 147
    succinate, 153,155,156
    superoxide detection, 147

Brain, aging of *(cont'd)*
   synaptic mitochondria, 154, 155,156
   total iron versus age, 154
   basal ganglia; possible effects in, 160
   factors contributing to
     cytochrome versus age, mitochondrial, 158
     dopamine, 159
     and ferritin, 157
     and five bars' pressure, effects, 159
     lipid peroxidation, 157
     O_2^-, decrease with age, 159
     striatum, 159
     versus temperature, 159
   iron, 160
     ascorbate, 145
     binding to nucleotides, 145
     and Mg^{2+}, 145
     moving of atoms, 145
     and $OH^.$, 144
     oxidation of Fe(II) to Fe(III), 145
   lipofuscin accumulation, 143–144
   and malondialdehyde, 144–145
     adenosine phosphates, 145
     ADP, 145
     nucleotides, 145
     and ferrous ion, 144
     and H_2O_2, 144
     ligands, 144
     $OH^.$, problems with formation of, 144
     spin-trapping technique, 144
     and superoxide, 144
   mitochondria versus age
     cytochrome *b*, 160
     oxygen consumption, 159
     synaptic, drop in cytochrome species, 159
     synaptic, O_2^- drop, 159
     ubiquinone, 160
   O_2^-, generation of, versus age
     cytochrome, 156,157
     cytochrome oxidase, 157
     oxygen consumption, mitochondrial, 156,157
     SOD to antimycin A-inhibited respiration, 155
     SOD to complete respiration, 155
     total electron flux for O_2^- generation, 156
     types of cytochrome, 157
   oxygen consumption, 160
   susceptibility to free radicals, 160

C

Ca^{2+}, 106,107,109,111,112,113,192
Cancer; *see also* Aging, phenomena in; Tissue injuries
   and anomalous ESR signals, 32
   antioxidants against, 33
   cirrhosis, 33
   diabetes, 33
   error catastrophe theory, 33–34
   eye, diseases of, 33
   as free radical disease, 7
   general hazards for, 293
     chemicals, early work on, 293
   methionine residues, 34
   osteoarthritis, 33
   and α-1-PI, protein, 34
   polynuclear aromatic hydrocarbons, 32
   retrolental fibroplasia, 33
   superoxide, 33
   role of free radicals in carcinogenesis by chemicals, 287
Carbon-centered radicals, 18,19
Carotenoids
   and aging, 252
   beta varieties, 252
   carotenoid deoxygenase, 252
   ratio of, to vitamin A, effect on lifespan, 253
   serum concentration of, versus lifespan, 253
   and vitamin A, 252
Catalase, 258–259
   brain, 146
   diabetes, 310,313
   iron toxicity, 258
   lung injury, 384
   reduction of H_2O_2, 80
   retina, 328
   scavenging, 89
Cataracts, *see* Lens, of eye
Catecholamines, as antioxidants, 259
Cell, free radical sources in, 46

Cell, numbers in culture, increase
    per passage formula, 218
Cell cultures, see Adrenocortical cells,
    bovine, as cell aging model
Ceroids, see Lipofuscin; Lipopigments
Chelators, 150,310
Chimney sweeps, 293
Cholesterol as antioxidant, 259
Choline as antioxidant, 259
Copper, see Superoxide dismutase
Cyclooxygenase, 44,358
Cytochrome c
    in brain, 158
    and diabetes, 311
    in endoplasmic reticulum, 49
    and lung injuries, 382
    in mitochondria, 149
    and superoxide anion, 359
Cytosol, soluble components of
    hydroperoxides, 45
    iron, 45
    list, 45

D

Dehydroepiandrosterone, 259
Diabetes, alloxan-induced, oxyradical
        production in
    alloxan, 307
    and beta cells, loss of, 307–308
    dialurate, 307
    Haber-Weiss reaction, 308-310
        catalase, 309
        chelates in, 310
        coupled oxidative reactions, 309
        DTPA, 310
        EDTA, 310
        Fenton reaction, 309
        iron-catalyzed, formulas, 309
        iron chelation, 310
        perhydroxyl, 308
        protonated superoxide radical, 308
        superoxide, 309
    in vitro studies, 310–311
        cytochrome c, 311
        dialurate, OH· radicals from, 311
        ethylene, 311
        hydrogen peroxide, 310
        6-hydroxydopamine, neurotoxin,
            311
        methional, 311
        superoxide, 310–311
    in vivo studies, 313–315
        and alloxan, 313
        and catalase, 313
        DTPA, benefits of, 314–315
        glucose elevation, 314
        inactivated SOD, uselessness
            of, 314
        polyethylene glycol, protector
            for SOD, 314
        protection by OH· scavengers
            against diabetes, 313
        red blood cells, peroxide in, 313
        SOD protection, 314
    isolated pancreatic islets
        alloxan toxicity to, protection
            against, 312
        ascorbate recycling of alloxan, 312
        and DTPA, 312
        methods, 312
    oxidative cycle, 307
    redox pair, 307
    and vitamin E, red blood cells' loss
        of, 308
Diene conjugation
    albumin, 271
    bonds, absorption of, 270
    linoleic acid, 270
    lipid peroxidation, 270
    liquid chromatography, 271
    nature, 270
    and oxygen free radicals, 270
    problems with, 270–271
    and proteins, 271
Dietary restrictions
    adult-initiated benefits, 195
        body weights versus survivor-
            ship, 196
        energy dissipation, lifetime,
            195–196
        immune system, 197
    dehydroepiandrosterone, 186
    diets, materials in, 182
    hepatic drug metabolizing
        enzymes, 186
    immunity system, 194
        cytotoxic T lymphocytes, 194
        natural killer cells, 194
        splenic T lymphocytes, 194
    insulin, 186, 187
    lens proteins, 194–195
    lifetime energy dissipation, 189–190

Dietary restrictions *(cont'd)*
  lifespan extension, 187–189
    benefits, weaning-initiated, 187–188
    diets by groups, 188
    genetic limits, 189
    hepatoma, 188–189
    lymphoma, 187,188,189
    lipid peroxidation, retardation of, 185–186
    lipofuscin accumulation, 185–186
    lysosomal enzyme activity, 186
    male Fischer 344 rats
      calories/gram body weight/lifetime, 183
      controls, 183
      duration versus energy expenses, 183
      effects of dietary restriction, 182–183
      per animal consumption, 184
      problems with analysis, 183–184
    male Sprague-Dawley rats
      females, maturation of, 185
      prolongation of breeding time, 185
      testosterone output, 185
    male Wistar rats
      collagen diseases, 185
      dopamine, 184
      every other day feeding, 184
      receptors, 184
    metabolic rates
      cytochrome *c,* 193
      drop of O_2 consumption, 192
      liver mitochondria respiration, with malate and pyruvate, 193
      mitochondria, 192–193
      recovery of mitochondria, diet effects, 193
  mice, 5
  mitochondria
    adenine translocation, 191
    Ca^{2+}, 192
    escape of free radicals from quenching, 190
    inner membranes, 191
    lipid fluidity, 191–192
    lipid peroxidation, 190
    loss, 191
    muscle homogenates, 192
    and O_2 consumption, 190
    O_2 radicals, formation of versus age, 190
    phospholipids, 191
  NADPH-generating enzymes, 186
  pancreas, effects, 186-187
  under- versus malnutrition, 181
  species tested, 181
  types, 4
  weaning-initiated rodents, 181
DHEA, *see* Dehydroepiandrosterone
Diethyldithiocarbamate, 125,393
Diethylenetriaminepentaacetate, 310
Diffusion of free radicals versus reactivity, 297
Dismutation reaction, 80,359;
  see also Superoxide dismutase
DNA
  free radical reactions with
    bases, 288
    ionizing radiation, 288
    metal-bleomycin complexes and, 288–289
    psoralen, 289
    sugar-phosphates, 288
    in tumor promotion, bleomycin, 278
  oxidative damage to
    effects on, of agents, 327
    in epithelial cells, 327
    mitotic rates, 327
    psoralen, 328
    pyknosis, 327,328
    repair synthesis initiation, 328
    temperature effects, 327–328
    two-strand breaks, 327
    ultraviolet light, 328
    x-rays, 327
  scission, *see* Somatic mutation theory of aging
DTPA, *see* (Diethylenetriaminepentaacetate)

E

EDTA, *see* (Ethylenediaminetetraacetate)
Electron pair chemical bonds
  heterolysis, 14
  homolysis, 14
  and spin flips, in magnetic field, 14
  termination of free radical chains, 14

SUBJECT INDEX

Electron spin resonance studies of cancer; *see also* Cancer
  bleomycin, possible mechanisms of, 278–279
  concepts, problems with, 275
  as experimental tool in studies, 289–290
  free radicals, possible lack of role in cancer, 280
  free radicals in tumor promotion, 281–282
    possible mechanism, 281,282
    and phorbol myristine acetate, 281
    and superoxide, 281
  hydroxyls, 276
  importance of free radicals versus speed of disappearance, 279
  ionizing radiation, 277
  and iron ions, 279
  less-reactive radicals generated by more-reactive radicals, 279
  location of radicals, 277,278–279
  occurrence versus importance of intermediates, 276–277
  and oxygen radicals, 282
  quinone-hydroquinone complex, 282–286
    adriamycin, 284,285
    ESR spectra, 284
    metabolic intermediates, 284
    and metal ions, 283
    redox couple to target, 283
    research with, 283–284
    semiquinone, 283
  reactivity, versus key reactions, 277
  superoxide anion reaction with iron, 279
  singlet oxygen, 276
  tallysomycin, 278
  terminology, 275–276
  tyrosinase, functions of
    and analog of melanin monomers, 285
    reactions of substrates with, 285
    research with, 285–286
  water breakup, 277
Endoplasmic reticulum and nuclear membrane
  azide, and OH$^{\cdot}$ increase, 49
  cytochromes in, 49
  flavin-containing oxidases, 49
  and O_2^-, 49
  reductases, 49
Ethylenediaminetetraacetate, 310
Eye, *see* Lens; of eye; *entries under* Retina

F

Fanconi's anemia, 6
Fenton reactions, 309
Ferric and ferrous ions, *see* Iron
"Free Radical Mechanisms in Tissue Injury," 397
Free radicals; *see also* Oxygen, species of free radicals of, and biological reactivity
  chemistry of, 14–15
    alkoxyl, 15
    in cigarette tar, 15
    compared to OH ion, 15
    diffusion *in vivo*, 15
    halflife, 15
    hydroxyl from linoleate, H removal reaction constant, 14
    peroxyl, 15
    reasons for reactivity, 14–15
  dietary studies, early, 2–3
  early work, 1
  lifetime table, 16
  nature, 1
  in vivo production, 25
    in eye, 26
    α-1-protease inhibitor, 26
    SOD, 26
  oxygen-centered
    from activated phagocytes, 43
    and antineoplastic agents, 43
    intermediates, yield of, 43
    as mediators, 43
    primary radicals, reactions of, 43
    radiation, effects, 43
    reduction of oxygen, 43
  pathology, 397–398
  present support of aging theory, 3
  stable in soluton, 1
  theory, history and development of, 9–10
  20th Century progress, 1

G

Glucose, as antioxidant, 259

Glutathione, 256–257
  concentration in tissue, 257
  lack of correlation with lifespan, 256
  toxicity of, Ames test, 256
Glutathione peroxidase, 257
  adrenocortical cells, 204,211–212, 215
  forms of, 258
  H_2O_2, 388
  versus lifespan, 256, 258
  lung membranes, 387
  and metabolism, 125
  neurons, 226
  retina, free radicals in, 328
  scavenging, 89

H

H^+, 107,308
Haber-Weiss reaction, metal-catalyzed,
  see Diabetes, alloxan-induced,
  oxyradical production in
Hepatocytes, and free radical effects
  on surfaces of
  agents toxic to, 103
  and aging,116
  bleb formation, possible mechanisms
    of, 106–107
    cytoskeleton, 106
    H^+, 107
    and ions versus ATP, 106
    Mg^{2+}, 107
    microfilaments, 106
  blebbing, and altered calcium
    homeostasis
    calcium controls, 110
    ionophore A23187, 111
    isolated perfused rat liver, 110
    loss of Ca^{2+} with incubation, 11
    loss of blebs in incubation
      experiments, 111
    menadione, 111
    t-BH, 111
    and toxins, 110
  blebbing, and ATP depletion
    actin, dependence of, on ATP, 107
    antimycin A, effects on ATP, 108
    calcium release, 107,109
    uncoupler, 107
  blebbing, membrane in, 103
  calcium, regulation in hepatocytes, 109
  calcium ions, 106,109,111,112, 113

  and cytoskeletal changes, 115
    actin microfilaments, 115
    villin, 115
  dithiothreitol, effects, 105
  and glutathione, 103
    isolated, appearance of, 103
    mechanisms, proposed, 114
    menadione, effects, 104,111
    mitochondria, mobilization of
      calcium from, 111–112
    Ca^{2+}, sequestration in
      microsomes, 112
    Ca^{2+}-ATPase, 112
    GSH in, 112
    preincubation effects, 112
    pyridine nucleotides, 112
  point of no return in, 106
  pyridine nucleotides, fluorescence of, 114
  'quin 2', 114
  tert-butylhydroperoxide, 103,110,111
  thiol oxidation, 106
  toxins, mechanisms of, 112–113
  translocase, 113
    protection of, 113
  vesicles, calcium sequestration
    versus toxicity, 113
Homolysis, uncatalyzed
  ethyl linoleate hydroperoxide, 26
  lipid hydroperoxides, rates, 26
Hydrogen peroxide
  in brain, 144,146
  catalase, 80
  chemistry of, 359–362
  in diabetes, 310
  glutathione peroxidase, 388
  lens, 326,327
  membranes, 44,45,386
  mitochondria, 46,47
  and oxygen concentration, 387
  sources, 78,79
4-Hydroxyalkenals, 369-374
  adenylate cyclase, 371
  aldehydes, 371
  alkenals, 371
  Ca^{2+}, 371
  and chemokinesis, 372
  and chemotaxis in neutrophils, 372
  4-hydroxyalkenals, in leucocytes, 372
  4-hydroxyenals, 373
  4-hydroxynonenal, 371

SUBJECT INDEX

laser killing of erythrocytes, and free radical release, 371
and leucocytes, 370
leukotrienes, 373–374
Na^+, 371
tripeptides, 373
Hydroxyl; *see also* Oxygen, free radical species of, and biological reactivity
in cells, 47
from dialurate, 311
endoplasmic reticulum, 49
formation, 144
inflammation, 98
iron, 144
lipopigments, 132,133
melanin, 73
membranes, 44
prostaglandins, 363
scavengers, 150

I

Inflammation
activated inflammatory cell, 90–91
aerobic survival, 87
biradical, defined, 88
cell membranes and proteins, 94–95
articular cartilage, possible damage mechanism, 95
invasion of cell by free radicals, 94
propagating lipid free radical chain reactions, 94
proteoglycan aging degradation, possible mechanism, 95
electron spin parallelism, 88
enzyme scavenging
catalase, 89
cytochrome oxidase, 89
glutathione peroxidase, 89
hydrophilic, 90
hydrophobic, 90
lipid peroxidation, 90
localization of enzymes, 90
peroxidases, 89
SODs, pathway, 89
vitamin A, 90
vitamin E, 90
free radicals, nature of, 88
inversion of electron spin, 88
iron/lactoferrin, 92
metal chelates, reduction schemes, 92
myeloperoxidase, 91
and O_2 ($'\Delta$)g, 92
polymorphonuclear leukocytes, 91
microvascular envelope
arachidonic acid, 96
cheek vasculature, 97
chemotactic compounds, 96
granulocytes, 96
hypoxanthine-xanthine oxidase, 96
leakage sites, effects of xanthine oxidase, 98
and O_2^-, 96
and $OH^{\cdot}$, 96
permeability control, 96
molecular oxygen, excited states, 88
molecular pathway for univalent reduction of oxygen, 88
oxygen singlet, 88
prerequisites for free radical injury
connective tissue, components, 93
electron flux from initiation reactions, 93
extracellular space, 92–94
granules, 91
and hyaluronic acid degradation by O_2^-, 93,94
respiratory burst, 91
Inflammatory diseases, free radicals in
aldehyde production, rat hepatocytes
medium polar, 357
nonpolar, 357
arachidonic acid metabolism, 356
and cyclooxygenase, 358
endoperoxides, 358
4-hydroxyalkenals, 356
leukotriene, synthesis pathway, 358
and lipid peroxidation mechanisms of
aldehydes, 356
liver problems, 356
malonaldehyde, 356
phases, 356
TBA test, 356
membrane peroxidation, stimulation of, 358
peroxidases, 358
PGG_2, 358
prostaglandin cascade, 358
unsaturated fatty acid breakdown, 357

Iron
  brain, 145,146–147,149–150
  chelation, 310
  and cytosol, 45
  and electron spin resonance studies, 279
  Haber-Weiss reaction, 47,309
  hydroxyls, 363
  lactoferrin, 92
  retina, 331,332,341–346
  and SOD, 79
  and vitamin C, 151,153,256

L

"La Science Experimentale," 397
Laboratory tests, indications for free radical reaction, 267–286; see also specific test
Lens, of eye
  capsule, 318
  diagram, 318
  differentiation in, 318
  epithelial cells, germinative zone, 318
  fibers in, 318
  lens sutures, 318–319
  lipids, 319
  metabolism, 319
Lens of eye, free radical damage in
  browning (brunescence), 323
  cataract, 323
  lipids oxidation
    3-aminotriazole, 326,327
    cell membranes, 326–327
    and H_2O_2, 326,327
    malondialdehyde, 326
    percents weights, 326
    riboflavin, 326,327
    vitamin E, 326
  lipid peroxidation, 323
  proteins, oxidation of
    brown material, 324–325
    and cataracts, 324
    fluorescent materials, 325
    insolubilization, 324
    photosensitizers, 325
    research problems, 325
    and sulfhydryl groups, 324,325
    tryptophan irradiation, 325
    ultraviolet, 324
    x-rays, 324
  sources, 323
Leukocytes, 91,370
Life, origin and evolution of
  cells, self-defenses of, 3
  DNA repair, 3–4
  early Earth, 3
  enzymatic reactions, 4
  free radicals effect, 4
  nonenzymatic reactions, 4
  somatic cells, 3,4
  sun, radiation of, 4
  zygotes, DNA of, 4
Linoleic acid, free radicals in autoxidation of
  autoxidation of, 23
  and carbon-centered radicals, 18
    molarity, 19
  chain length, 19–20
  diene conjugation, 270
  micelles, 18
  rate constants, kinetic calculations, 19
  steady-state peroxyls, 18
  termination rate, 18
Lipids, peroxidation of
  alkoxy radicals, 55
  arachidonic acid metabolism, 62,63, 64–65
  autoxidation reactions, 59
  autoxidation reactions, 59
  bilayer, 57–58
    peroxidation in, 58
  diene, autoxidation of, 54
  bulk fatty acids, possible hydro-epoxide mechanism, 62
  eicosenoids, 62
  epoxides, 57, 59–60
    of cholesterol, 59–60
    of fatty acids, 59–60
    formation of, 60
  ethane production mechanism, 56
  fatty acids on silica model, 58, 59
  and free radical initiation, 53
  homolysis of, 54
  hydroxyketones, 57
  isomerization, 54
  isomers, geometric, formation of linoleate hydroperoxides, 55
  ketones, 57
  linoleic acid monolayer, hydro-epoxides in, 61
  lipofuscin, 60
  versus metabolism, 123
  oleic acid, 53–54
  oxidized fatty acids in phospholipids, 60
  oxyradicals, escape from control

of, 61-62
pentane production, 56
and peroxy radicals, 53
phosphatidyl choline liposome, 59
and phospholipase, 60,63
   A2 activity, 63
   C activity, 63
phospholipids, hydrolysis of, 60
radical initialization, short-term
   experiment, 59-60
rancidity, 56
β-scission, 55,56
termination reactions, 54
and tocopheral, 58,59
Lipofuscin; see also Vitamin E
accumulation, decoupling of,
   from aging
     double bonds in fatty acids, 169
     problems with, 168-169
     RPE, lipofuscin spectrum in, 169
     and vitamin E, 168
and cell function
   centrophenoxine, 175
   photoreceptor disk membranes,
     restoration by vitamin E,
     176-177
   research problems, 175
   and senescent rats, 176
   slowdown of accumulation by
     vitamin E, 176
general topics
   aging, 6,122
     marker for aging, 122
   antioxidants against, 167-168
   autoxidation, 177
   dietary restriction, 185-186
   neurons, 224-230
   photoreceptors, 171
   properties of, 223-224
   in retina, 164-166,167,329
     damage to, by, 329
   and vitamin A, 173
Lipooxygenase pathway, 63,64,368
Lipopigments, free radicals in formation
   aging, overall effects with lipid
     peroxides, 135
   aging dog, 129
   in Batten's syndrome, 129
   blue light fluorescence, 129
   canine ceroid lipofuscinosis, 129
   ceroids, 129
   chain length, 131
   classes, 129

inclusions, 129
lipid damage to ceroid, 131
lipopigment formation, inhibition of,
   by scavengers
   and ascorbate, 135
   centrophenoxine, 135
   cithiolone, 135
   and vitamin E, 135
lipid globule to lipofuscin, possible
   pathway, 131
lipopigments, factors associated with
   brain, 132
   and melanin, 131-132
   old heart, human, 131
metals, and free radicals
   in lipid peroxidation
   copper chelation, 134
   copper ion, 132
   ESR signals, 134
   ferric complexes, 132
   hydroxyls, 132,133
   lipid peroxides, 133
   malondialdehyde, 133,134
   and oxygen, 132
   pathways, 132,133
   peroxide, and hydrogen
     abstraction, 132
   PUFA, 133
   superoxide, 132
peroxide regulating system
   enzymes, 135
   and peroxidases versus age, 136
   retina, 136
   retinal pigment epithelium, 136
   rod outer segments, 136
pictures of examples, 130
polyunsaturated fatty acids, 131,133
possible pathways, 137
in 70-year-old human, 129
syndromes associated with, 129
Longevity, biological nature of
   cancer, oncogenes for, 242
   dysdifferentiation, 242
   and human evolution, 242
   number of genes determining,
     242-243
   regulating genes for, 242
Lung injury, and free radicals
   alveolar area, 381
   catalase, 384
   cyanide blocking, 382
   and cytochrome c, 382
   defense mechanisms, 384

Lung injury, and free radicals *(cont'd)*
  and DNA, 383
  and fatty acids, 384
  and glutathione peroxidases, 384–385
    system, 385
  glutathione reductase, 384
  H_2O_2
    versus glutathione peroxidase, 388
    release, by oxygen concentration, 383
  hyperoxia, 384
  lipid peroxidation, 383
  liposomes, 385
  and membranes, 386–387
    arachidonic metabolites, 386–387
    G6PD, 387
    glutathione peroxidases, 387
    H_2O_2 in, 386
    inflammatory cells, influx of, 386–387
    lipids, 386
  morphological changes
    alveolar I, 386
    alveolar II, 386
    endothelium, 386
    interstitial, 386
  NADPH, 384–385
  oxygen
    85%, 385
    100%, 383,385
  and Priestley, work of, 381
  reduction, 382
  and SOD, 384
  xenobiotic free radical reactions in, 388–389

M

Magnesium, 106,107,145
Malondialdehyde
  lens of eye, 326
  lipids, oxidation of, 326
  lipopigments, 133,134,144–145
  TBA test, 268
Manganese, *see* Superoxide dismutase
Melanin formation, and free radicals
  derivation, 67
  eumelanin, 67
  generation from tyrosine and DOPA, 68
  neuromelanin, 67
  radical in formation, 67–70
    autoxidation, 69,70
  enzyme oxidation, 68
  dopaquinone, 69
  free radicals' range of reactivity, 67–68
  one-electron oxidations, 69
  and precursors, free radicals from, 69
  primary free radicals, 69,70
  secondary free radicals, 70
  semiquinones, 69
  toxic free radical species, 70
  tyrosinase, 68
  reactions of melanins, free radicals in
    above 300nm, 72
    in air, 73
    concentration, factors that affect, 72
    decay components, 72
    and light, 72
    and oxygen consumption, 72–73
    OH, 73
    photoreactions, mechanisms of, 73
    as quenchers, 73
    reversible reactions, 72,73
    types, 73
  structures, free radicals in
    aqueous suspension at pH 7, free radicals in, 72
    and ESR spectrum, 70–71
    metal ions, involvement of, 71
    semiquinones 71
    types of melanin, 71
Metabolic activation of chemical carcinogens
  compounds yielding free radicals, 294
  cycle, 294
  electron transport system, 293
  flavoproteins, 293,294
  NADPH-cytochrome P_{450}, 293
  species, 293–294
Metabolism, rate of, versus free radicals' effects on aging
  3-aminotriazole, glutathione inhibitor, 125
  catalase, 125
  diethyldithiocarbomate, SOD inhibitor, 125
  and cupro-zinc SOD, 125
  early diet experiments, 124
  energy dissipation
    in nonprimates, 119
    in primates, 119

and escape of free radicals, 123
exercise, exhaustive, in rats, 123
flies, enzymes in, versus
    age, 124–125
flies, experiment with, 120–121
glutathione peroxidases, 125
hibernation, 119
insects
    flight of, versus temperature, 120
    oxygen consumption versus
        temperature, 120
lipid peroxidation, 123
and lipofuscin, as aging marker
    accumulation of, in flies versus
        metabolic rates, 122
    fluometry for, 121
    hamsters, compared, 122
    by physiological age, 122
    and soluble fluorescent
        material, 122
malondialdehyde, 123,124
    analogs, 123,124
organelles, free radicals in, 123
other controls on, 119
oxidative stress, 125-126
    and thiobarbituric acid, substances
        reacting to, 126
and oxygen consumption
    free radicals from, 122,123
    hydroperoxides, 122,123
    versus temperature, 119–120
rate of living theory, 120
Schiff base substances, 123
SOD versus metabolism rates, 124
threshhold theory, 120
Metal ions, *see specific element*
Mitochondria; *see also* Cytochrome *c*
    adrenocortical cultures, 208–210
    aldehyde oxidase, 47
    calcium ions from, 111–112
    and cytochrome *c*, 46,149,158
    dietary restriction and, 191–193
    dioxygenases, 47
    electron carriers, enzymes as, 48
    free radicals, 47, 78
    H_2O_2 release, 46,47
    hydroxyl in cells, 47
    intact, 46–47
    iron-catalyzed Haber-Weiss
        reaction, 47
    nonsynaptic, in brain, 154–156,159
    O_2^-, 47

    oxygen consumption, 48
    particles of, 46,47
    reduction of O_2, 46
    respiratory chain components, 48
    synaptic, 154–156,159
    ubiquinone-cytochrome *b*, 48
    xanthine oxidase, 47

N
NADPH
    enzymes generating, 186
    lung injuries, 384-385
Neurons, lipofuscin in, antioxidants on
    antioxidants, 224–230
    diet, 226,228
    diet restriction
        versus age of animal, 230
        effects, 229
        and protein deprivation, 230
    dissolution of, 224
    distribution in neurons, 223,224
    flourescence, 224
    formation versus diet, 224,226
    frontal cortex, 227
    glutathione peroxidases, 226
    hepatocytes, 228
    learning deficit, 228
    lipid peroxidation, 224,228
        versus age, 228
    malonaldehyde, 224,226,227,228
        cross-linkages of, 224
    percents versus age, 223
    properties of lipofuscin, 223–224
    Purkinje cells, 225
    pyramidal cells versus vitamin E, 225
    retinal pigment epithelium, 227
    selenium, 226
    staining versus age, 223,224
    vitamin C, 228,230
    vitamin-E-deficiency-associated
        diseases, 224
    word, derivation of, 223

O
Olefins, oxidizability of
    experimental values, 21
    factors, 20
    free radicals, 20–21
    rate formulas, 20

Oxygen, species of free radicals of, and biological reactivity
  activated phagocytes, 368
  aldehydes, 367–368
  arachidonic acid, 363,365
  arthritis, aldehydes in, 366,367
  granulocytes, 368
  hydrogen peroxide, 359–362
    action, 360
    chloramines, 361–362
    cytotoxicity, 362
    GSH, reactions with, 360–361
    and GSH redox unit scheme, 360
    halide complex, 362
    halides, oxidation, 361
    hypohalous acids, 362
    iodination, 361
    myeloperoxidase, 361
    neutrophils, 362
    oxidative decarboxylation of amino acids, 361,362
    peptide cleavage, 361,362
    phagocytic cells, 361
  4-hydroxyalkenals, 369–371
  hydroxyl radical
    and ferric ions, 363
    and ferrous ions, 363
    Haber-Weiss reaction, 362–363
    hydroperoxy acids, 363–364
    and lipids, 364
    and prostaglandins, 363
    and superoxide, 363
  12L-hydroxy-5,8,10-heptadecatrienoic acid, 369
  hypochlorite pathway, 366
  lipoxygenase pathway, 368
  malonaldehyde, 369
  membranes, 367
  monocytes, 367
  myeloperoxidase-H_2O_2-halide system, 365
  neutrophils, 367
  oxygen-mediated injuries, 368
  perhyrdoxyl radical, 359
    protonation, 359
  phagocytes, 367,368
    activated, 368
    products, 368
  phlogogens, 365
    malonaldehyde, 365
    prostaglandins, 365
  polymorphonuclear cells, 368
  polyunsaturated fatty acids, 366
  prostaglandins, 365,369
  synthesis pathway, 365
  respiratory burst, 368
  singlet oxygen
    bonds sensitive to, 364
    guanine, 364
    Haber-Weiss reaction, 364
    hypochlorous acid, 364
  superoxide anion, 358–359
    cytochrome c, 359
    in cytosol, 359
    dismutation reaction, 359
    impaired electron, 358,359
    NADPH-dependent enzymatic line, 358,359
    production, 358
    and prostaglandins, 359
    quinol dehydrogenase, 359
  thromboxanes, 369

P

Peroxisomes, 49
  and H_2O_2, 49
Peroxyl; see also Lipopigments; Oxygen, species of free radicals of and biological reactivity
  action of, 17
  atherosclerosis, 8
  autoxidation, 22
  diabetes, 308
  formation, 16
  H^+ abstraction, 132
  red blood cells, 313
Phagocytes, 44,361,367,368
Phosphatidyls
  choline, 340
  ethanolamine, 340
  serine, 340
Phospholipids, 60
Photoreceptor cells; see also entries under Retina
  bridge in, 322
  cones, 321
  discs, 321–322
  inner segment, 322–323
  metabolism, 323
  outer segment, 321–322
  rods, 321,322
  and RPE, 322
Plasma membrane and free radicals
  arachidonic acid, 44
  cyclooxygenase metabolism, 44
  diffusion, 45
  H_2O_2, 44,45

membrane-bound NADPH
    oxidase, 44
myeloperoxidase, 44
  and OH·, 44
  perfusion across, 44,45
  phagocytes, 44
  and prostaglandins, 45
  and superoxide, 45
Polyunsaturated fatty acids, 131,133
The Prolongation of Life, 2
Prostaglandins
  cascade, 358
  and hydroxyl reactions, 363
  and oxygen, 365,369
  and plasma membrane, 45
  superoxide anion, 359
Protein fluorescence, 271-272

Q

Quinones
  hydroquinones, complex of,
      with, 282-286
  system, 286

R

Radiation, cancer hazards of, 293
Radicals, enzyme-catalyzed
      production of
  and alkoxyls, 27
  in vivo, 27
Red blood cells, 308,313
Respiratory chain, 48,78
Retina, autoxidation in
  aging of eye, versus lipofuscin
    blood supply, 164
    choriocapillaris, 164
    cones, 164,165
    disks, turnover of, in rods, 164-165
    neural retina, 164
    photosensitive proteins, 164,165
    retina, structure of, 164,165
    rods, 164,165,166
    RPE, functions of, 164,165
    vitamin A, 164,165
  lipofuscin accumulation, universality of, 163-164
  lipofuscin versus age in
      RPE, 166,167
  senescence, conditions for, 163-164
Retina, cellular organization of
  capillaries, 319,321
  choroid, 321
  layers in, 319
  light absorption, 321

light path, 321
nutrition of, 319, 321
Retina, experiments with degeneration
  arachidonic acid, 338
  conjugated diene assay, 334
  constant illumination
    dienes, lipid conjugated, 341
    fatty acids, 339
    lipid content, 338
    phospholipids, 340
    protein content in ROS, 338
  diglyceride acetates, 333,339
    1,2- variety, 333
  discs, 334,335
  electroretinography, 332
  examinaton procedures, 332
  fatty acids, 338
  ferrous ions, 332,341-346
    a-wave changes, 343
    a-wave loss, 341,342
    b-wave amplitude, 344
    b-wave loss, 341,342
    chemical changes, 343
    effects, 345
    ERG, 342
    and frog, ROS effects in, 346
    morphological changes, 341,343
  lipid analysis, 333
  photoreceptor membranes
    isolation of, 333
    purity, 333
  phosphatidyls, 338
  results
    cell loss, 334
    degeneration, 334
    discs, 334,335
    examples, 335,336
    inferior, 334
    ONL, 334
    opsin, 334-335
    superior, 334
  ROS, 335-337
  SDS gel electrophoresis, 337
  sodium sulfate, effects, 343,344,346
  substances not affected by constant
      illumination, 338-339
  2-dimensional thin layer chromatography, 333
  22:6ω3, 338,339
Retina, free radical damage in
  adriamycin, 331
  albino rats, 329-330
  blue-blinding, 330
  catalase, 328

Retina, free radical damage in *(cont'd)*
  circulation, 328
  degeneration, 331
  dienes, 330
  docosahexaenoic acid, 328
  factors in, 328
  glutathione peroxidase, 328
  hydroperoxides, 330-331
    injection of, 330,331
  iron ions, 331
  light damage, 330
  lipid peroxidation, 328-329
  lipofuscin, 329
  macula lesions, 329
  MDA, 331
  and oxygen on premature babies, 329
  rhodopsin regenerability, 331
  selenium, 329
  singlet oxygen, 330
  SOD, 328
  vitamin A, 329
  vitamin E
    deficiency, 329
    location, 329

S

Selenium
  cultures, 204,214,215,216,217,218
  deficiencies, 203-204
  growth rate, 210,213
  lipofuscin, 226
  mitochondria, 209
  neurons, 226
  retina, 329
  and vitamin E, 213
Semiquinones, 69,71
Senescence; *see also* Aging, phenomena of
  autoxidation, 177-178
  criteria, 178
Singlet oxygen; *see also* Oxygen, species of free radicals of, and biological reactivity
  inflammation, 88
  retina, free radical damage to, 330
  scavengers for, 150
Slater, 397
SOD, *see* Superoxide dismutase
Somatic mutation theory of aging
  aminotriazole, 394
  benzo(a)pyrene, 391

  metabolites of, 391
  benzoyl peroxide, 395
  catalase, 394
  diethyldithiocarbamate, SOD inhibitor, 393
  DNA scission, 391
  excess oxygen, radiation equivalent of, 394-395
  γ radiation, 392,393
  hamster embryo fibroblasts, oxygen susceptibility of, 392
  HGPRT locus, mutation of, 392
  hyperoxia, 394
  liposome-encapsulated catalase, 394
  liposome-encapsulated SOD, 394
  N-methyl-N'-nitro-N-nitroso-guanidine, 392
  oxygen reduction, 391-392
  phorbol ester tumor promoters, 395
  reduced oxygen species
    DNA effects, 395
    and metal ions, 395
  (^{3}H)-thymidine incorporation, 393
Spin, electron, 13
Superoxide dismutase, 392,394,395; *see also* Oxygen, species of free radicals of, and biological reactivity
  in brain tissue, 246
  in congenic mice, 246-247
  Cu/Zn variety, 246
  and cytotoxicity of O_2^-, 79
  energy values, 245
  factors, 77
  functions, 244-246
  H_2O_2, sources of, 78,79
  histocompatibility loci of genes, 10
  and iron, 79
  versus lifespan, 244,245
  liver tissue content, 246
  in mice liver cells, 248
  mitochondria, concentrations of free radicals in, 78
  Mn variety, 246
  nature of SODs, by type
    Cu/Zn variety, 80,246
    dismutation, mechanism of, 80
    distribution, 80
    metals in, 80
    Mn/Fe varieties, 80
    rate constant, 80
    reaction catalyzed by, 80
  O_2^-, 77-78

sources of, 78
univalent reduction of, 77
peroxide, control of, 78
physical function
  in cataracts, 82
  decline with age, 82
  in eukaryotes, 82
  oxygen concentration, 81–82
  and O_2^- scavengers, 81
  and oxygen tolerance, anaerobic organisms, 81
range of effects, 79
ratios by age, 246
respiring cells, generation of radicals in, 78
similarity of structure across species, 245
species versus lifespan and SOD content, 245
specificity, possible, 79
synthesis
  glucose, 81
  Mn variety, induction of, 81
  and oxygen concentration, 80–81
  and superoxide anion, 81
Superoxide radical; *see also* Oxygen, species of free radicals of, and biological reactivity; Superoxide dismutase
  adrenocortical cell cultures, 206–207
  in brain versus age, 156–157
  diabetes, 309,310–311
  dismutation reaction, 80,359
  electron flux in generation, 156
  endoplasmic reticulum, 49
  with H^+, 308
  hyaluronic acid degradation, 93,94
  inflammation, 96
  lipopigments, 132
  mitochondria, 47
  prostaglandins, 359
  toxicity, 79

T

TBA tests, *see* Thiobarbituric acid tests
Thiobarbituric acid tests, 267–270
  cascade reactions, 268
  cross-reactions, 268
  limits, 268–269
  malondialdehyde, 268
  promoters, 269–270
  rate-limiting steps, 269
  red blood cells, 269
  and SOD, 269
  thiobarbituric acid, 268
  vitamin E, 269
  uses, 268
Tissue autoxidation, versus age and mean life span; *see also* entries *under* Autoxidation
  of brain homogenates, 260
  experiment, 259–260
Tissue injuries
  cervix, samples and ESR signals from, 294
  damage paths, 296–297
  flash photolysis, 296
  and free radicals, diffusion of, versus reactivity, 297
  kinetics, and free radicals in solution, 295
  lipid peroxidation
    aldehydic products, 299
    and cancer, 300–301
    and cell division, 302
    hepatoma cells, 301–302
    4-hydroxyalkenals, 299
    4-hydroxypentenal, 300
    and lipooxygenase, 299
    in liver microsomes, 297
    normal hepatocytes, 301–302
    species, 298
  phenyl-t-butyl-nitrone, 294
  pulse radiolysis, 296
  spintrapping, 294
  trichlormethyl, 296
  and water, 294
α-Tocopherol, *see* Vitamin E
Tumor promoters, 281–282
Tyrosinase, 285,286

U

Ultraviolet light, 324,328
Urate; *see also* Uric acid, as antioxidant
  in brains, primate, 251
  evolutionary increase of, mechanisms, 252
  versus lifespan, 250
  and intelligence, 249
  in plasma, nonprimates, 251
  removal of Fe^{2+} by, 256
Uric acid, as antioxidant
  behavior, 249
  and gout, 250

Uric acid, as antioxidant *(cont'd)*
  kidney function, 249
  membranes, 249
  as nervous system stimulant, 249
  plasma levels, primates, 250
  and uricase, 249
  and vitamin C, loss of ability to synthesize in primates, 249

V

Vitamin A; *see also* Carotenoids
  and autoxidation, 169
  and inflammation, 90
  and lipofuscin, 173
  loss of, in eye, 172
  in retina, 164,165,329
    and free radical damage, 329
  and vitamin E, 171
Vitamin C; *see also* Ascorbic acid; Ascorbyl radicals in cancer
  in brain, 146
  iron-ascorbate, 151,153
  loss of ability to synthesize in primates, 249
  and neurons, lipofuscin in, 228,230
Vitamin E
  antitoxin, 209
  autoxidation, 23,24,25
    kinetics of, 23,24
  in brain, 146
  in culture medium, 214,216,217
  deficiency diseases, 224
  deficiency in eye, 170,173
  in eye lens, 326
  inflammation, 90
  versus lipofuscin accumulation, 167,171
  lipopigments, 135
  and natural antioxidants, 261–262
  photoreceptor disk membranes, 176–177
  plasma levels by species, 254
  pyramidal cells, 225
  red blood cells, 308
  retina, 329
  serum deficiency, 203
  slowdown of lipofuscin accumulation, 176
  TBA, 269
  and vitamin A, 171,172

X

X rays, 324,327
Xanthine oxidase, 47
Xenobiotics, and free radical production, 27–28
  carbon tetrachloride, 28
  cigarette smoke, 27–28
  mechanisms, 27
  and nitrogen oxides, 27
  without oxygen, 28
  ozone, 28
  quinoid species, 28
  tar, 28

Z

Zinc, *see* Superoxide dismutase